Leo Corry

Modern Algebra and the Rise of Mathematical Structures

Second revised edition

Birkhäuser Verlag
Basel · Boston · Berlin

Author's address:

Dr. Leo Corry
The Cohn Institute for the History
and Philosophy of Science and Ideas
Tel Aviv University
Ramat Aviv 69978
Israel

The first edition of this book was published in 1996 in the series *Science Networks – Historical Studies* Vol. 17.

A CIP catalogue record for this book is available from the Library of Congress, Washington D.C. U.S.A.

Bibliographic information published by Die Deutsche Bibliothek:
Die Deutsche Bibliothek lists this publication in the Deutsche Nationalbibliografie; deatailed biblio-graphic data is available in the Internet at <http://dnb.ddb.de>

ISBN 3-7643-7002-5 Birkhäuser Verlag, Basel – Boston – Berlin

© 2004 Birkhäuser Verlag, P.O. Box 133, CH-4010 Basel, Switzerland
Part of Springer Science+Business Media
Cover design: Micha Lotrovsky, CH-4106 Therwil, Switzerland
Printed on acid-free paper produced from chlorine-free pulp. TCF ∞
Printed in Germany
ISBN 3-7643-7002-5

9 8 7 6 5 4 3 2 1   www.birkhauser.ch

To the memory of my father,
Ricardo Corry Z., ז״ל

Preface to the second edition

Not many recent books on the history of mathematics have seen a second edition, I believe. Thus, I was greatly surprised and pleased when the editors of Science Networks approached me, more than a year ago, asking if I would be interested in publishing a second edition of mine. I was eager to accept, of course, though in the end it took me much longer to complete the task than I had initially assumed.

Since its publication, I have received many positive reactions about the book, and I hope that it has had some significant impact on the current historiography of mathematics. In the present edition, I have not made considerable changes in the overall conception or in the details of the text, except where necessary. I have tried indeed to update the footnotes, by incorporating, wherever possible, references to relevant works that have appeared over the last five years. I have also made a great effort, hopefully to a certain degree of success, to simplify and improve the prose of some linguistically intricate passages, particularly in Chapter 2.

Friends and colleagues have called my attention to many typos and other minor errors in the first edition. I am grateful to them all. Some went even further and were kind enough to provide me with a detailed list of those places that needed to be corrected. I thank especially José Ferreirós, Colin McLarty, and John Allen, for their attentive reading and useful comments. I owe an enormous debt to Walter Purkert for systematically indicating problems with some German quotations scattered throughout the text. Norbert Schappacher provided me with new, important information on the work of Rudolf Fueter on number theory. Miriam Greenfield dispelled some of my doubts concerning the English prose of the text. If there still remain some errors, especially typos or any kind of language oddities, in the present edition, I take full responsibility for them.

Preface to the first edition

This book grew out of a doctoral dissertation submitted to Tel Aviv University in 1990. Initially, I had intended to concentrate on a detailed account of the origins and early development of category theory. Although category theory is not usually included under the label of "metamathematics", it is obviously a metamathematical theory in the etymological sense of the word. Much as proof theory, for instance, is a mathematical theory whose subject matter is a well-determined aspect of mathematical knowledge and of mathematical practice—namely, mathematical proof—so does category theory involve an attempt to characterize and analyze, from the perspective of an elaborate mathematical theory, the organization of mathematical knowledge within different disciplines as well as the interconnections among those disciplines. Many disciplines usually included under the heading of "metamathematics" have been the object of considerable historical and philosophical research; category theory has until now received too little attention.

Metamathematical theories are particularly interesting subjects for historical and philosophical research, since they evince a peculiar trait of mathematics, namely, the possibility of discussing certain meta-disciplinary issues within the body of the discipline itself. Proof theory, which is a rather modern discipline, set out to elucidate in strict mathematical terms what has been a constitutive element of mathematical knowledge at least since the time of the Greeks. Category theory, on the other hand, considers a much more recent idea—the idea of a mathematical structure. During a considerable portion of the present century mathematics has been perceived as a science of structures. Students have made their way in the discipline essentially by becoming gradually acquainted with the different structures that constitute contemporary mathematics. Along the way, they learn similar conceptual schemes and conceptual tools which are equally applied in the study of diverse structures. Category theory may well be characterized as the mathematical theory that seeks a systematic analysis of these different structures, and the recurring mathematical phenomena that arise in them.

Category theory was not the first attempt to formalize the idea of a mathematical structure. An well-know, earlier, one was advanced by Nicolas Bourbaki, a group of mathematicians whose name is typically associated with the idea of mathematical structure. Bourbaki formulated a theory of structures as part of a multi-volume treatise, and one often reads that category theory constitutes an improvement of Bourbaki's theory. Less well-known is a third attempt, advanced by Oystein Ore from around 1935, to provide a general concept of algebraic structure in terms of lattice-theoretical ideas. Ore's program stressed the fact that a general theory of structures must ignore the existence of elements in each individual mathematical system, and that instead it must concentrate on the inter-relations among them. This line of thought, more systematically developed, emerged as a leading principle of category as well.

The existence of these three theories, together with the well-known pervasiveness of the idea of structure in contemporary mathematics, suggests— or at least so it seems to me—that in order to describe the origins of category theory, one could start from the stage at which the conception of mathematics understood in terms of mathematical structure was widely adopted in mathematics. From this vantage point of view, one should then discuss the rise and development of Ore's and Bourbaki's theories, and their relationship, if any, to the rise in category theory.

I had originally planned to follow this line of argumentation. However, it soon appeared to me that the very idea of mathematical structure was a rather fuzzy one, and that the parallel between category theory and other metamathematical theories cannot shed much light on the deeper questions involved. It is not only that mathematical structures are a much more recent feature of mathematical knowledge and practice than, say, mathematical proofs. The idea of a mathematical structure, in spite of it ubiquity in twentieth century mathematics, is an idea whose nature, meaning and role are not clearly understood and have seldom been systematically discussed. Thus, the more one studies texts in which the term "mathematical structure" appears, the clearer it becomes that this notion has been used in different contexts by different persons, with different meanings in mind, or with no clear meaning at all.

This being the case, I decided to broaden the initial scope of my research in order to try to answer some more general historical questions, such as: How can the structural approach to mathematics be characterized? What is a mathematical structure? When, and as a result of which processes, did structures begin to be adopted in mathematical research and practice? When did mathematicians become conscious of this adoption? How did they respond to it?

The present book is an attempt to answer some of these questions, at least partially, and also to suggest a framework for a more comprehensive study of them. Thus, the book is divided into two parts. Part One describes the development of ideal theory from Richard Dedekind to Emmy Noether. This particular development transformed a theory, initially conceived in the context of accounting for factorization properties of algebraic numbers, into a paradigmatic structural theory of modern algebra. I have tried to ascribe a more precise meaning to the term "structural approach" in this development, while explaining the specific differences between Dedekind's and Noether's theories that spell out why the former is "less structural" and the latter a "more structural" mathematical theory. In particular, I have tried to show that the structural approach to mathematics did not come about simply because it became clear that mathematical concepts can be formulated abstractly. This was certainly an important factor in the overall process. However, it was also necessary for mathematicians to reflect on the interesting questions to be asked about such questions as well as on the legitimate, interesting answers to be expected when addressing those questions while using particular kinds of conceptual tools. I have chosen to pursue here these questions in the context of ideal theory. The rise of structural approach to mathematics in general involved a much more complex process, that took place in different disciplines at different paces, motivated by diverse kinds of mathematical concerns.

Part Two deals with three different mathematical theories that may be understood as attempts to elucidate the concept of mathematical structure, namely, those advanced by Ore and Bourbaki, together with category theory. Drawing on the developments presented in Part One. I describe here the roots of three attempts, and their historical and conceptual interrelations.

My initial plan for dealing with the rise of category theory was thus essentially transformed: I describe in much less detail than originally thought the direct roots and the initial stages of category theory itself. On the other hand, I have attempted to provide a wider perspective from which the detailed account should be considered. Such a detailed account thus remains for a future undertaking.

In writing this work—first as a doctoral thesis and later as a book—I have received help and support from many persons to whom I am deeply grateful. First and foremost, I owe a great debt to Sabetai Unguru, who introduced me into the study of the history of mathematics and supervised the writing of my dissertation. His wise guidance and constant encouragement have been a stimulating source of support.

In the long and winding process of transforming my dissertation into a finished book, I have enjoyed the invaluable editorial help of David Rowe. He generously put at my disposal much more of his time, professional expertise, and of his uncompromising critical abilities than I could have deservedly asked for. His many suggestions and criticisms made me rethink considerable portions of the book, as well as its overall structure. The present form of the book, however, should not be taken in any way as representing his own views on the issues discussed. In fact, he explicitly expressed his disagreement with the options put forward in some sections of the book. Still, where he agreed and where he disagreed, his comments always helped me to improve my arguments and presentations.

I owe special thanks to Professor Saunders Mac Lane for insightful remarks on a draft of chapters 8 and 9, as well as for useful material he sent to me, and to Professor Samuel Eilenberg, for an illuminating conversation in Tel Aviv in 1986, on the occasion of his being awarded the Wolf Prize.

Several people have been kind enough to read and comment upon portions of earlier versions of the manuscript. I thank them all for their helpful remarks: Liliane Beaulieu, Pierre Cartier, Newton C.A da Costa, Andrée Ch. Ehresmann, José Ferreirós, Catherine Goldstein, Ivor Grattan-Guinness, Ralf Haubrich, Giorgio Israel, Colin Mclarty, Herbert Mehrtens.

I have benefited from information and opinions kindly communicated to me, in answer to my letters, by Professors Garret Birkhoff, the late Jean Dieudonneé, Harold Dorwart, Harold Edwards, Solomon Feferman, Peter Hilton, Jean-Pierre Serre and Pierre Samuel.

Naturally, none of the persons mentioned above should be held responsible for the mistakes, excesses or shortcomings that the reader may still find in the book.

Professor Andrée Ch.Ehresmann, at Amiens, kindly placed at my disposal original, unpublished material belonging to the estate of her late husband, Professor Charles Ehresmann. I thank her very much for her help, and for having granted me permission to quote from these documents.

Portions of Chapter 7 and 8 appeared originally in my article "Nicolas Bourbaki and the Concept of Mathematical Structure", *Synthese* 92, 1992, pp. 315-348. I thank Kluwer Academic Publishers for allowing their reproduction (with considerable changes) here.

I also thank Dr.Helmuth Rohlfing from the Handschriftenabteilung, Niedersächsiche Staats- und Universitätbibliothek Göttingen, for his helpful advise and for permission to quote from David Hilbert's *Nachlass* (Cod.Ms.Hilbert 558, *Mechanik*, WS 1898-99). Likewise, I thank Mr. Mathees, librarian of the Mathematisches Institut, Universität Göttingen, for permission to quote from a manuscript of Hilbert's lecturers *Logische Principien des Mathematisches Denkens*, SS 1905.

Ayelet Raemer helped me revising the English prose of the final version. I enormously appreciate her kind help. Any Hispanisms, Hebraisms and other linguistic oddities that may have remained in the text are all of my exclusive choice and responsibility. All translations into English that appear in the text are mine, unless otherwise indicated.

In preparing the camera-ready copy of the book, I benefited from the kind help of Yonatan Kaplun, Ido Yavetz and Studio Itzuvnik, and from the technical guidance of Doris Wörner from Birkhäuser Verlag. I thank them all.

Most of the research and the writing of this book was done in the environs of the Cohn Institute for the History and Philosophy if Science and Ideas, Tel Aviv University. During the years that I have been connected with the Cohn Institute I have had the opportunity to meet many people, who contributed to shaping my own ideas and approach to the history of mathematics in ways that cannot be acknowledge in footnotes. I wish to thank especially Yehuda Elkana for constant support and encouragement, and Gabriela Williams for her devoted help. Shmuel Rosset, of the School of Mathematics, Tel Aviv University, has been faithful interlocutor on mathematical, as well as on other issues over many years. His critical attitude towards my work has been of great help.

Last, but certainly not least, I want to thank all those who constitute my non-academic environment, and who have provided over the years a congenial atmosphere for carrying out my work: all my friends in Kibbutz Nirim, my family, and, above all, my dear wife Efrat.

Tel Aviv – Nirim, October 1995

Table of Contents

Introduction:
Structures In Mathematics

It is commonplace for mathematicians and non-mathematicians alike to refer to the structural character of twentieth century mathematics, or at least of considerable parts of it. Such references can be found in mathematical texts,[1] as well as in historical accounts[2] and in philosophical or semi-philosophical debates[3] concerning the discipline. However, when one attempts to discern the meaning attached to the term "mathematical structure" in the various places where it appears, one soon realizes that, while most writers using it take its meaning for granted and feel no need to add further clarifications, they often in fact ascribe diverging meanings to the term. In some cases, they ascribe to it no clear meaning at all.

The idea of a mathematical structure, its meaning, its origins and its development constitute the main themes of the present book. What is indeed a mathematical structure? Or, more precisely, how can the structural approach to mathematics be characterized? How did the idea of structure arise and develop within mathematical research and what was its actual influence? In which ways have mathematicians attempted to elucidate this idea and its place in mathematics? These are the main questions addressed here.

In order to answer these questions, the present account will focus—in the first place—on the evolution of algebra between 1860 and 1930, especially in Germany. Existing historical accounts have stressed the existence of several significant trends at the center of this evolution. As the most important among

---

1. For instance, from among books appearing in the bibliography, Hall 1966, 2; Kurosh 1963, 5; Lawvere 1966, 1; Safarevich 1974, v.

2. Bell 1945, 245-246; Bos 1993, 178; Dieudonné 1979; Dieudonné 1985, 59 ff.; Dieudonné 1987, 160-166; Mac Lane 1980, 362; Mehrtens 1990, 319-326; Scholz (ed.) 1990, 418-424; Wussing 1984, 15; Wussing 1989, 243-245.

3. Aspray & Kitcher 1988, 14; Grattan-Guinness 1992; Lane (ed.) 1970, 20; Mac Lane 1987a, 34; Maddy 1980; Resnik 1981, 1982; Shapiro 1983.

them the following are usually mentioned: (1) the penetration of methods derived from Galois's works into the study of polynomial equations; (2) the gradual introduction and elucidation of concepts like group and field, with the concomitant adoption of abstract mathematical definitions (a trend associated with the works of Richard Dedekind, Heinrich Weber, Georg Ferdinand Frobenius, Ernst Steinitz and others); (3) the improvement of methods for dealing with invariants of systems of polynomial forms, advanced by the German algorithmic school of Paul Gordan and Max Noether; (4) the slow but consistent adoption of set-theoretical methods and of the modern axiomatic approach implied by the works of Georg Cantor and of David Hilbert; (5) the systematic study of factorization in general domains, conducted by Emmy Noether, Emil Artin and their students.

The intense activity developed in the framework of these trends by an ever increasing number of researchers, not only in German universities but also (though to a lesser extent) in France and in England, contributed to the unprecedented enlargement of the body of accepted algebraic knowledge. Thus, towards the end of the 1920s, one finds a score of recently consolidated theories, usually aimed at investigating the properties of abstractly defined mathematical entities, as the focus of interest in algebraic research: groups, fields, ideals, rings, and others.

The individual trends and theories involved in this process have to various degrees been the subject of detailed historical research. Nevertheless, it seems that the evolution of the notion of a *mathematical structure itself*, as well as its connection with the above-mentioned process and with its outcome, have not been yet properly elucidated. The reason for this lies perhaps in the noteworthy, though sometimes overlooked, differences between this idea of a mathematical structure and most other mainstream mathematical ideas that have played central roles in the development of mathematical knowledge and practice during the last two centuries. While historical research has traditionally focused on the latter kinds of ideas, this book attempts to explain in what sense was the evolution of the idea of a mathematical structure different and to describe this evolution. In order to do so, however, it is first necessary to begin with a brief methodological introduction that provides a general background and clarifies some terms that will be used in the forthcoming chapters.

Scientific disciplines give rise to two, more or less discernible sorts of questions. The first sort includes questions about the subject matter of the discipline. The second sort comprises questions about the discipline *qua* discipline, or meta-questions. To answer questions of the first sort is always among the aims of any given discipline and, obviously, practitioners of that discipline are usually engaged in such an activity. Concerning the questions of the second sort, however, whereas one may certainly find some investigators consciously attempting to answer them, one may also find others only implicitly or tacitly answering them. Still others ignore those questions altogether. One may even encounter individuals who try to avoid dealing with them.

There are some statements which can easily be classified as being answers to either of the two above mentioned sorts of questions. For other statements, however, it is harder to establish whether they are answers to questions about the subject matter, or about the discipline *qua* discipline. Each of Newton's three laws, for instance, clearly belongs to the first category; all three are statements about how bodies move. The claim that Copernicus's system is "simpler" than Ptolemy's clearly belongs to the second: it is a claim about astronomical theories rather than a claim about heavenly bodies. Gödel's theorems are deep results *within* a specific branch of mathematics, but they may also be taken as claims *about* mathematics—the discipline.

Two domains of discourse can accordingly be tentatively identified when speaking about any scientific discipline; they can be described schematically as the "body of knowledge" and the "images of knowledge." The body of knowledge includes statements that are answers to questions related to the subject matter of the given discipline. The images of knowledge, on the other hand, include claims which express knowledge about the discipline *qua* discipline.

The body of knowledge includes theories, 'facts', methods, open problems. The images of knowledge serve as guiding principles, or selectors. They pose and resolve questions which arise from the body of knowledge, questions which are in general not part of and cannot be settled within the body of knowledge itself. The images of knowledge determine attitudes concerning issues such as the following: Which of the open problems of the discipline most urgently demands attention? What is to be considered a relevant experiment, or a relevant argument? What procedures, individuals or institutions have authority to adjudicate disagreements within the discipline? What is to be taken as the legitimate methodology of the discipline? What is the most efficient and illuminating technique that should be used to solve a certain kind of problem in the discipline? What is the appropriate university curriculum for educating the next generation of scientists in a given discipline? Thus the images of knowledge cover both cognitive and normative views of scientists concerning their own discipline.

In conjunction with the analysis of their interaction, this schematic separation of scientific knowledge into two domains of discourse may provide a useful perspective for the study of the history of science in general.[4] In the particular case of mathematics, it brings to the fore a peculiar trait of this discipline, which will be of special interest for the discussion advanced in the present book: the possibility of formulating and proving metastatements about the discipline of mathematics, from *within* the body of mathematical knowledge.[5] This capacity of mathematics of studying itself mathematically will be referred to in what follows as "the reflexive character of mathematics."[6]

The distinction between body and images of knowledge should not be confused with a second distinction that has often been admitted, either explicitly or implicitly, by historians of mathematics: the distinction between "mathematical content" and "mathematical form."[7] Such a distinction has been criticized on the grounds that a clear separation between mathematical form and content is impracticable, historically unilluminating, and indeed misleading.[8] The distinction between body and images of knowledge, however, is of a different kind. The borderline between the two domains is blurred and historically conditioned. Moreover, one should not perceive the difference between

---

4. The centrality of meta-issues for the history of science and the terminology "body of knowledge" and "images of knowledge" have been taken from the work of Yehuda Elkana. These concepts arose in the framework of an ambitious program aimed at an anthropological characterization of scientific knowledge as a cultural system (Cf. Elkana 1981; 1986). They are introduced in this book, however, with much more limited aims in mind.

5. The implications of applying this scheme for mathematics have been elaborated in Corry 1989.

6. A somewhat similar analysis has been advanced by Mario Bunge (Bunge 1961) concerning physics. Bunge focuses on a particular class of statements which he calls 'metanomological'; these are 'lawlike statements about scientific laws', and comprise factual as well as normative ones. Some of Bunge's claims could be reformulated to make them relevant to the case of mathematics and to the present argument.

7. The distinction has been explicitly stated, for instance, by Michael Crowe (1975, 19) who claimed that "there are no revolutions *in* mathematics", thus separating "mathematics proper" from "incidental factors" such as "nomenclature, symbolism, metamathematics."

8. Mehrtens 1976; Rowe 1993, 321.

the body and the images of knowledge in terms of two layers, one more important, the other less so. Rather than differing in their importance, these two domains differ in the range of the questions they address: whereas the former answers questions dealing with the subject matter of the discipline, the latter answers questions about the discipline itself *qua* discipline. Thus, unlike the rather artificial distinction between the form and content of mathematical ideas, the relationship between body and images of mathematics resembles that between a text and its context. The body and the image of mathematics appear as organically interconnected domains in the actual history of the discipline. Their distinction is undertaken for analytical purposes only, usually in hindsight. It is a task for the historian of mathematics to characterize the images of knowledge of a given period and to explain their interaction with the body of knowledge, on the one hand, and the subsequent transformations in both the body and the images of mathematics, on the other.

The reflexive character of mathematics and the interaction of reflexive mathematical thinking with both the body and the images of mathematics have manifested themselves variously throughout history. Passages discussing the proper division of mathematical science into subdisciplines are found in ancient mathematical texts from the Pythagoreans in the fifth century B.C. and up until Pappus and Proclus in the fourth and fifth centuries A.D., respectively. The changing opinions on these issues had direct repercussions on various aspects of mathematical practice, including the kinds of problems addressed by Greek mathematicians, the techniques they applied to solve them, and the answers they were able to obtain. In the seventeenth century one finds a classical instance of meaningful reflexive mathematical thinking, with significant consequences on mathematical practice and knowledge, in Descartes's writings on the classifications of curves and permissible methods in geometry.[9] But it was in the early decades of the present century that reflexive mathematical thinking witnessed an impetus of unprecedented scope and consequences.[10] In fact, it was in this period of time that reflexive thinking became "canonized", as it were, through the establishment of "metamathematics" (in its various manifestations) as an autonomous domain of mathematical research. The achievements of metamathematics not only directly enhanced

---

9. This issue is discussed in Bos 1981. Bos obviously does not use the terminology introduced here.

our understanding of many basic aspects of mathematics as a system of knowledge, but they also shaped many of the dominant images of twentieth century mathematics. To a large extent, they account for the dominance attained since the first decades of the present century by foundationalist positions in the philosophy of mathematics. These positions are characterized by the tendency to reduce philosophical debates about mathematics to foundational research, namely to *mathematical* research in branches that happen to deal with problems related to "foundations."[11]

The mathematical success of reflexive theories since the beginning of the twentieth century promoted a new and influential image of mathematics, namely, the view that meta-issues in mathematics are best elucidated by creating "reflexive" mathematical theories aimed at providing clear-cut answers. Actually, only relatively few meta-issues have been settled in this way, especially those concerning decidability and effective calculability of mathematical propositions. Yet, the impact of such results has gone well beyond their relatively limited scope. They have underscored the very possibility of settling once and for all certain open questions about mathematics and have thus set a standard to be met by other pronouncements on similar issues. An extreme version of this view even suggests that *the only* meaningful assertions about mathematics should be gained through technical mathematical research.[12] In fact, certain twentieth century mainstream trends in Anglo-American philosophy, especially those connected with the analytical approaches and with logical positivism, can be seen as parallel to the dominance of this image of mathematics.

Certain trends in the historiography of mathematics during the present century have also been influenced by the idea that meaningful debate about the body of mathematics should be done only through strict mathematical argumentation. These trends suggest that history of mathematics should be written

---

10. A thorough and ambitious study of the development of early twentieth century mathematical thinking, with special emphasis on thinking *about* mathematics appears in Mehrtens 1990. Many of the central issues discussed by Mehrtens touch upon the interaction between body and images of knowledge, and upon the issue of reflexive mathematical knowledge. Although Mehrtens's discussion throughout his book is not couched in those terms, he explicitly states that the debate between moderns and countermoderns in mathematics (one of the central issues addressed in the book) is a controversy about images of knowledge (Mehrtens 1990, 420).

11. This issue is discussed in detail in Tymoczko (ed.) 1985 (especially in the introductory section and in pp. 1-8) and in Kitcher 1988, esp. 294-298. See also Mehrtens 1990, 421-424.

12. On this point, see Bishop 1975, 515.

only by mathematicians (or, at least judged only by mathematicians) and that historical research about mathematics is meaningful only inasmuch as it throws new light on present mathematical research, by examining ideas elaborated by mathematicians in the past with current mathematical tools.[13] The consequence of this view has been that the historiography of mathematics has been relatively dominated by a positivistic perspective, especially when compared with recent trends in other scientific disciplines. This perspective has tended to present the history of the discipline as an uninterrupted accumulation of successful discoveries leading inexorably from one theory to the next,[14] up to the present state of a particular mathematical discipline.[15]

Now, both recognizing the interaction between body and images of mathematics as an important factor in the development of the discipline and studying the peculiar changes that separately affect both domains of discourse may open a useful perspective for historical research. Concerning particular episodes in the history of mathematics, this perspective may direct the attention of historians to details and trends which might otherwise perhaps have been overlooked. In particular, they may underscore the influence of the images of knowledge characteristic of a discipline at a certain period of time on the creation, growth, relative evaluation, and eventual oblivion of parts of the body of mathematics.[16] This scheme turns out to be especially useful in studying the rise and development of the notion of a mathematical structure, as will be seen in the chapters of the present book.

---

13. This position is articulated in Weil 1980. See also Askey 1988. For a different opinion see Grabiner 1975, May 1975, and, more recently, Dauben 1994, each of which could have been titled "A Historian-of-Mathematics' Apology." Their main point is that there is room for independent historical research of mathematics which is at least as important as that carried out by mathematicians themselves. This issue is considered in similar terms in Lorenzo 1977, 33.

14. Mehrtens 1990, 140, explicitly establishes a direct connection between the reflexive capabilities of mathematics, the success of Hilbert's metamathematics and the positivistic historiography of mathematics: "Die Möglichkeit der Mathematik, sich selbst zu thematisieren, von Hilbert mit der axiomatischen Methode und der Metamathematik ins Bewußtsein gehoben, erlaubt ihr auch, ihre Geschichte rekurrent zu rekonstruieren und als den einem großen Strom von Problemen und Lösungen darzustellen, der Diskontinuitäten und Sackgassen nicht kennt—als eine große Erfolgsgeschichte der mathematischen Vernunft."

15. As Ivor Grattan-Guinness (1990, 157) has put it: "[T]hey usually view history as the record of a 'royal road to me' ... In other words, they confound the question, 'How did we get here?', with the different question, 'What happened in the past?.'"

16. For a more detailed discussion on historiography, in terms of body/images-of-knowledge see Corry 1989, 434-436; Corry 1993.

This book elaborates upon the idea that the nature and the evolution of the structural approach to mathematics are best understood when this approach is seen as a specific image of mathematical knowledge. Like any other image of scientific knowledge, this one evolved from a particular historical process. In this case, the process involves the evolution of algebra, and the interaction between the body and the images of knowledge in this discipline between 1860 and 1930, mainly in Germany.

As already said, algebraic research in this period produced an unprecedented growth in the body of knowledge. This growth in the body of knowledge raised many pressing questions concerning the very nature of algebra. Mathematicians involved in algebraic research ought to come to terms in their works with questions such as these: What are the relevant mathematical entities that should be studied, and how are they interrelated? What is the proper way to define these entities and what are the relevant questions that should be asked about them? What kind of methodological and conceptual tools should be used in solving these questions? What kinds of answers are to be expected in answering the main research questions? What branches of mathematics are to be most urgently investigated? How should the textbooks on these issues be written?

During the late 1920s an innovative and coherent way of answering the above questions gradually crystallized mainly in Germany, and especially among Emmy Noether's Göttingen students and colleagues. In essence, a new image of knowledge evolved. This image was put forward and made known to a wider mathematical audience in 1930 in the form of a new textbook of algebra: Bartel L. van der Waerden's *Moderne Algebra*. Like most mathematical textbooks that eventually become classics, *Moderne Algebra* assembled the important results that had been obtained during the last decades of research in its domain of concern and exposed them in a systematic, and didactically clear, fashion. But beyond that significant achievement, this textbook put forward a new image of the discipline that implied in itself a striking innovation: the structural image of algebra. In the forthcoming account, it is this specific, historically conditioned image of mathematical knowledge that will be considered as implicitly defining the idea of a mathematical structure. In order to understand the nature of this image and the process that brought about its creation and widespread adoption, it is necessary to examine thoroughly the kinds of answers that the structural image provides to the disciplinary questions posed in the preceding paragraph. It is also necessary to describe the process leading to the consolidation and implementation of precisely those answers, rather than any alternative ones. Such an examination constitutes the main subject of the first part of the present book.

The structural image of algebra construes the self-identity of this discipline in terms of the implicit idea of an algebraic structure. The essence of the structural approach to algebra lies in the recognition that it is mathematically enlightening to conceive a handful of concepts (groups, rings, fields, etc.) as individual "varieties" of the same mathematical "species" (both words intended here with their biological, not mathematical, connotation), namely, the species of algebraic structures. With the adoption of this approach, the study of algebraic structures gave algebraic research a new focus, subsuming under it the traditional tasks of the discipline, namely, the study of polynomial forms and polynomial equations, and the problem of solvability of polynomial equations. Moreover, under the new approach to algebra, this discipline came to cover under its unified scope, the study of other related, but theretofore separated domains of research, particularly algebraic number theory. The study of algebra in terms of algebraic structures thus opened a new kind of mathematical horizon, which an ever increasing number of leading mathematicians found to be worth pursuing.

Moderne Algebra had an enormous impact on mathematicians not only in Germany but also in France and in the USA. This impact led to a widespread adoption of the structural approach in algebraic research and to a thoroughgoing reelaboration of the existing body of knowledge of the discipline. It turned out that by pursuing research under the view promoted by the structural image of algebra, many open problems were solved in an economic and elegant way. Moreover, the structural approach to algebra enabled mathematicians to present problems which had been solved previously, in a new and interesting light. It also led to formulate new interesting problems, and, more generally, it seemed to open new perspectives for research before the mathematicians who adopted the approach. Moreover, parallel developments that took place in other mainstream mathematical disciplines, such as topology and functional analysis, led to the adoption of what was considered to be a structural perspective in these disciplines.

In the mid-1940s, the gradual adoption of the structural image in various central mathematical disciplines yielded a new overall conception of the nature, the aims and the organization of the *whole* of mathematical science. Under this new conception the idea soon arose that mathematical structures are the actual subject matter of mathematical knowledge in general. Now, while research in many central branches of mathematics began to be conducted under the structural image, certain mathematicians addressed a more specific question: What, in fact, is a mathematical structure? Several attempts to answer this question in terms of a reflexive mathematical theory were pursued, with varying degrees of success. These attempts are analyzed in the second part of this book.

The two parts of this book jointly describe the rise of a specific image of mathematical knowledge, at the center of which stands the implicit idea of an algebraic structure, and the reflexive theories elaborated in order to elucidate this latter idea. One should hasten to add, however, that the structural approach to mathematics has had its own evolution during the twentieth century. It has been embraced with varying degrees of enthusiasm and of effectiveness at different times, at different places, and in different mathematical disciplines. This more recent evolution will not be studied in detail here. However, it would be interesting to pursue the question to what extent have the achievements and drawbacks of the reflexive mathematical theories of "structure" analyzed in the second part of this book affected the actual adoption of the structural approach (intended in the general, non-formal meaning) in increasing or decreasing degree in the various branches of mathematics during the twentieth century. Our analysis of the rise of the structural approach is meant to put forward the claim that the structural character of mathematics can only be referred to and understood if one carefully separates the two meanings of the term. On the one hand, the non-formal, usually implicit, idea of a structure as the focus of research of a mathematical discipline that has adopted the structural approach. On the other hand, several possible ways to formulate elaborate mathematical theories that presumably encapsulate the mathematical essence of the former idea.

The remaining pages of this introduction describe cursorily how the various chapters of this book contribute to fulfilling its aim. Realizing the possibility to define central concepts in mathematics, and in particular in algebra, by means of systems of abstract postulates defined on a non-empty, abstractly conceived set, was an important, and indeed necessary stage in the way to the consolidation of the structural approach. In fact, this stage has received considerable attention in existing historical accounts of the rise of the individual domains that together constitute modern, structural algebra. But the historical depiction of the development of the idea of an algebraic structure remains incomplete if it considers only this particular feature of the more involved, actual historical process. As will be seen, the technical *possibility* of defining a concept in abstract axiomatic terms was attained much earlier than the understanding of the *usefulness* of placing this kind of definition at the focus of attention in mathematical research. Thus, it is not enough to describe the gradual adoption of abstract formulations for the main concepts of algebra. A complete account of the process must also explain how the main kinds of questions, conceptual tools, and expected, legitimate answers that characterize the structural approach to algebra came to take their central position in algebraic research. Moreover, it will be shown that the consolidation of the structural approach hinged heavily upon the recognition of significant common traits in the conception of various domains that were previously seen as separate. It was this recognition that suggested the convenience of investigating these domains from a unified perspective. Therefore, both the *evolution of the individual instances* of algebraic structures, and the gradual *interplay between them* should be conceded special attention. The five chapters that constitute the first part of the book are intended to provide an outline of the rise of the structural approach following these guidelines.

Chapter 1 explains more in detail what is meant by the structural approach to algebra. This is done by analyzing the images of knowledge characteristic of algebraic research in the late nineteenth century, and by comparing it with the images of knowledge embodied in *Moderne Algebra*, the first truly structuralist presentation of the discipline. The rise of the structural approach, and its gradual adoption as the canonical conception of algebra is discussed in Chapters 2 to 5, by focusing on the development of ideal theory between the works of Richard Dedekind and those of Emmy Noether. This particular development transformed a theory initially conceived as a tool to investigate factorization properties of algebraic numbers (i.e., a non-structural theory), into a paradigmatic structural theory of modern algebra. This transformation was accompanied by a process, through which the theory merged organically with what then constituted a separate domain, namely, the theory of factorization of polynomial forms. In our account we consider two prominent milestones of this transformation: the impact of David Hilbert's contributions (Chapter 3) and the early definitions and research of abstract rings by Abraham Fraenkel (Chapter 4).

Part Two discusses various reflexive attempts to formulate mathematical theories of structure. The first of these attempts was led by Oystein Ore at Yale between about 1935 and 1945. Ore developed together with several collaborators a general concept of algebraic structure, based on lattice-theoretical ideas. This research program is discussed in Chapter 6. Chapter 7 discusses a second such attempt: the one advanced by Nicolas Bourbaki. Bourbaki's name has usually been associated, more than any other, with the idea of structure in mathematics. As will be seen in some detail, Bourbaki's work brings to the fore a remarkable interplay between the general, non-formalized idea of a mathematical structure and the axiomatically formalized one. The third attempt discussed in Part Two is the one connected with the theory of categories and functors. The initial stages of the theory of categories crystallized in the USA between 1943 and about 1960. These stages are analyzed in Chapter 8. Category theory is the most elaborate and successful instance of an axiomatized theory allowing for a systematic characterization and analysis of the different structures, and the recurring mathematical phenomena that come forward in the latter.

Although these three theories are usually not included under the label of "metamathematics", they are obviously metamathematical theories in the etymological sense of the word. Metamathematical theories have often proven particularly interesting for historical and philosophical research. Perhaps one main reason for this is the fact that they bring to the fore the reflexive capabilities which are peculiar to mathematics. In fact, most disciplines that are usually included under the heading of "metamathematics" have been the subject of recent detailed analysis. It is thus remarkable that this has not been the case for the reflexive theories of structure. The present book is intended as a contribution to the elaboration of this direction.

The account presented in this book can be thus characterized as an outline of the rise of the structural approach to mathematics. This outline comprises three stages. First, we elucidate the nature of the structural approach and ascribe a more precise meaning to the term. Second we describe the way in which a particular branch of algebra evolved until attaining its structural character. Third, we describe the early stages of three attempts advanced after the consolidation of the structural approach, meant to engulf within reflexive mathematical theories belonging to the body of knowledge, the idea that the disciplinary concern of mathematics is, in fact, the research of mathematical structures.

Part One: Structures in the Images of Mathematics

The first part of the present study outlines the origins and development of the conception according to which algebra is the mathematical discipline dealing with algebraic structures. This is done by focusing on the evolution of the concept of ideal and its application to the research of factorization properties, beginning in the context of algebraic number theory with the work of Richard Dedekind, and culminating in the context of an abstract, structural theory with the work of Emmy Noether. It must be stressed from the beginning, however, that the actual process of gradual adoption of the structural approach was not confined to ideal theory. More generally, the adoption of a structural approach and the rise of the idea of structure as a central concern of mathematics was in fact not restricted to algebra. The reader should thus keep in mind that the developments described here are part of more comprehensive changes affecting other branches of mathematics as well. In fact, accounts similar to the one advanced here seem to be equally possible, first for other domains of algebra, and then, for other mathematical disciplines, particularly topology and functional analysis.[1]

Moreover, the larger space and the considerably detailed attention conceded in our account to the works of Dedekind and of Noether, at the expense of their contemporaries, should not in itself be considered as a value judgement. Central as their respective contributions to algebra certainly were, there were many other major figures that played a key role in the development of algebra in the period of time considered here. Among them one could mention Heinrich Weber, Paul Gordan, Max Noether, Georg Ferdinand Frobenius, Otto Hölder, Issai Schur (whose contributions are only tangentially referred to in this book). Selecting Dedekind and Noether as focus of the account corre-

---

1. For existing accounts on the evolution of these latter disciplines see Bernkopf 1966; Birkhoff & Kreyszig 1984; Browder 1975; Fingerman 1981; Monna 1973; Siegmund-Schultze 1981.

sponds more to the high visibility of structural elements gradually introduced into the research of factorization properties in their works, than to any assessment of their relative value as compared to others that receive less attention here. The present study examines thus a cross-section of mathematics in the period considered, and it is not intended as a comprehensive account of it.

Although a parallel account could have been given, which focuses on a different concept or theory among those which were later to be considered as the core of "modern algebra" (groups, fields, hypercomplex systems, etc.), or upon the works of mathematicians other than those selected here, there are several reasons justifying the choice of ideal theory, and the relevant works of Dedekind and of Noether on this subject, as the focus of the present story. The most immediate reason is that, while detailed studies on the history of the rise and development of other algebraic concepts already exist,[2] a parallel account of ideal theory has not yet been written. To be sure, important research has been done on the beginnings of ideal theory, especially by Harold Edwards and by Ralf Haubrich.[3] There are also scattered accounts of Emmy Noether's work on ideals, appearing as part of general accounts of her work[4] or of the rise of modern algebra.[5] But a more focused investigation of the evolution of ideas leading from Dedekind's work to Noether's, including the subsumption of the concept of ideal under an abstract theory of rings, appears here, to the best of our knowledge, for the first time.

The second reason for the choice of ideal theory as the focus for the present study concerns the general character of our inquiry. The main aim of this first part of the book is to describe the rise of the structural approach to algebra in terms of the adoption of a new image of algebraic knowledge. Central to the view advanced here is to describe the rise of the structural approach, not merely as the gradual adoption of abstract, axiomatic formulations of the concepts introduced in the various theories involved. Rather, stress is laid upon the gradual recognition of those theories as individual, interrelated, varieties of a more general species, namely, the species of algebraic structures. In the structural approach, these varieties deserve to be investigated from a common perspective by asking similar questions about them and attempting to answer those questions with similar conceptual tools. In order to develop such

---

2. See, e.g., Crowe 1967, Mehrtens 1979, Purkert 1971, Wussing 1984.
3. See especially Edwards 1980; Haubrich 1992, 1996.
4. For instance, Jacobson 1983, Kimberling 1981.
5. For instance, Scholz 1990 (ed.), 403-407.

claims, it is necessary to stress the varying kinds of interplay between the various kinds of algebraic structures. The theory of ideals seems especially well suited for this purpose.

Most types of algebraic structures developed within particular contexts which stressed the specific role played by them within individual theories, rather than their interrelations with other kinds of structures. Thus, for instance, quaternions and related hypercomplex systems developed in close connection with physical theories. The concept of group, to take another example, emerged in connection with Galois theory, on the one hand, and with the study of geometrical transformations, on the other. Fields arose in the framework of research on algebraic numbers, where they were seen as distinguished subdomains of the system of complex numbers. The close interplay and the conceptual affinities between these and other kinds of theories became a focus of interest only gradually, at various stages beginning from the first decade of the twentieth century, and often as a direct consequence of the increased adoption of the structural approach rather than as a trigger for it. The case of ideals is somewhat different in this respect. From the beginning of its development, its interplay with other, newly introduced algebraic concepts, such as fields, modules, groups, lattices, and polynomial rings, was clearly manifest. In fact, this interplay was central to the main concerns of the theory. Thus the theory of ideals had a rich pre-history, characterized by strong interconnections with other incipient structures, even before turning into a useful tool of central importance for the theory of abstract rings.

The abstract concept that provided the framework for the axiomatic study of ideals, and that enabled their incorporation into the picture of algebra as a hierarchy of structures, was the concept of ring. This latter concept did not arise from within the theory of ideals itself, or even in direct connection with it, but from rather different concerns (as will be seen in Chapter 4). Thus, in studying the evolution of the theory of ideals as a tool for research of factorization properties, one sees intrinsic reasons for a shift of interest, which are independent of any autonomous attempt to provide an abstract axiomatic formulation of a particular concept. The evolution of the theory of ideals thus helps stress the intrinsic interest gradually posed by structural questions, even when formulated within more classical, "concrete" mathematical entities.

A further reason for focusing on ideals concerns the prominence of the mathematicians involved in its development and the strong influence that they exerted on each other. The meaning of "the structural image of algebra" is explained in Chapter 1 by examining the picture of the discipline as presented in van der Waerden's *Moderne Algebra*. It is well-known that the most decisive influence upon van der Waerden's conception came from the algebraic ideas of Emmy Noether (as well as from Emil Artin), and thus, indirectly, from Dedekind. David Hilbert (whose influence is discussed in Chapter 3) played a decisive role in promoting Dedekind's view as the established approach to algebraic number theory. Emmy Noether came to Göttingen in 1915 to work with Hilbert. Obviously, her work was strongly influenced by his conceptions, although, as will be seen in later chapters, Hilbert's own image of algebra was quite different from the structural one. Thus a coherent picture is advanced, beginning with Dedekind's work, moving to Noether, through several intermediary works, and culminating in van der Waerden's textbook.

The choice of ideal theory as the primary focus of our account thus seems to be well justified, in spite of the obvious limitations of any delimited account of a process affecting such a wide range of issues in the history of nineteenth-century mathematics. In addition to its intrinsic justification, our analysis of ideal theory will enable a natural passage from the first to the second part of the book. First, in Chapter 6 we analyze Oystein Ore's program to unify algebra using lattice-theoretical ideas. Ore collaborated with Emmy Noether in the edition of Dedekind's work, and one can identify a clear influence of the latter on Ore's program. Second, there is a natural path leading from van der Waerden's textbook to the work of Nicolas Bourbaki, examined in Chapter 7. To a considerable extent, one of the main influences leading to Bourbaki's conception was the impact that van der Waerden's book exerted on a group of young French mathematicians during the early 1930s, driving them away from the then dominant French mathematical traditions and towards the German approach epitomized in that book. Bourbaki's work will thus be described as an extension of van der Waerden's image of algebra to the whole of mathematics. Thus, in spite of the basic differences between the issues considered in the two parts of the present study, there is also an essential continuity among them.

It also seems necessary to add here a general clarification concerning the terminology used in the first part of the book. One of the central aims of our analysis here is to show that the conceptual proximity attributed nowadays to the mathematical entities included under the heading of "algebraic structures" was actually absent from nineteenth-century images of mathematics. Therefore the very use of the unqualified term "algebra" throughout the book might be seen as somewhat problematic in the present context, since the discipline it denotes in the first part, differs from that in the second part. Moreover, the central figures discussed in the forthcoming chapters (Dedekind, Weber, Hilbert, Noether, etc.) were mathematicians with an astonishingly wide mathematical background and spectrum of interests. They cannot be called algebraists in the modern sense of the word or even in the sense attributed to the word in the nineteenth century. In fact, most of classical algebra was strongly connected with other parts of mathematics, and so are the particular works considered here. Nevertheless, by bearing in mind these qualifications, necessary in fact for any analysis in which present disciplinary categories are projected into the past, the use of the term algebra should not raise any problem in following the line of argumentation without confusion. On the one hand, we stress the shift of interests in a certain area of mathematical research. On the other hand, we describe how certain concepts and theories, especially those connected with factorization problems and ideal theory, came to be considered as special instances of the more general idea of "algebraic structure".

Part One comprises five chapters. Chapter 1 describes the images of knowledge that dominated algebraic research in the second half of the nineteenth century, as they came forward in the leading textbooks of the period, and contrasts them to the structural image of algebra manifest in van der Waerden's book. This analysis is meant to explain the nature of the structural approach to algebra. Chapter 2 describes in some detail Dedekind's work on ideals, while placing it in the wider context of his research at large, which also included Galois theory, algebraic function theory, Dedekind's early research on lattices, and his foundational research on the systems of natural and real numbers. This analysis allows understanding the peculiarities of Dedekind's conceptions and techniques, which are in many respects similar to those characteristic of twentieth-century research. At the same time, however, it shows how the different algebraic concepts introduced by Dedekind in his works (ideals, fields, groups, modules, lattices) belong, in his images of mathematics, to different conceptual levels and play differently conceived roles in the theories in which they appear. Chapter 3 discusses the impact of David Hilbert's work on early twentieth-century mathematics, and analyses with special attention the almost complete absence of specifically structural concerns in it. At the same time, however, it also analyses its indirect, yet significant contribution to the shaping of the structural view of algebra. Chapter 4 describes the rise and the early study of the abstract concept of ring, its roots in the study of *p*-adic systems, and the influence of Ernst Steinitz's early research on abstract fields. Contemporary parallel development in the theory of polynomial forms are also discussed here, in order to prepare the background for the work of Emmy Noether. This is discussed, finally, in chapter 5, laying special stress on her motivations and in the way that structural concerns already dominate her work.

Of the five chapters of Part One, chapters 2 and 5 include some more technically detailed passages concerning factorization of ideals (especially in §§ 2.2, 5.2 and 5.3). These passages are meant to allow a clear comparison between the less structural character of Dedekind's proofs and Noether's more structural one. They will perhaps appear as more appealing to readers with a stronger interest and background in algebra and number theory.

Chapter 1   Structures in Algebra: Changing Images

In 1930, the Dutch mathematician Bartel Leendert van der Waerden published his now famous *Moderne Algebra*. The book's publication date is mentioned in many places as one of the central milestones in the development of modern abstract algebra.[1] Thus, for instance, according to Bourbaki:

> Van der Waerden's treatise, published in 1930, assembled for the first time the works of Noether, Artin, Steinitz, etc. in a joint exposition, thus opening the way and guiding many later research-works in abstract algebra.[2]

Almost sixty years after the publication of the book, abstract algebra has grown into one of the most highly developed and pervasive domains of contemporary mathematics. Many excellent textbooks have since been written in order to keep pace with these developments. Nevertheless, *Moderne Algebra* could still satisfactorily be used today in any introductory university course in this field.[3] This reason alone (and there are in fact many others) amply justifies the prestige the book possesses as the classical text of modern structural algebra.

Much of Van der Waerden's later writings were dedicated to the history of mathematics. In one of his latest books on the history of algebra he wrote:

> Modern algebra begins with Evariste Galois. With Galois, the character of algebra changed radically. Before Galois, the efforts of algebraists were mainly directed towards the solution of algebraic equations... After Galois, the efforts of

---

1. Cf., e.g., Birkhoff 1973, 771; Dieudonné 1970, 136; Guèrindon & Dieudonné 1978, 122-123; Kurosh 1963, 5; Mac Lane 1988, 328-329; Mal'cev 1971; Mehrtens 1979, 152-157; Novy 1973, 223; Scholz (ed.) 1990, 407.

2. Bourbaki 1969, 77: "Le traité de van der Waerden, publié en 1930, a réuni pour la première fois [les] travaux [de Noether, Artin, Krull, Steinitz, etc.] en un exposé d'ensemble, ouvrant la voie et servant de guide aux multiples recherches d'Algèbre abstraite ultérieures."

3. In later re-editions (1937, 1950), the contents of the book remained essentially unchanged, although the *Moderne* was later dropped, leaving the title simply as *Algebra*.

the leading algebraists were mainly directed towards the structure of rings, fields, algebras, and the like. (van der Waerden 1985, 76)

This claim, however, immediately raises several questions. On the assumption that mathematicians who followed Evariste Galois (1811-1832) indeed turned their attention towards the structure of the various abstractly defined algebraic systems, why did it take so long for a textbook (van der Waerden's own book) to be published in which algebra was presented systematically from this perspective? Is it that *Moderne Algebra* was not so innovative as it is sometimes claimed? If so, how is it that such a wide consensus exists regarding the book's novelty and historical importance? If one accepts van der Waerden's historical claim, it would seem hard to explain the professional activity of many first-rate mathematicians in the algebraic domain after 1830. In fact, from the perspective of the increasingly generalized algebraic theories presented in *Moderne Algebra*, many problems which the leading algebraists between Galois and van der Waerden tackled with great efforts are particular applications of subsidiary importance.

These questions exemplify the difficulties inherent in the indiscriminate use of the term "mathematical structure" in historical accounts. In fact, even in those places where the unique historic contribution of *Moderne Algebra* is explicitly pointed out, the term "mathematical structure" is taken for granted and used without further qualifications. Clearly, then, it is necessary to make these terms more precise.

The present chapter explains why, and in what sense, *Moderne Algebra* should indeed be considered a real turning point in the history of algebra, marking the consolidation of the structural trend in algebra. Such an explanation will, however, imply disagreeing with van der Waerden's own historical account.

It is indeed the case, that the genesis of many of the central concepts of modern abstract algebra goes back to the mid-nineteenth century. The development of the particular concepts involved has been described in detail in several recent accounts.[4] Nevertheless, the real change in the overall conception of algebra, i.e., in the images of algebraic knowledge, was to come much later. Throughout the eighteenth century algebra was the branch of mathematics

---

4. Several such accounts are listed in the bibliography. See, e.g., Crowe 1967; Edwards 1980; Hawkins 1971, 1974; Kiernan 1971; Mehrtens 1979; Purkert 1971; Scholz (ed.) 1990; Silvestri 1979; Wussing 1984.

dealing with the theory of polynomial equations, including all the various kinds of techniques used to solve them and to analyze the relationships among roots and coefficients of a polynomial. During the nineteenth century these problems remained a main focus of interest of the discipline, and some new ones (algebraic invariants, determinants, etc.) were added. Parallel to this, number theory dealt with problems of divisibility, congruence and factorization. The new "structural algebra" eventually came to provide a common framework within which several of these problems, originally arising in these two disciplines, algebra and number theory, could be addressed in generalized formulations, and from a new, unified perspective. This new perspective gradually redefined the aims of research in algebra, its main tools, its interesting questions and legitimate, expected answers, i.e., its images of knowledge.

The process of transformation from the old image of algebra to the new image, beginning in the mid-nineteenth century and up into the late 1920s, is discussed in Chapters 2-5 from the perspective of the development of ideal theory. Before entering that discussion, however, the present chapter is devoted to analyze the image of the discipline as it was generally accepted during the nineteenth century and to compare it with the structural one. This is done by examining the way in which both images are reflected in the leading textbooks of algebra of their respective periods. In fact, an ideal way to understand the images of knowledge characteristic of any discipline at a certain period of time is to examine its textbooks. Some textbooks are instrumental in producing new images of knowledge and promoting them further, while others simply reflect dominant images of knowledge.[5] The archetypal textbooks of the classical nineteenth-century and of the structural images of algebra were Weber's *Lehrbuch der Algebra* and van der Waerden's *Moderne Algebra*, respectively, which are analyzed in §§ 1.2-1.3 below. Section 1.1 discusses developments in algebra prior to the publication of Weber's book. Section 1.4 discusses other textbooks which, although roughly contemporary to *Moderne Algebra*, present an image of algebra that is different from the structural one.

---

5. This claim suggests some kind of connection between the scheme body/images of knowledge and Kuhn's account of the development of scientific theories in terms of scientific revolutions and paradigms. This relation is analyzed in detail in Corry 1993.

1.1   Jordan and Hölder: Two Versions of a Theorem

Cours d'algèbre supérieure[6] by Joseph A. Serret (1819-1885), and *Traité des substitutions et des équations algébriques*[7] by Camille Jordan (1838-1921) were the two leading textbooks of algebra in France since the mid-nineteenth century. Comparing the first three editions of the *Cours*, published in 1849, 1854 and 1866 respectively, one notices that Serret gradually introduced significant changes, so as to incorporate the latest innovations of current research in algebra. These changes involved important additions to the body of algebra. In spite of these additions, however, throughout the successive editions of the *Cours*, the image of the discipline remained unchanged: "algebra" was the discipline dealing with the theory of polynomial equations and polynomial forms.[8] Serret's conception of the discipline remained essentially unchanged even in the third edition of his book, which included for the first time a treatment of Galois theory. However, from a more general perspective, the gradual adoption of Galois's methods and point of view for the study of polynomial equations and the related problem of solvability signified a major change of perspective for the whole domain. This change will concern us in the forthcoming chapters. At this point, in order to understand the image of knowledge that dominated research on polynomial equations and, more generally, on what should be considered as the discipline of algebra during the nineteenth century, it is necessary to describe briefly the early stages of the development of Galois theory and the changes it brought with it.[9]

The publication of *"Réflexions sur la théorie algébrique des équations"*[10] by Joseph-Louis Lagrange (1736-1813) marked the beginning of a truly new period in the development of the theory of polynomial equations. Like many other mathematicians before him, Lagrange addressed the problem of finding a solution by radicals for the general polynomial equation of fifth degree, or higher. At variance with the somewhat naive and straightforward attempts of his predecessors, Lagrange based his own research on a detailed analysis of

---

6. Serret 1849.

7. Jordan 1870.

8. More detailed accounts of the *Cours* appear in Kiernan 1971, 110-114; Novy 1973, 218-222; Toti Rigatelli 85-89; Wussing 1984, 129-135.

9. More detailed descriptions of the developments discussed briefly here appear in Kiernan 1971, 13 ff.; Scholz (ed.) 1990, 365 ff.; Wussing 1984, 118 ff.

10. Lagrange 1770-71.

the existing algorithms for the equations of the second, third and fourth degree, in order to determine the principles behind their workings.

Lagrange's important insights in this context came from the analysis of the effect of the permutations of the roots of a given equation on the values of its coefficients. More significantly, he analyzed the effects of these permutations on the values of the coefficients of the auxiliary, "resolvent" equations, that appear in the intermediate steps of the known algorithms. Since the coefficients of the equations are symmetric functions of the roots, any permutation of the roots leaves their values unaffected. The coefficients of the resolvents happen to be also rational functions of the roots, and Lagrange asked what permutations leave these coefficients invariant. Thus, for instance, in the known algorithm for solving the cubic equation, the cubic radicals that appear in the final expression of the roots may be expressed as $\frac{1}{3}(x + \alpha x' + \alpha^2 x'')$ where x, x', x'' are the roots of the original equation (in a certain order), and where the quantity α is a cubic root of unit, $\alpha^3 = 1$. Now, the six possible permutations of the three roots of the equation yield only two different values of the function $(x + \alpha x' + \alpha^2 x'')^3$. A similar analysis for the algorithm of the quartic involves the expression $x_1 x_2 + x_3 x_4$ which remains invariant under permutations that interchange the first two or the last two roots, but not under permutations that interchange, e.g., x_1 with x_3 or with x_4. Thus, among the twenty-four possible permutations of the four roots only eight leave this relation invariant, and the expression takes only three different values under all these permutations. The value of the expression is therefore a root of a third-degree equation, whose coefficients are rational functions of the coefficients of the given equation. Lagrange concluded that this is the reason why the original equation is solvable by the given algorithm. He thus connected the success of the existing methods with the existence of resolvents, whose roots are rational expressions of the roots and of the coefficients of the original equation, and such that the degree of the resolvent equals the number of different values that a root of the resolvent takes under all possible permutations of the roots of the original equation.[11]

Lagrange used these ideas to study the possibility of developing an algorithm for the quintic. Although he attained no definite conclusions, his work

---

11. See Kiernan 1971, 45-56.

fostered the belief that no general solutions by radicals exist for the quintic, and thus provided a new starting point for his followers. But before Galois was able to develop his own general theory of solvability, it was still necessary to further elaborate some of the indispensable, basic tools. Thus, for instance, when speaking of permutations of roots, Lagrange himself concentrated on the effect of the permutation on a certain function of the roots, rather than on the properties of the permutations considered in themselves. The development of an independent calculus of permutations and of an elaborated theory of its properties, an essential requisite for Galois's work, was left to Augustin Louis Cauchy (1789-1857), who developed it starting only in 1815.[12]

The Italian Paolo Ruffini (1765-1822) was the first mathematician ever to assert the impossibility of an algebraic solution for the *general* polynomial equation of degree greater than four, in papers published in 1799 and 1813. In his work, he elaborated on Lagrange's ideas and adumbrated the notion of a group of permutations, working out some of its basic properties. Ruffini's proofs, however, contained several, significant gaps and they were never widely read and understood.[13]

Parallel to this, and in a somewhat contrary direction, between 1796 and 1801, in the framework of his seminal number-theoretical investigations, Carl Friedrich Gauss (1777-1855) systematically dealt with the so-called cyclotomic equations, $x^p-1 = 0$ ($p>2$, prime), and developed new methods for solving these *particular* cases of higher-order polynomial equations.[14]

The Norwegian mathematical star of the early nineteenth century, Niels Henrik Abel (1802-1829), completed in 1824 the first clear and accepted proof of the impossibility of solving by radicals equations of degree five or above. His work was basically a direct elaboration of Lagrange's line of thought. He proved that if an equation is solvable by radicals, then its roots may be expressed in such a way, that the radicals that appear in the final expression of the roots and involve the coefficients of the equation may be expressed as rational functions of the roots of the equation and of certain roots of unity. Using this result he was able to prove, in a rather complicated way, the impossibility of solving by radicals the general equation of degree five.[15]

---

12. See Dahan 1980; Wussing 1984, 86-95.
13. See Ayoub 1980; Kiernan 1971, 56-60 & 67-72; Wussing 1984, 80-84 & 96-101.
14. See Kiernan 1971, 61-64.
15. See Kiernan 1971, 67-72.

In view of this result, but aware also of the progress made by Gauss, he suggested in 1828 a research program for polynomial equations based on two main points: (1) to find all equations of a given degree solvable by radicals; (2) to decide if a given equation can be solved by radicals. His early death in complete poverty, two days before receiving an announcement at home that he had been appointed professor in Berlin, prevented Abel of undertaking this program, as well as many other research plans he had conceived, especially in analysis.

This is the background against which Galois started his own study of the solvability by radicals of any given equation of degree five or more. However, rather than establishing for specific equations if they can or cannot be solved by radicals, as Abel had suggested, Galois pursued the somewhat more general problem of defining necessary and sufficient conditions for the solvability of any given equation. It is not possible here to give a detailed analysis of Galois's contribution or describe his short and turbulent, life, which has been the subject of books and films and has inspired countless myths.[16] It will suffice to point out those aspects of his works in which he departed from the line of thought described in the preceding paragraphs, thus opening new perspectives for the research of polynomial equations and eventually for algebra at large.

Galois started from the group of permutations of the roots of a polynomial equation, and from the domain K, obtained by adjoining to the rational numbers all the rational expression involving the coefficients of the given equation. This is a field, and Galois, like Abel before him, used the concept in a meaningful way, although without giving it a specific name. Galois considered all the permutations that leave invariant all the rational expressions of the roots of the equations, with coefficients in K. These permutations happen to be closed under composition, and therefore they form a group in themselves: the invariant group of the equation (later also known as the Galois group of the equation). Galois showed how the Galois group of any equation can be determined, even if one does not actually know the roots. He then developed distinct group-theoretical techniques to expand successively the initially given field K, through adjunction of roots that can be rationally expressed in terms

---

16. Toti-Rigatelli 1996 is a short, very-well documented biography of Galois, that includes a detailed analysis of his work as well as a comprehensive bibliography.

of elements of the given field. He also formulated a criterion to determine whether or not the given equation is solvable by radicals, by referring to the properties of the group of the equation and of successively built subgroups.

In finding criteria for the solvability of particular equations, Galois's methods did not enable one to *calculate* the roots of those equations, but rather they allowed one to analyze the conditions for the *existence* of a rational expression of the roots in terms of the coefficients of the given equation.[17] Galois's work, with its insistence on the properties of the group of permutations, the invariant group of the equation, and the study of its sub-groups, introduced many basic concepts that were later developed in the framework of the autonomous theory of groups. In fact, Galois conceived his own contribution to be the formulation of a rather general theory based on the concept of group, a particular application of which was the problem of solvability of a particular equation of arbitrary degree.[18] It thus offered a new field of interest encouraging further research, closely related to the study of polynomial equations, but at the same time it promoted an intrinsic interest on its own. Interest in techniques based on the direct study of the resolvents, such as those introduced by Lagrange and later elaborated by mathematicians like Abel, did not immediately disappear with the introduction of Galois's point of view.[19] Thus, posterior developments in algebra would have to take into account the existence of these two parallel, and strongly connected perspectives for research in algebra.

Galois's work were virtually unknown to his contemporaries, and they were first published in 1846, fifteen years after his death, by Joseph Liouville (1809-1882).[20] Their publication implied the necessity of adapting the existing theory of polynomial equations to the new perspectives that Galois's ideas had opened. However, Galois's writings were highly obscure and difficult, and his proofs contained many gaps that needed to be filled. Moreover, his ideas were in many respects so innovative that it took considerable time and effort to assimilate them properly into the existing body of knowledge. Thus,

---

17. In theory, once the Galois group of an equation is determined, and if the equation is solvable by radicals, then the roots can be found, although not without difficulty. However, determining the group of an equation, in the first, place, is practically impossible in general.

18. Kiernan 1971, 42.

19. Cf. Wussing 1984, 118, ff.

20. The developments leading to, and following, Liouville's publication of Galois's manuscripts are described in Lützen 1990, 129-132, 559-580.

Galois theory was incorporated into algebra in different ways. For example, Dedekind followed closely both Galois's results and his approach. Dedekind stressed the parallel relationship between the Galois group and its subgroups, on the one hand, and the field of rationals and its successive extensions by addition of roots, on the other hand. Dedekind developed an early and very idiosyncratic version of the theory, which had no immediate influence on teaching and research, but which would eventually evolve into the standardly accepted one, through the work of Emil Artin (1898-1962) in the 1930s.[21] By contrast, Serret incorporated many of the new techniques developed by Galois without however modifying his own images of algebraic knowledge in any essential way. Serret's images of algebra are manifest in the various editions of his *Cours*.

The third edition of the *Cours* (1866) was the first university textbook to publish an exposition of Galois theory.[22] Serret's book presented it essentially as an auxiliary tool for actually determining the roots of a given equation. He did not discuss the concept of group on its own, and he formulated the main theorems of Galois theory in the traditional language of solvability theory, going back to the works of Lagrange and Abel.[23] The following editions of the *Cours* basically repeated the contents of the third one; they did not introduce the important advances in Galois theory, group theory and field theory that took place in the last part of the nineteenth century. Nevertheless the *Cours* remained an oft-quoted reference book up to the turn of the century. As will be seen, the first thirty years of abstract algebra in the twentieth century concentrated in Germany, while French mathematicians remained far behind their colleagues in this field. The decisive influence of Serret's book, whose approach and content was to be out of date by 1900, may partly account for this gap.[24]

Somewhat different was the approach followed by Jordan in his *Traité*. This was the first textbook to deliberately and explicitly connect the theory of

---

21. See Kiernan 1971, 129-132. Dedekind's work on Galois theory is discussed in § 2.1 and § 2.2.4 below.

22. In fact, Serret had wanted to deal with Galois's ideas already in the first, and then in the second, edition of his book. He refrained from doing so until the third one, however, because Liouville, his influential mentor, expressed his intention to publish a commentary of his own, which he actually never did. See Lützen 1990, 196-199.

23. Cf. Toti Rigatelli 86-88; Wussing 1984, 129-131.

24. On this point, see Kiernan 1971, 114.

equations and the theory of permutations. It constituted, in fact, a comprehensive exposition of all fields of mathematics which, by that time, could be approached with the techniques and results of the theory of permutations. Although it still presented algebra as the discipline concerned mainly with solvability of polynomial equations by radicals, the theory of groups also appeared in it as a domain worthy of interest in its own right. Many of the central tools and results presented in the book had been obtained by Jordan himself, but had appeared only scattered around in many journals. Here they were compiled and organized, for the first time, as a unified and systematic exposition. Jordan introduced many of the concepts and nomenclature that would later become standard for the theory: conjugation, normalizer, normal (or self-conjugate) subgroup, simple group, etc. At the same time, however, groups were defined throughout Jordan's book only as groups of permutations. An abstract definition of groups had already been proposed by the British mathematician Arthur Cayley (1821-1895) back in 1854. Cayley had also shown that every finite group can be realized as a permutation group. Jordan was acquainted with Cayley's work, yet his own groups were never defined, either in his research work or in his textbook, following Cayley's model. In this respect, Jordan might be seen as "delaying" the adoption of the new, structural, approach to algebra. But this latter claim could be only accepted as true regarding one particular feature of that approach, namely, the adoption of the abstract formulation of concepts. In fact, the kind of questions on which Jordan focused his research implied an innovative approach that was to open new directions and eventually become central to the structural conception of algebra. This change in perspective, introduced by Jordan and elaborated in works that drew on his, can be understood by examining one of his important contributions to the study of groups: the concept of "composition series."

Jordan had originally defined "composition series" in an article published in 1869. In this article a composition series was defined not in terms of groups of permutations, but rather in terms of algebraic equations alone.[25] In the *Traité* the concept already appears as a purely group-theoretical one.[26] Given a group G, which is not simple, and a sequence of subgroups $I, I_1, \ldots$ such that every subgroup in the sequence is a normal subgroup of its predecessor, this sequence is called a "composition series" whenever no other subgroup satis-

---

25. Jordan 1869.
26. Jordan 1870, 42-48.

fying that condition may be inserted between two members of the sequence. Jordan called the numbers representing the quotients of the orders of two successive groups in one such sequence "composition factors." For any group, there are many possible composition series, but Jordan proved that the number and values of the composition factors (except perhaps for their order) is an invariant of the group.

Jordan thus formulated and proved a general theorem without mentioning the equation to which the permutation group in question is related. At the same time, the connection of the theorem with equations was mentioned in a different place as a particular application of the general theorem. But still, Jordan's definition of group was far from general: in his book a group is a collection of permutations closed under composition (Jordan uses the terms *groupe* or *faisceau*).[27] Further assumptions regarding the properties of this operation were either tacitly accepted or introduced as special conditions.[28] This theorem is known today in very general formulations. Since Jordan's conception was limited to permutation groups, he could only formulate the theorem in a somewhat restricted way, i.e., in terms of "composition factors", rather than in terms of quotient groups as will be later the case. The task of providing the wider, more abstract setting for the theorem had to be postponed almost twenty years until the German mathematician Otto Hölder (1859-1943) completed the second part of the proof of the theorem named today after both mathematicians.

But although Jordan's formulation of the theorem may appear in retrospect as somewhat limited, for Jordan himself the information it provided was more than helpful in indicating the kind of significant questions that should be pursued in investigating groups. Jordan concluded from the theorem that "one can legitimately classify the composite groups by the number and value of their composition factors."[29] The search for this kind of intrinsic invariant is as central to the structural approach as the axiomatic formulation of concepts.

---

27. Jordan 1870, 22.

28. Cf. Silvestri 1979, 324-327; Kiernan 1971, 114-125; Wussing 1984, 154-163. There is, however, some disagreement among these authors regarding Jordan's conception of groups. While Kiernan states that for Jordan, the group is the central concept and the solvability of equations, only an application of the theory as Galois had originally conceived his theory (p. 117), Silvestri emphasizes precisely the fact that Jordan's group concept was still tied to permutations and the solution of algebraic equations (p. 323). Wussing's view is closer to Kiernan's.

29. Jordan 1870, 48: "Le théorème que nous venons d'établir montre qu'on peut légitimement classer les groupes composés d'après le nombre et la valeur de leurs facteurs de composition."

Jordan's example shows that the former is not logically conditioned by the latter and that, from an historical point of view, both features did not always go hand in hand.

Hölder reconsidered Jordan's theorem in an article of 1889, entitled "Reduction of an arbitrary algebraic equation to a sequence of equations". This article comprises two sections bearing significant subtitles, "group theoretical part" and "algebraic part", the former being considerably shorter than the latter. Thus, groups are of interest in Hölder's article inasmuch as they provide a tool for dealing with the "algebraic" problem which constitutes its main issue: the solution of polynomial equations.

Hölder defined groups in a more or less axiomatic way, and explicitly emphasized, like Cayley before him, the irrelevance of the nature of the elements, or "operations", that constitute the group. Hölder's definition was based on an earlier article of 1882 by Walther von Dyck (1856-1934). This article represented a decisive step towards the definite formulation of the abstract concept of group, since it explicitly acknowledged for the first time the role of groups as a synthesis of trends that had arisen separately in number theory, in the theory of continuous transformations, and in the theory of polynomials.[30] In the first, group-theoretic section Hölder reformulated Jordan's result on the invariance of the factors of composition of any group, but taking the factors to be themselves groups rather than numerical coefficients. For this purpose, Hölder introduced the concept of a "quotient group" defined by a normal subgroup of any group, and explained its basic properties.[31] With the concept of quotient group at hand, Hölder restated Jordan's theorem in a more general version: given a group and a composition series in it, *the collection of quotient groups* determined by the series (rather than the quotients of the orders of the groups) is an invariant of the given group. Hölder also formulated a similar theorem (without proof) for principal series, namely, maximal series in which the subgroups are normal in the larger group, and not just in the immediate predecessor group of the series. Thus, by formulating the concept of group in more abstract terms Hölder was led to a deeper understanding of the structural problem initially posed by Jordan. Not only could one find a certain invariant of the given group; the invariant itself could also be expressed

---

30. An account of von Dyck's career appears in Faber 1935. On von Dyck's work on groups, see Chandler & Magnus 1982, 5-10; Wussing 1984, 238-243.

31. The trend of ideas leading from Jordan's work to Hölder's definition is described in Nicholson 1993, 75-81.

in terms of groups. Here the abstract formulation of the concept, though not necessary for the formulation of the problem itself, turns out to be illuminating in solving it.

Moreover, in addition to the improved formulation of Jordan's theorem, the newly introduced concepts raised new, interesting problems. For instance, one could show that any group may be factorized into a product of two groups: one is the epimorphic (*meroedrisch isomorph*) image of the group, while the other is a normal subgroup of it. The reciprocal problem could then also be formulated: given two arbitrary groups, to build a product group such that one of the factors is epimorphic to it and the second, a normal subgroup.[32] This problem will be later called the "extension problem for groups", and Hölder rightly claimed that it may have more than one answer.[33] In his article Hölder had promised to deal in detail with that problem at a later opportunity, something which he never did.

Hölder's approach and his clear conception of the kinds of important questions that should be considered came more explicitly to the fore in his next article, published in 1893.[34] This article was published as an explicit answer to the problem, proposed by Cayley in 1878, of characterizing all the groups of a given order. As early as 1854, Cayley had published a study on groups whose main ideas and nomenclature had been directly taken over from Galois. The work evinced a thorough mastery of the contents of Galois's work, and, at the same time, the clear influence of the British school of abstract algebra. Thus in 1854 Cayley himself had already defined a group in the most abstract way, while introducing for the first time the tables of permutation groups.[35] His paper, however, had no direct repercussion on the works of contemporary mathematicians. The total absence of Cayley's point of view in a book like Jordan's is but one clear testimony to this fact. Cayley returned to groups in 1878 and set forward the task of finding all the groups of a given order, as a central open question of the discipline. Von Dyck and Hölder separately took up the challenge and began to work in that direction.

In his paper of 1893, Hölder attempted to characterize all the groups of orders p^3, pq^2, pqr and p^4 (with p,q and r prime numbers). He began his article

---

32. Hölder 1889, 33.

33. The extension problem for groups is one of the central problems of the theory of group-cohomology. For a general discussion of this problem in modern terms see Jacobson 1980, 363-369.

34. Hölder 1893.

35. Cf. Wussing 1984, 232.

by defining groups in an abstract formulation similar to the one used in his previous paper. He emphasized that two isomorphic (*holoedrisch isomorphe*) groups are essentially one and the same object. These sorts of remarks may be found in many works of the period and even later, indicating the rather slow pace of adoption of the abstract point of view by the mathematical community. The fact that many abstract definitions of mathematical concepts had already been formulated by that time seems not to have in itself sufficed to re-orient research in the abstract direction. Moreover, as will be seen in Chapter 2 below, even after the abstract formulation was thoroughly adopted and recognized for groups, it was not adopted in every other branch where such formulation was feasible. Hölder's adoption of a more abstract formulation had turned out to be very fruitful, but it was yet a far cry from an overall adoption of the abstract formulation of concepts as the standard approach in algebra.

The different treatments of the concept of group by Serret, Jordan and Hölder illustrate the kind of processes undergone in the last third of the past century by ideas that eventually came to be related to modern algebra. In his book Serret treated groups within the framework of Galois theory, but he presented them only as an auxiliary tool for addressing the problem of solvability. Jordan and Hölder addressed an invariance property of groups, independently of any equation that the group is related to, but their respective analyses yielded two different kinds of results: the first was formulated in terms of numerical coefficients and the second in terms of quotient groups. None of these three mathematicians saw the concept of group as part of a more general idea of "algebraic structures", of which it was a particular instance. For them, a group was an important tool, useful for the problem they were trying to deal with.

Textbooks like Jordan's, and especially Serret's, were soon to become outdated in view of the important advances initiated by the research of Hölder and von Dyck. By the turn of the century a new textbook appeared which was to close the existing gap. It soon became the standard text of the field. This was Heinrich Weber's *Lehrbuch der Algebra*. The *Lehrbuch* included most of the recent advances in algebra missing from its predecessors. Yet it, too, preserved the main traits of the old image of algebra as the science of polynomial equations and algebraic forms.

1.2   Heinrich Weber: *Lehrbuch der Algebra*

When the first edition of volume one of the *Lehrbuch der Algebra* appeared in 1895, Heinrich Weber was well-aware of the latest advances in algebra, and in particular of the possibility of formulating concepts in purely abstract terms. As a matter of fact, in 1893 he had been the first to provide abstract definitions of both groups and fields within the framework of a single article. Moreover, his research on algebraic functions, in collaboration with Richard Dedekind, shows that Weber was deeply acquainted with the latter's theory of ideals, which, as will be seen in the following chapters, played a central role in the rise of the structural approach to algebra. Weber's *Lehrbuch* was the last great textbook of algebra of the nineteenth century and it may be considered as the foremost representative of the classical nineteenth-century image of algebra. It is thus worth examining here in some detail the images of algebra manifest in it. These images depart in certain senses from those in Serret's and Jordan's textbooks. However, at the same time, they remain true to the picture of algebra as the discipline of polynomial forms, whose basic properties are derived from those of the systems of real and complex numbers.

Heinrich Weber (1842-1913) studied in Heidelberg and Leipzig. He habilitated in Königsberg in 1866, and taught there until 1883, except for the years 1870 to 1875, when he was professor at the ETH Zürich. He later spent several years at the Charlotenburg technological institute (Berlin), and in Marburg, and was *Ordinarius* in Göttingen between 1892 and 1895. Finally he moved to Strasbourg, where he remained until his death in 1913.[36] Weber's Königsberg years were the most productive of his successful career. The tradition of analysis and mathematical physics developed in this university, under the leadership of Carl Gustav Jacobi (1804-1851), Franz Neumann (1798-1895) and, somewhat later, Friedrich Richelot (1808-1875), was a main force behind the increasing dominance attained by Germany in the mathematical world over the nineteenth century.[37] Weber was but one of the outstanding mathematicians whose names came to be connected with that school. The early careers of Adolf Hurwitz, Hermann Minkowski and David Hilbert were also later associated with this institution and their works were decisively influenced by its tradition (see § 3.1 below). Weber's output covered many different domains of mathematics, that had been especially active over the

---

36. See Schappacher & Volkert 1997; Schoeneberg 1981.
37. On the Königsberg school see Klein 1926-7 Vol. 1, 112-115 & 216-221.

preceding decades of research.[38] His textbook of algebra was only one among several important works he published and that reached a wide mathematical audience.[39]

One of the active domains of research in which Weber contributed important innovations was the theory of polynomials. In 1893, during his Göttingen period, Weber published an important article entitled *"Die allgemeinen Grundlagen der Galois'schen Gleichungstheorie."* This article contained an exposition of the theory of Galois in the most general terms known to that date. Weber elaborated a direction of inquiry introduced by Dedekind, which saw the theory not just as an analysis of the problem of solvability, but rather as a more general examination of the interplay between specific groups and certain, well-defined fields. Under this view, the study of solvability was a particular application of the more general theory. In many respects, Weber's article represents the first truly modern published presentation of Galois theory. In particular, it introduced all the elements needed to establish the isomorphism between the group of permutations of the roots of the equation and the group of automorphisms of the splitting field that leave the elements of the base field invariant.[40] This presentation of Weber suggests—in a natural way—the convenience of adopting an abstract formulation of the central concepts of group and field. Moreover, it stresses the importance of the interplay between structural properties of both entities. It therefore implied an important move towards the understanding of the idea of an algebraic structure, and

---

38. See Klein 1926-27 Vol. 1, 275. Klein described Weber as the most versatile representative (*der vielseitigste Vertreter*) of the nineteenth-century trend, of which Klein himself so proudly felt part, that sought to elaborate the interconnections between mathematical domains such as the theory of invariants, the theory of polynomials equations, the theory of functions, geometry and the theory of numbers (see *ibid*, p. 327).

39. For instance, together with J. Wellstein, Weber published the widely-read *Enzyklopädie der Elementarmathematik* (Weber & Wellstein 1903). Together with Dedekind, Weber edited Riemann's mathematical papers. See below Chapter 2, note #3. On Weber's important contributions to the early development of Class Field Theory, see Hasse 1967, 266-267.

40. See Weber 1985, 543 ff. Weber's theorem is not in itself formulated in a complete and explicit fashion, in the terms used above. The more modern formulation of this result and, in particular, the identification of the central role played by the splitting field, is due to Emil Artin, who introduced it relatively late, namely, in his 1938 lectures. Even van der Waerden's original presentation, based to a large extent on Artin's earlier lectures, do not contain this latter, crucial detail. For an account of the significance of Weber's work for the development of Galois theory see Kiernan 1971, 135-141. For Artin's formulation see Kiernan 145-147.

towards the adoption of the structural approach. It is thus worth examining in some detail how these concepts appear in Weber's article.

In the opening sections of the paper Weber wrote:

> The theory appears here as a direct consequence of the concept of field, itself an extension of the concept of group. It appears as a formal law without any regard to the numerical meaning of the elements involved.... The theory is thus conceived as a pure formalism, which acquires life and content only when the elements are assigned with numerical values.[41]

Weber both defined and examined the basic properties of groups in an absolutely abstract fashion. In this article—unlike in Serret's and Jordan's presentations—permutation groups appear always as particular instances of the more general concept. The properties of permutation groups are never discussed separately. Likewise, Weber considered finite and infinite groups as subjects of a single theory. Although he was certainly not the first to do so,[42] Weber's 1893 article was probably the first place where the two concepts received an extended, *abstract* common analysis. The concept of field was not only defined here for the first time as an extension of the concept of group (namely, by adding a second operation), but it was also stressed that by approaching the concept in this way, it may be perceived in its full generality, thus enabling the concomitant treatment of finite and infinite fields. Like groups, finite and infinite fields were subsumed here for the first time under a single, general definition. Significantly, however, Weber did not consider in this article (or elsewhere) the problem of the characteristic of the field.[43]

Weber defined a group as a system of finite, or infinitely many elements endowed with a composition rule that is assumed to be associative, right- and left-cancellable, but, in general, not commutative. The existence of a unit for the system and of an inverse element for each element of the system were proven as consequences of the axioms. Weber also defined the notion of iso-

---

41. Weber 1893, 521: "[Die Theorie] ergiebt sich hier als eine unmittelbare Consequenz des zum Körperbegriff erweiterten Gruppenbegriffs, als ein formales Gesetz ganz ohne Rücksicht auf die Zahlenbedeutung der verwendeten Elemente.... Die Theorie erscheint bei dieser Auffassung freilich als ein reiner Formalismus, der durch Belegung der einzelnen Elemente mit Zahlwerthen erst Inhalt und Leben gewinnt."

42. Wussing 1984 gives a detailed description of the separate evolution of these two concepts and their late combination into a single one. See especially pp. 230-251.

43. The characteristic of any field is the smallest positive integer p such that any element of the field added to itself p times equals zero. If there is no such finite integer, then the field's characteristics is said to be zero. See below § 4.2 on Steinitz's theory of fields.

morphism and stated that two isomorphic groups may be considered as representing a generic concept (*Gattungsbegriff*). Thus it makes no difference whether one considers one or the other specific representative while studying the properties of the group.[44] This fundamental idea of modern algebra, and in fact of modern mathematics at large, was already seen to have been emphasized by Hölder in his group-theoretical work. In fact, it was already clearly stated in Felix Klein's famous Erlangen Program of 1872,[45] and it may be gradually found in an increasing number of contemporary works. Eventually this idea became an obvious principle in no need of explicit mention; this is the case, for instance in van der Waerden's book, as will be seen below.

A commutative group is transformed into a field if a second operation (multiplication) is added and an additional new condition (distributivity) is demanded on the relationship between addition and multiplication. Among the properties of fields, Weber mentioned that for a product to equal zero, it is necessary that at least one of the factors be itself zero. Weber went on to mention a number of possible directions in which the concept of field could be further expanded. For instance, one could consider congruences modulo a composite number n, instead of a prime, and then one would obtain divisors of zero, namely, null products of two non-zero factors. One could also consider cases of non-commutative addition, or the definition of a third operation. However, Weber asserted that the logical possibility or fruitfulness of such ideas was yet to be investigated.

The innovative approach adopted by Weber in this paper and the issues considered in it may indeed strike the modern reader by their novel appearance. But more striking perhaps is the fact that all these innovations had minimal direct influence on the algebraic research of other, contemporary mathematicians—even in Germany. Thus, in an article on Galois theory written by Hölder for the *Encyklopädie der Mathematischen Wissenschaften* in 1898, that can certainly be considered as a faithful indicator of contemporary conceptions in Germany, Weber's innovations are not even mentioned.[46] Some twenty years were still needed until Ernst Steinitz would publish his work on abstract fields, where he systematically developed for the first time the ideas introduced here by Weber (§ 4.2 below). More significantly yet, even

---

44. Weber 1893, 524.
45. See Rowe 1989, 189-190.
46. See Hölder 1898.

in Weber's own textbook, published barely two years thereafter, groups would be neither defined nor analyzed as an abstract concept, nor would fields make their appearance within his general approach. As late as 1924, a new textbook of algebra was published under the explicit influence of Weber's textbook, in which the "new abstract approach" of Weber's article is totally absent.[47]

Weber, then, published the first volume of his textbook on algebra in 1895. In the preface to the first volume, he explained that the development of algebra during the last decades had rendered the existing textbooks obsolete (including Serret's, whose excellence by previous standards Weber explicitly pointed out) and had brought about the need for a new coherent presentation of results and their applications. This volume presented the "elementary parts of algebra", namely all that may be subsumed under the designation of "formal algebraic manipulation" (*Buchstabenrechnung*), beginning with the rules for the determination of the roots of an equation and finishing with an exposition of Galois theory. Weber explicitly acknowledged Dedekind's influence in consolidating his long-standing interest in algebra, an influence that came mainly through the notes of Dedekind's 1857-58 Göttingen lectures on Galois theory, the manuscript of which Weber had had the opportunity to read.[48]

The book opens with a thorough discussion of the different systems of numbers. This was meant to provide the necessary, natural starting point for dealing with the problem of finding the roots of polynomial equations, which dominates a considerable portion of the book. At the same time, it is in line with the arithmetizing trends so widespread and dominant at the time in German mathematics.[49] Thus, Weber started with the properties of the integers, including the existence of a *g.c.d.* (and how to obtain it through the so-called Euclidean algorithm) and of unique prime factorization within the domain. Next, the notion of set (*Mannigfaltigkeit oder Menge*) was introduced—following Dedekind—in what we would call today a naive formulation: a system of objects or elements of any kind, such that for any given object one can always say whether it belongs to the set or not. This was followed by the definition of concepts such as ordered sets, discrete and dense (*dicht*) sets (exemplified by the integers and the rationals), cuts (*Schnitte*) and continuity (*Stetigkeit*)—all of them as previously defined by Dedekind. Weber's remarks

---

47. Namely, Fricke 1924. See § 1.4 below.
48. On Dedekind's lectures on Galois theory, see below § 2.1.
49. See Ferreirós 1999, 34-38.

evidently constitute an echo of Dedekind's views: density and continuity—wrote Weber—are properties which our senses cannot perceive, and one cannot rigorously prove their existence in the outside world. And he added: "It is however possible to construct abstract conceptual systems satisfying the density property but not the continuity property, or the continuity property but not the density property."[50] In Dedekind's definition of the real numbers this formulation appears almost verbatim.[51]

Within this framework of ideas, the rational numbers are introduced as a dense, but discontinuous set and the real numbers as the set of cuts of the rationals. Like Dedekind before him, Weber showed here a definite tendency to formulate the definitions of number systems in abstract terms. However, this kind of abstraction is essentially different from that commonly used in twentieth-century mathematics. The systems of numbers considered by Weber are *particular entities* (albeit admittedly free creations of the human spirit) whose properties are given once and for all in advance. Here, one confronts one of the central images which dominated both nineteenth-century algebraic research, and Weber's own conception of algebra: in spite of the technical ability to formulate mathematical concepts in abstract terms, *algebraic knowledge in its entirety derives from the fundamental properties of the basic systems of numbers*. The rise of the structural image of algebra will imply a change in the conceptual hierarchy: abstract structures will be defined and studied in advance, so that the number systems may be introduced as specific instances of them. In the image of algebra embodied in Weber's book, the algebraic properties of number systems do not derive from those of some more basic, underlying, abstract algebraic structures. Rather it is the other way around: algebra is based on the given properties of the number systems. Moreover, from a modern perspective, Weber's examination of properties such as density and continuity is in fact irrelevant to the algebraic problems discussed in his book. Thus when the conceptual order of study eventually changed in algebra, these "topological" considerations disappeared from algebraic discourse.

The introduction of the *Lehrbuch* closes with a remark on the formal manipulation of symbols. One can distinguish two main forms of the latter:

---

50. Weber 1895, 6: "Es lassen sich aber sehr wohl reine Begriffsysteme construiren, denen die Dichtigkeit ohne Stetigkeit oder auch Stetigkeit ohne Dichtigkeit zukommen."

51. Cf. Dedekind *Werke* Vol. 3, 320-321. On Dedekind's influence on Weber's *Lehrbuch* see Ferreirós 1999, 114-115.

identities and equations. Algebra, wrote Weber, is the discipline whose aim is the resolution of equations. This statement is not mere lip-service to the prevailing views. As will be presently seen, the contents of the book faithfully reflected this declared central role of equations. Other issues, such as the study of groups, appear as conceptually subsidiary to this aim and, needless to say, no other abstract algebraic concept is systematically investigated in the *Lehrbuch*.

The first volume of Weber's *Lehrbuch* comprises three books. The first two of them deal with the classical theories of polynomial equations. Chapters III and IV provide evidence of the attachment of the book to nineteenth century images of algebra. In Chapter III, the concept of root of an equation is discussed in terms that may be classified as "analytic": limits, continuity, ε–δ arguments etc. Chapter IV deals with "symmetric functions." Symmetric functions had been used by Lagrange in his early research on solvability of polynomial equations (§ 1.1 above) Later, the gradual development of Galois theory as the main tool for studying solvability of polynomial equations eventually rendered symmetric functions a rather dispensable tool, yet Weber included a treatment of them in the *Lehrbuch* as part of a tradition of which his approach to algebra was part and parcel. Under the conception of algebra characteristic of this tradition, a treatment of the theory of polynomial equations should include every particular technique devised to deal with their solvability.

In Book II one finds additional discussions that are analytic in character, and that would be excluded from later textbooks of algebra. This is the case, for instance, of the theorem of Sturm discussed in Chapter VIII. Sturm's problem concerns the question of how many real roots of a given polynomial equation lie between two given real numbers.[52] This, and further similar problems, are solved with the help of derivatives and other analytical tools. Likewise, Weber treats approximation techniques: interpolation and Newton's method are mentioned among others in Chapter X. Chapter XI deals with roots of unity: no mention whatsoever is made of their group-theoretical properties.

Galois theory is finally introduced in Book III, after nearly five hundred pages of discussion on the resolution of polynomial equations. First, a field of numbers is defined as a set of numbers closed under the four operations.

---

52. For a detailed account of the development of Sturm's theorem and its significance for the history of algebra see Sinaceur 1991, esp. 33-141, 246-261 & 355-363.

Indeed, the concept is extended to fields of functions or to any set closed under the four operations of addition, multiplication, subtraction and non-zero division.[53] Weber referred here to his article of 1893. But if then he had insisted upon the potential interest involved in studying finite fields, now, in the *Lehrbuch,* he considered only (infinite) fields of characteristic zero.

Groups are mentioned for the first time on page 511. But even here one does not find a general treatment of groups; this is left for later chapters. At this stage, Weber considered substitutions of one root of a function with another. These substitutions may be composed with one another; the collection of substitutions together with the compositions defined by them are said to form a group and, more specifically, a finite group (p. 513). At this stage, Weber defined a group of permutations—a concept which he used in the next chapter—and the Galois group of a given field.

Chapter XIV shows the application of groups of permutations to the theory of equations. First, Weber showed that any permutation may be decomposed into transpositions and cycles. He then defined some central concepts: subgroups (*Teiler*), and the cosets (*Nebengruppen*) determined by a given permutation, as well as the index of a subgroup of permutations. Weber explicitly stated that the aim of this whole section was to improve our understanding of the issues dealt with in the preceding chapter (p. 529). Thus the focus of interest does not lie in the study of the properties of the group of permutations as such, but only insofar as it sheds light on the theory of equations.

In the following chapters, Weber analyzed particular cases of equations using the insights provided by the already developed theory. The exposition culminates towards the end of the book, in Chapter XVII, where the algebraic solution of equations is discussed. Weber acknowledged the centrality of this problem for the contemporary development of algebra, and the important contribution of group theory to its better understanding. Thus, he wrote:

> One of the oldest questions which the new algebra has preferentially addressed is that of the so-called algebraic solution of equations, meaning the representation of the solution of an equation through a series of radicals, or their calculation through a series of root-extractions. The theory of groups sheds much light on this question.[54]

---

53. Weber 1985, 491-492.

It is in this section that Weber proves that the alternating group is simple—a result needed for the proof of the irresolubility of the general fifth degree equation.[55]

A definition of group in terms similar to those of Weber's own 1893 article appears only in the second volume of the *Lehrbuch*. After the basic concepts of the theory of groups were introduced in the first four chapters of the second volume in a general and abstract way, Weber stated the object of the abstract study of groups:

> The general definition of group leaves much in darkness concerning the nature of the concept.... The definition of group contains more than appears at first sight, and the number of possible groups that can be defined given the number of their elements is quite limited. The general laws concerning this question are barely known, and thus every new special group, in particular of a reduced number of elements, offers much interest and invites detailed research.

Weber also pointed out that the determination of all the possible groups for a given number of elements was still an open question. It had been recently addressed by Cayley, but only for the lowest orders.[56]

In the following chapters, Weber discussed special instances of groups (groups of characters, groups of linear substitutions, polyhedric groups) and presented some applications of group theory, such as Galois theory, invariants, and others. The last part of the second volume dealt with algebraic number theory. Following Dedekind, the fields considered are only fields of numbers.

---

54. Weber 1895, 644: "Eine der ältesten Fragen, an der sich vorzugsweise die neuere Algebra entwickelt hat, ist die nach der sogennanten algebraischen Auflösung der Gleichungen, worunter man eine Darstellung der Wurzeln einer Gleichung durch eine Reihe von Radicalen, oder die Berechnung durch eine endliche Kette von Wurzelziehungen versteht. Auf diese Frage fällt von der Gruppentheorie das hellste Licht."

55. Weber 1895, 649-652.

56. Weber 1895 Vol. 2, 121: "Die allgemeine Definition der Gruppe... lässt über die Natur dieses Begriffes noch manches im Dunkel.... In der Definition der Gruppe ist mehr enthalten, als es auf den ersten Blick den Anschein hat, und die Zahl der möglichen Gruppen, die aus einer gegebenen Anzahl von Elementen zusammengesetzt werden können, ist eine sehr beschränkte. Die allgemeinen Gesetze, die hier herrschen, sind erst zum kleinsten Theile erkannt, so dass jede neue specielle Gruppe, namentlich bei kleinerer Gliederzahl, ein neues Interesse bietet und zu eingehendem Studium auffordet.

Welche Gruppen sind zwischen einer gegebenen Zahl von Elementen, d.h. bei gegebenem Grade überhaupt möglich? Das ist die allgemeine Frage, um die es sich handelt, von deren vollständiger Lösung wir aber noch weit entfernt sind. Cayley hat diese Aufgabe zuerst für die niedriegsten Gradzahlen in Angriff genommen."

Our account of Dedekind's theory of ideals in Chapter 2 below could also serve as a detailed description of this section of Weber's book.

The third volume of the *Lehrbuch* appeared in 1908. This volume represents in fact a second edition of Weber's book on elliptic functions and algebraic numbers, first published in 1891.[57] It deals with the reciprocal interrelations between problems and techniques of the theory of fields of algebraic numbers and of the theory of elliptic functions. A detailed account of its content is beyond our interest here. However, a faithful description of contemporary images of algebra cannot fail to stress the importance of the connection established in this third volume between these two domains, algebra and the theory of elliptic functions. Our present account tends to stress the rise of concepts that were later to become part of modern algebra as well as the interplay among them. However, the kinds of conceptual and technical interconnections that were pursued by mathematicians like Weber during the second half of the nineteenth century in relation to algebraic problems cover a much broader spectrum; the third volume of Weber's book touches upon some important portions of that spectrum. The existence of these kinds of interconnections underscores the difficulties inherent to the use of one and the same term (algebra) to denote the disciplines known by this name in the nineteenth and in the twentieth-century.[58]

Our account of Weber's article and book shows a complex picture of his images of algebraic knowledge, comprising elements of both classical as well as more modern conceptions. The central issue of the first volume of the *Lehrbuch*—first published as late as 1895 and re-published without essential changes in 1898— was the resolution of polynomial equations, and its presentation remains similar to those of previous ones in textbooks of algebra. All the concepts and techniques related to Galois theory (in particular the concepts of group and field) are introduced, to a large extent, only as ancillary to that central issue. The treatment accorded to groups in the book is especially illustrative of this situation. By the end of the century, group theory was the paradigm of an abstractly developed theory, if there was any. It was, perhaps, the only discipline in contemporary mathematical research that may be really qualified as "structural." Research on groups had increasingly focused on

---

57. Weber 1891.

58. Some additional details about this interrelation between algebraic number theory and the theory of elliptic functions may be gathered from the discussion in the opening passages of § 2.2.1 below.

questions that we recognize today as structural, and, at the same time, the possibility of defining the concept abstractly had been increasingly acknowledged. More importantly, the idea that two isomorphic groups are in essence one and the same mathematical construct had been increasingly adopted. Weber's 1893 article exemplifies clearly this trend. Yet, in Weber's book, group theory plays a role which, at most, may be described as ambiguous regarding the overall picture of algebra. For, although in its second volume, the theory of groups is indeed presented as a mathematical domain of intrinsic interest for research and many techniques and problems are presented in an up-to-date, structurally-oriented fashion, the theory appears in the first volume as no more than a tool of the theory of equations (albeit, it is now clear, a central one). Weber's textbook, and much more so his 1893 article, bring to the fore the interplay between groups and fields abstractly considered more than any former, similar work. However, in spite of this, the classical conceptual hierarchy that viewed algebra as based on the essential properties of the number systems is questioned in neither of these two works.

Weber's *Lehrbuch* became the standard German textbook on algebra and underwent several reprints. Its influence can be easily detected, among others, through the widespread adoption of a large portion of the terminology introduced in it.[59] Thus, the image of algebra conveyed by Weber's book was to dominate the algebraic scene for almost thirty years, until van der Waerden's introduction of the new, structural image of algebra. But obviously, influential as the latter was on the further development of algebra, it did not immediately obliterate Weber's influence, which can still be traced to around 1930 and perhaps even beyond. We will return briefly to comment on this point after describing van der Waerden's *Moderne Algebra* and the image of knowledge conveyed by it.

1.3   Bartel L. van der Waerden: *Moderne Algebra*

Bartel Leendert van der Waerden was born in Amsterdam in 1903, where he entered the university in 1919. After graduating there, he travelled to Göt-

---

59. Cf. Mehrtens 1979, 145; Novy 1973, 228; Wussing 1984, 251. Wussing 1989, 245, attributes to the modern, structural character (in his words) of Weber's book the source of its success and influence, and also points out that the next stage in the evolution of algebra was marked by van der Waerden's book. Wussing does not, however, explain here the differences between the two texts.

tingen and Hamburg, where he worked with the then leading German algebraists: Emmy Noether and Emil Artin respectively. In 1926 he received his
doctorate in Amsterdam, in 1927 habilitated in Göttingen, in 1928 he was
Ordinarius at Groningen and in 1931 professor in Leipzig.[60] By that time, the
German school of algebra was at its apogee, with many brilliant, fruitful
researchers continually contributing hosts of new results. As frequently happens in the history of mathematics, an urgent need was felt for a compilation
and systematic exposition of those advances.[61] Van der Waerden was not the
only one to feel the need to publish a textbook of algebra by that time. One can
also mention three additional contemporary examples: *Modern Algebraic
Theories* (1926) by Leonard Eugene Dickson, *Höhere Algebra* (1926) by Helmut Hasse and *Einführung in die Algebra* (1929) by Otto Haupt. Van der
Waerden and these three other mathematicians were faced with similar
advances in algebra when they came to publish their respective books. It was,
however, van der Waerden's book that turned into the new classical text of
algebra, while simultaneously algebra became the discipline presented in van
der Waerden's book. The book was translated into many languages, being
widely used in universities all around the world. The success of *Moderne
Algebra* was responsible for the widespread absorption of the idea of "algebraic structure." It was also probably responsible for the adoption of the idea
that the rise of this concept was a natural, unavoidable outcome of the growth
of algebra. An examination of the contents of *Moderne Algebra* in the present
section will give us a better understanding of the meaning of its "structural"
character. Likewise, our examination of other textbooks of algebra published
in 1924 or thereafter (in § 1.4 below) will show that van der Waerden's structural presentation of algebra was not the only conceivable alternative in the
late twenties. Van der Waerden's exposition of existing results was an original
contribution in itself, and a highly important one at that, since it presented a
new image of mathematical knowledge: the idea of "algebraic structure" in
full strength.

Roughly stated, the aim and contents of van der Waerden's book may be
characterized by a formulation which may seem rather trivial today: *to define*

---

60. For biographical details on van der Waerden and an account of his important mathematical
contributions—which, beyond *Moderne Algebra*, we don't discuss at all here—see Frei 1993. For a
more recent biography written after his death in 1997, see Dold-Samplonious 1997.

61. Accounts of the immediate background to the publication of *Moderne Algebra* appear in
Benis-Sinaceur 1987, 11-16; van der Waerden 1975.

the diverse algebraic domains and to attempt to elucidate fully their structure.
What is meant by "defining an algebraic domain"? Van der Waerden defined
algebraic domains in two different ways: (1) by endowing a non-empty set
with one or two abstractly defined operations (such as in the case of groups,
rings, hypercomplex systems, etc....) or (2) by taking an existing algebraic
domain and constructing a new one over it, by means of a well-specified procedure (field of quotients of an integral domain, rings of polynomials of a
given ring, extension of a field, etc.).

What is meant by "elucidating the structure of an algebraic domain"?
There is no clear-cut answer to this question and, in fact, it is never explicitly
discussed in *Moderne Algebra*. Chapter V, for instance, opens with the statement: "The aim of this chapter is to give a general view of the structure of
commutative fields, and of their simplest subfields and extension fields."[62] A
field is defined in Chapter III of the book, following the first of the two possible ways described above, namely by endowing a non-empty set with two
operations which satisfy certain conditions. But what must be known of a field
(or of any other "algebraic domain", for that matter) in order to claim that we
know its "structure"? Although this latter question is not explicitly asked in
van der Waerden's book, an implicit answer to it may be found by examining
the kinds of more specific, technical questions and answers that repeatedly
arise in the actual discussion of the different domains.

In order to understand in more precise terms what van der Waerden's analysis of the structure of the various algebraic domains consists of, it is necessary to stress from the outset that at no place in the book does he state what an
algebraic structure is, either at the general, non-formal level or by means of the
introduction of a suitable, rigorously defined, mathematical concept. Van der
Waerden's approach is based on the recurrent use of several common fundamental concepts and questions. Among the most salient of such recurring concepts, one finds isomorphisms, homomorphisms, residue classes, composition
series and direct products, etc. None of these concepts are defined in a general
fashion so as to be a-priori available for each of the particular algebraic systems. Rather they are simply put to use repeatedly in each of those systems,
and they are redefined in the corresponding setting as the need arises. The
example of homomorphisms illustrates this situation.

---

62. All the following quotations appear in the English translation of the second edition (see bibliography).

In the opening chapters, van der Waerden gives a general definition of homomorphisms. A homomorphism is defined as a bijection "preserving a relation defined on the set." But this non-formal definition cannot actually be used for any practical purpose. Van der Waerden thus defined homomorphism again separately in the relevant chapter for each of the particular structures (groups, rings, etc.), without mentioning its connection with the more general idea. He specifically showed, e.g., that the homomorphic image of a ring is itself a ring. The idea of a homomorphism is thus *actually used* as a useful, unifying concept for investigating all algebraic domains, even though it is not *formally defined* as a general algebraic concept. Likewise, isomorphisms are defined separately for groups and for rings and fields. Van der Waerden showed *in each case* that the relation "being isomorphic to" is reflexive, transitive and symmetric. Thus the important notion, constitutive of the structural image of algebra, that two isomorphic constructs actually represent one and the same mathematical entity, appears only implicitly in the book. This notion plays, of course, an instrumental role in the elucidation of the structure of each and every domain considered, yet it is accompanied by no explicit declaration concerning its central role.

There is, indeed, an incipient attempt in the book to formulate a general formal concept of structure, namely, "Groups with Operators" which are considered in Chapter VI. Groups with operators had been first defined by Wolfgang Krull in a paper of 1925 (§ 6.1 below); van der Waerden used this concept in a similar way to that proposed by Krull and proved some theorems that are simultaneously valid for several different algebraic domains. Among them, we may mention the first theorem of isomorphism, the Jordan-Hölder theorem and some other results concerning the relation of the latter theorem to direct products. These results are not presented, however, as general theorems necessary for elucidating the structure of all algebraic domains. Moreover, van der Waerden's structural image of algebra does not hinge upon this concept, and in fact, it would remain unchanged, even if this particular chapter were suppressed from the book.

The concepts recurrently used by van der Waerden in his analysis of algebraic domains are instrumental in building the conceptual hierarchy that stands at the basis of his image of algebra, and which constitutes one of his central departures from previous textbooks of algebra. The conceptual framework of the book and, more particularly, the role of the various systems of numbers in the general fabric of algebra are set out in the introductory chapter.

Van der Waerden introduced the natural numbers through a cursory review of Peano's axioms. He then extended them to the integers through the (informally presented) construction of pairs of natural numbers. But then, neither the rational nor the real numbers are mentioned in this introduction, since they are defined later within a strictly algebraic setting in which only their relevant (namely, algebraic) properties are discussed in the following chapters. Thus whereas previous textbooks of algebra (and above all Weber's *Lehrbuch*) assumed the basic properties of the rational and real numbers as foundational for algebra, these systems are no longer considered in van der Waerden's book as given entities whose properties must be mastered before beginning with algebraic research. Rational and real numbers have no conceptual priority here over, say, polynomials. Rather, they are defined as particular cases of abstract algebraic constructs. Thus, in Chapter III, the concept of a field of fractions is introduced for integral domains in general and the rationals are then obtained as a particular case of this kind of construction, namely, as the field of quotients of the ring of integers (p. 42).

Characterizing the real numbers in purely algebraic terms necessitates a somewhat more elaborate procedure than the rationals. The problem lies in the possibility of defining—in abstract fashion and in purely algebraic terms— properties such as absolute value, "being a pure real" (or a pure imaginary) number, positiveness, etc., within an algebraic number field. Consider for example the equation $x^4 = 2$. Let w be a purely real, and iw a purely imaginary root of this equation. Define an isomorphism from the extension of the field of rationals by w, $\Gamma(w)$ into the extension $\Gamma(iw)$ by mapping w into iw. This isomorphism preserves the main basic algebraic properties, but the real number w is mapped into the imaginary number iw, the positive number $(w)^2 = \sqrt{2}$ into the negative number $(iw)^2 = -\sqrt{2}$, and, accordingly, the number $1 + \sqrt{2}$, whose absolute value is greater than one, into $1 - \sqrt{2}$, whose absolute value is less than 1.

The possibility of characterizing the basic arithmetical properties of the field of real numbers in "purely algebraic terms" (i.e., independently of considerations of continuity, or similar, "analytical", ones) had been systematically explored by Emil Artin and by his collaborator Otto Schreier (1901-1929), in their mathematical seminar at the university of Hamburg between 1924 and 1927. Van der Waerden attended this seminar during the critical

years of 1926-27, when Artin and Schreier published their path-breaking articles on "real fields."[63] In these articles a real field was defined as an abstract
field having the additional property that -1 (i.e., the opposite—with respect to
addition in the field—of the field's unit) cannot be written as a sum of squares
of other elements in the field. Artin and Schreier were also able to define, in a
similar fashion, additional properties such as order and closure. More importantly, they were able to prove many significant theorems concerning the systems of real and complex numbers, and concerning polynomials over these
fields, based exclusively on this kind of algebraic considerations. In particular,
Artin and Schreier were able to characterize in purely algebraic terms the differences between the main fields of numbers: rationals, algebraic, real and
complex.[64]

In *Moderne Algebra* (§ 70), van der Waerden followed Artin and
Schreier's work very closely when he gave his purely algebraic definition of
the reals as a specific case of an "Ordered Field."[65] It is significant, however,
that in doing so, he did not give up completely the old image of algebra, and
felt the need to include a standard "analytic" definition of the real numbers
using Cauchy sequences (and adopting the term originally introduced by Cantor to designate them: "fundamental sequences"). This is the only section of
the book where an ε–δ argument appears. As we saw, in Weber's *Lehrbuch*,
several chapters had been devoted to analytic questions concerning polynomials, including interpolation and approximation techniques. After van der
Waerden, on the other hand, textbooks of algebra tended to completely
exclude arguments of this sort.[66] But in 1930, van der Waerden apparently still
felt that a book on algebra cannot completely do without such considerations,
even though they are not actually needed for the theories developed in it.
Indeed, van der Waerden explicitly acknowledged that ε–δ arguments are
extraneous in character to his approach to algebra. At the same time, however,
he justified their inclusion, claiming that the purely algebraic character of the

---

63. See Artin & Schreier 1926; Artin & Schreier 1927. A very detailed account of the significance of the contribution of Artin and Schreier to the rise of modern algebra appears in Sinaceur
1991. See especially pp. 145-254. For a shorter account, see Sinaceur 1984. According to van der
Waerden's testimony (Sinaceur 1991, 27) he attended the seminar and assisted Artin and Schreier,
day after day, in their elaboration of the theory of real fields.

64. See Sinaceur 1991, 241 ff.

65. See van der Waerden 1975, 38.

66. For instance in Chevalley 1956, fields "are of necessity left outside" and the real numbers
are not even mentioned.

book is not altered by doing so, since the definition of the real numbers in which those arguments appear is not used later on:

> The reason for this procedure is not so much the fact that it is a logical necessity to begin with it, but because the problems involved in the purely algebraic theory become clearer as soon as we know what the real and complex numbers actually are, and because we can at the same time discuss the important basic concepts of ordering and of a fundamental sequence. (p. 208)

Thus the analytic definition is included, but it plays only a heuristic role within the book.

The task of finding the real and complex roots of an algebraic equation—the classical main core of algebra so far—was relegated in van der Waerden's book for the first time to a subsidiary role. Indeed, three short sections in the chapter on Galois theory deal with this specific application of the theory, and at this point no previous consideration of the properties of real numbers were necessary to develop it.[67] In this way, two of the most central building blocks of the edifice of classical algebra as conceived until the end of the nineteenth century, the systems of the rational and the real numbers, are presented here merely as final products of a series of successive algebraic constructs, the "structure" of which was gradually elucidated. At the same time, other, non-algebraic properties (continuity, density) of these systems were not considered at all.

This possibility of concentrating on a single feature of a classical object of mathematics is, doubtless, an important aspect of the study of algebraic "structures" and one of the most striking innovations reflected in the book. Naturally, this possibility is afforded by the abstract formulation of concepts, but the mere availability of such a formulation was in itself not enough. As we saw, Weber was fully aware of the abstract formulation, yet his conception of algebra was absolutely subordinated to the preliminary knowledge of the system of real numbers, and he considered it necessary to begin by elucidating *all* of the properties of the latter.

Within this conceptual setting, and together with the *concepts* which are repeatedly used in the book, one finds several *problems* which are repeatedly considered by van der Waerden in the study of each algebraic system. Giving an account of these problems is necessary to understand the essence of van der

---

67. On the situation in Galois theory around 1930, and its relation to van der Waerden's presentation, see Neumann 1997, 307-311.

Waerden's image of algebra. For instance, factorization and its behavior in a given domain is a central question in the research of structures. For every domain, some specific kind of subdomain is usually considered, which may be taken as playing a role similar to the role that prime numbers play for the ring of integers. Accordingly, the elucidation of the structure of a domain typically involves the elucidation of the relationship between a given element of the domain and its "prime elements" and the problem of "prime factorization" is usually discussed for the different domains. Thus, simple groups play the role of such "prime elements" in group theory, prime ideals in ring theory, irreducible polynomials in rings of polynomials, prime fields in fields, and so on. For instance, Chapter XII—on the general theory of commutative rings—is entirely devoted to elucidating the role of the relevant "prime elements", a role played in this case by the prime ideals:

> In this chapter we shall investigate the divisibility properties of the ideals of commutative rings and determine the extent to which the simple laws that are valid in a domain such as integers may be carried over to more general rings. (Vol. 2, p. 18)

Chapter XIV, on integral algebraic quantities, is by definition devoted to the problem of factorization. Likewise, many of the theorems of linear algebra and, naturally, those of hypercomplex systems, deal with decomposition into more basic elements (nilpotent ideals, semisimple rings, etc.).

A further question which is central to the study of the structure of a domain is closely related to the problem of factorization, namely, the relation between a given domain and the system of its subdomains. In fact, the research of "prime elements" and factorization within a given domain is only one aspect of this broader question. Galois theory provides a classical example of this direction of research inasmuch as it "is concerned with the finite separable extensions of a field K... It establishes a relationship between the extension fields of K, which are contained in a given normal field, and the subgroups of a certain finite group" (p. 153). So central is this feature to the structural approach, that in the mid-thirties, Garrett Birkhoff and Oystein Ore independently attempted a formalization of the very concept of "algebraic structure" by studying the lattice of certain subdomains of a given algebraic domain (§ 6.3 below).

There is an additional kind of question which is also closely connected with the one discussed in the last paragraph and which appears throughout the book. It was said above that an algebraic domain may be created by taking a specific domain and by performing certain standard algebraic constructs on it (e.g., fields of fractions of an integral domain, etc.). There arises the question as to what extent specific properties of the original domain are reflected in the new one. For example, if a given ring is an integral domain, the same is true for its ring of polynomials (p. 47); if a given ring satisfies the *base condition* (namely, that every ideal in it has a finite generating set) then any quotient ring of that ring and also the polynomial ring associated with it satisfy the same condition (Vol. 2, p. 18). One may also ask which properties are passed over from a given algebraic domain to its subsystems or to its quotient systems. These kinds of questions are exemplified in van der Waerden's study of the structure of fields (Ch. V). This chapter was modelled directly on the seminal work of Ernst Steinitz, who had advanced a completely new approach for the treatment of abstract fields as an issue of intrinsic interest in 1910 (§ 4.2 below). Emmy Noether and the Göttingen algebraists had been deeply impressed by Steinitz's work (§ 5.1 below), and van der Waerden also studied it in detail under her guidance. In his book, van der Waerden pointed out the basic role played by "prime fields" as building blocks of the theory, and he claimed that after all the properties inherited from the prime field by its extension are known, then the structure of all fields is also known.

A last kind of recurring problem that can be mentioned as part of this inventory concerns the possibility to realize the whole structure of certain algebraic systems by examining only a limited set of data. Thus, in the case of a vector space, its structure is completely determined (up to isomorphism) by the field of coefficients and by the cardinality of the base. The structure of a hypercomplex system (or algebra), as a second example, is completely determined by n^3 constants that provide a multiplication table for the n elements of the base. In Van der Waerden's book we find repeated attempts to establish similar results for various algebraic systems.

The idiosyncratic "structural" character of van der Waerden's approach arises, then, from the combined and persistent use of all the above mentioned factors: the recurrent use of certain kinds of concepts, the focusing on certain, specific kinds of problems. Yet there is a further central component of his approach which has not been mentioned so far and which is usually identified, more than anything else, with the "structural approach", namely, the reliance on the modern axiomatic method. This component deserves separate attention. In *Moderne Algebra* van der Waerden indeed presents algebraic theories in a completely abstract way, which no other comprehensive exposition of algebra had ever done before. However, all the other factors just mentioned must help us realize that identifying the innovative character of the book with its use of the axiomatic method alone would be misleading. The real innovation implied by the book can only be understood by considering the particular way in which van der Waerden exploited the advantages of the axiomatic method in tackling the above mentioned issues.

The fact that the modern axiomatic method, on the one hand, and the structural concerns inventoried above, on the other, became the focus of interest of algebraic research simultaneously does not necessary imply that there was a *necessary* connection between the two. It is clear, that the modern axiomatic approach provides a very effective and economic way of dealing with those issues. Moreover, the relevance of the above mentioned questions becomes much more evident when formulated in abstract axiomatic terms. But as the following chapters will show, it was the internal development of the very theories presented in the book that led to a shift of interests from the classical to the new central problems of algebra and brought about the transition from the classical images of algebra to the new, structural ones. Such developments were not directly connected with the use of the modern axiomatic method or of abstract formulations in general, but rather they were dictated by the solutions proposed for the existing problems of the discipline. It was the fact that similar sorts of problems arose in the different disciplines now included under the heading of modern algebra, and that at the same time they were addressed from an increasingly common perspective, that led to the unified exposition of van der Waerden. The full adoption of the abstract axiomatic formulation was only a part of a wider development.[68]

The above account of the contents of *Moderne Algebra* may seem somewhat trivial and superfluous to anyone who received his mathematical training in the last fifty years; but it is this inventory and its comparison with the nine-

---

68. Some attempts to characterize modern trends in mathematics tend to focus almost exclusively on foundational issues, and in particular on the adoption of the axiomatic approach in the various branches of mathematics, while paying much less attention to the decisive role of the internal developments of the disciplines themselves in bringing about the adoption of modern approaches. A recent example of this appears in the already mentioned, ambitious attempt to characterize modern mathematics as a whole, advanced in Mehrtens 1990. Mehrtens discusses the role of algebra as a leading metaphor (*Grundmetapher*) in the rise of the modern trends in mathematics (especially in pp. 497-505).

teenth-century images of algebra that allows an understanding of the essence of the "structural approach" of *Moderne Algebra*.[69] Although it is commonly accepted that van der Waerden's *Moderne Algebra* was the first text in which the "modern structural approach to algebra" was fully implemented, the meaning of the latter expression is usually taken for granted. The above account is an attempt to indicate *precisely* what it is that makes this approach structural, and in what sense we cannot agree with van der Waerden's claim that modern algebra began with Galois. Although earlier versions of the central concepts discussed in *Moderne Algebra* were indeed known and increasingly investigated in the decades after Galois, important discoveries in the body of knowledge and basic changes in the images of knowledge were still needed before reaching the point where the interconnections among those concepts became a focus of interest. Moreover, the crucial change came about with the clear realization that those concepts may be seen as different varieties of a same species (namely, different kinds of algebraic structures), and that the central disciplinary concern of algebra is the study of those different varieties through a common approach. In fact, this fundamental realization appears in *Moderne Algebra* not only implicitly, as the above detailed account of the aims and scope of the various chapters of the book shows, but explicitly and even didactically epitomized in the "*Leitfaden*" appearing in the introduction to the book, showing the hierarchical, structural interrelation between the various concepts investigated in the book. Thus, van der Waerden's textbook may be considered, more than any other work, the first thorough presentation of the whole discipline of algebra under its newly consolidated image.

The last section of the present chapter discusses the images of algebra displayed by other textbooks of algebra published at about the same time as *Moderne Algebra*. This discussion is meant to show how, given a certain stage of development of the body of knowledge, two or more different systems of images of knowledge might be chosen to produce entirely different disciplines

---

69. It is worth comparing this account with Cartan 1958. In his expository lecture Cartan set out to explain to high school teachers the idea of an algebraic structure. Although he also defined the concept formally as a set on which one or two operations are abstractly defined (p. 8), his real explanation seems to be his subsequent account, similar in a certain sense to the one given here. Of course, Cartan does not claim, as we do here, that his account is *the* explanation of what an algebraic structure is. In fact, given the intended audience Cartan had to explain in greater detail some technical definitions (e.g., ring of polynomials, etc.) and he laid lesser stress on the kind of questions addressed in elucidating algebraic structures and techniques used to solve them.

or systems of knowledge. In particular this will underscore the specific contribution of van der Waerden in his textbook of 1930.

1.4   Other Textbooks of Algebra in the 1920s

The changes in the images of algebra implied by van der Waerden's book,
and the pervasiveness of the classical nineteenth-century conception of algebra as embodied in Weber's *Lehrbuch* can be better understood by examining
other textbooks of algebra, published during the 1920s. This period, as will be
seen in the following chapters, contributed many important additions to the
body of algebra. These, however, were not fully incorporated into the structural image of algebra before van der Waerden's text of 1930. By examining
other textbooks, it can clearly be seen that the organization of that enlarged
body of algebra did not necessarily convey the structural image, but rather,
that the latter was a specific choice of van der Waerden (and therefore a true
innovation in itself) among other possible alternatives.

First we examine the *Lehrbuch der Algebra* by Robert Fricke (1861-
1930), published in 1924. Fricke had collaborated with Felix Klein in the latter's work on the foundations of the theory of elliptic modular functions and
automorphic functions. His own contributions to the latter subject helped to
establish the central role played by groups in that discipline.[70] Later, he coedited the complete works of Klein (1921-23). Some years after the publication
of his textbook on algebra, he also coedited the complete works of Dedekind
(1930).

In 1920 Fricke published a book on elliptic functions which contained a
brief section on Galois theory. Afterwards he repeatedly expressed the desire
to publish an expanded version of that section. The opportunity presented
itself when F. Vieweg, the Braunschweig publisher of Weber, announced that
Weber's *Lehrbuch* had sold out, and asked Fricke to write a new book on the
same subject. Fricke's new textbook of algebra came out in 1924, bearing the
subtitle: *"Verfasst mit Benutzung vom Heinrich Webers gleichnamigem
Buche"*. This was Fricke's way of expressing his gratitude for the many
insights gleaned from Weber's book, although, in the preface he stated that he
would have preferred that Weber stayed closer to Dedekind's original point of
view.[71] In any case, Fricke basically reproduced the same image of algebra

---

70. Cf. Wussing 1984, 212-213.

advanced in Weber's *Lehrbuch*, while adapting the body of algebraic knowledge to some of the developments that had taken place in the field.

The first volume of Fricke's *Lehrbuch* is similar to that of Weber's in that its main aim is the thorough study of the theory of equations including such aspects as were later excluded from standard textbooks on algebra (e.g., approximations). It is different, however, in that groups are treated in a more systematic, abstract fashion. Thus, the theory of groups is first developed on its own in the early chapters, without mentioning its connection with the theory of equations. It is used later on, in those sections where the foundations of Galois theory are developed. Fricke's direct familiarity with the theory of groups certainly accounts for this approach. By contrast, being less acquainted with number theory, his treatment of fields is rather *ad-hoc*; the theory of fields is not developed in advance, but rather as part of the discussion on Galois theory. Fricke introduced here the concept of "field of numbers" to denote a set of numbers closed under the four operations (p. 354). He considered here only *fields of numbers*, like Dedekind before him and like Weber in his book, but *unlike Weber in his 1893 article*.

But the differences in the treatment accorded by Fricke to the concepts of group and of field are indicative of more than just his own mathematical background and preferences. In fact, when comparing both textbooks, Weber's and Fricke's, to that of van der Waerden, it is at this point that the main difference between the classical nineteenth-century and the structural image of algebra comes to the fore. As already stated, a central trait of the structural image of algebra is the reconsideration of several existing algebraic concepts (e.g., groups and fields) as *different instances of a single, underlying general idea, namely, the idea of an algebraic structure*. This point of view is absent from both Weber's and Fricke's images of algebra. Moreover, precisely the fact that Fricke's presentation of groups is much more abstract than Weber's, and that nevertheless both of them coincide in not recognizing the various algebraic systems as individual manifestations of a common underlying idea, helps stress a main thesis we have been discussing so far, namely, that awareness of the conceptual unity of the various algebraic structures is not a necessary consequence of their being formulated in abstract terms.

---

71. See Fricke 1924, *v*: "...für den Standpunkt eines 'Lehrbuches' der Algebra ein reinerer und engerer Anschluß an die Grundauffassungen Dedekinds rätlicher erscheinen wäre."

But Fricke's lack of a unified understanding of the notion of algebraic structure is manifest not only in his diverging approaches to the treatment of groups and fields. There is also more direct evidence for it. Fricke explicitly acknowledged (in a footnote) the unrealized possibility of improving his own approach, following a suggestion made to him by Emmy Noether. Noether had indicated to Fricke that one could define a field in a more general way than he had actually done, if the concept was approached in a manner similar to that adopted for groups: one could postulate two operations for the elements of an arbitrary set and demand some conditions to hold for these operations. Fricke remarked that he was aware of this kind of definition, since it had already been given by Weber in 1893. But evidently, he had not considered Weber's definition to be worthy of special attention until Noether's remark. And even after Noether's remark, Fricke did not change the approach adopted in his book. He explained that Weber's abstract definition had not produced significant changes regarding the relevance attributed to the concept of field, and he added in his footnote that:

> E. Steinitz approached the problem in detail in the years 1908-1910 in his work "Algebraic Theory of Fields" (1910) and showed that it involves a fruitful extension of the original concept of a field of numbers. (Fricke 1924, 354)

Steinitz's work and its influence on the development of field theory are analyzed in § 4.2 below. At this stage, however, it is important to bear in mind the attitude of a mathematician like Fricke who, while holding Weber's contributions to algebra in high esteem, did not even consider in 1924 the possibility of using the axiomatic definition of a field that the latter had already introduced in 1893. It took more than two decades after this article to finally show the convenience and fruitfulness of developing this approach for the whole of algebra; this happened only after Noether's research and the publication of *Moderne Algebra*, but not before. Were it not for Noether's suggestion, Fricke would not have even mentioned Steinitz's influential article and his innovative approach to fields.

In the already mentioned footnote, Fricke also mentioned the classification of fields according to the characteristic that may be either zero or p (p prime). He attributed this classification to Steinitz. He further commented, following Steinitz, that not all the theorems valid for fields with characteristic zero are also valid for fields with characteristic p (p prime). Fricke warned that all the fields considered in his text were of characteristic zero; therefore, he developed Galois theory only for fields arising within the sphere of his main interest, namely, the theory of polynomial equations. This approach remains faithful to the central concerns of Weber's image of algebra, and contrasts with the one adopted by van der Waerden (following Steinitz). For the latter the abstract field itself served as the real focus of interest, making fields of arbitrary characteristic equally interesting. The innovations of Steinitz's approach were not adopted in Fricke's book. They appear neither at the level of the technical definitions nor at the level of the images of algebra at large; they simply did not fit Fricke's overall image of algebra.

Weber's article of 1893, with its unified presentation of groups and fields, may today appear as having expressed a conscious understanding of both concepts as part of a more general idea; but in fact, as we have seen, Weber himself did not choose to organically incorporate this approach in his textbook of algebra in 1895 (nor in the following re-editions of it).[72] Fricke's treatment of groups in 1924 departed from that of Weber's, and followed the new, more mature abstract approach to the theory as it had been already presented in textbooks from the beginning of the century. But when it came to fields, even the definition (not to mention the exposition of the results) was far behind contemporary research, not only concerning technical results (as may be typically the case with textbooks), but also concerning images of algebra.

The second book we consider in this section is *Höhere Algebra* by Helmut Hasse, first published in 1926. In the introduction to his book, Hasse described algebra as "the theory of solving equations... with a number of unknown quantities." The two volumes of this book are entirely dedicated to that problem; some of the new developments of algebra, later included in *Moderne Algebra*, do appear here, but they are totally subordinated to the above-mentioned central aim. More specifically, in the introduction Hasse tentatively defined "the basic problem of algebra" in the following terms:

---

72. Novy 1973, 228, wrote on Weber's textbook: "Although the author wanted to connect various theories... [t]he study of algebraic structures was still mostly hidden in details not sufficiently generalized. Even the fundamental concepts, including the concept of a field and a group, remained illustrative and defined only, provided the theories which were being interpreted with their help required it [sic]." This historical assessment seems to me to imply a much clearer conception of "algebraic structure" than actually existed. I would say that the idea of algebraic structure is not "mostly hidden in details" here, but rather totally absent.

To develop general formal methods for solving equations between known and unknown elements of a field in terms of the unknowns, where the equations are formed by means of the four elementary operations. (Hasse 1926 Vol. 1, 6-7)

Fields, rings and integral domains are defined in the first chapter of the book, and some basic results are derived, but only in order to be able to formulate the "basic problem" in more precise terms. One thus observes a certain sloppiness in Hasse's use of this terminology.[73] Yet, after some technical definitions, Hasse restates the basic problem more precisely in the following terms (p. 48):

Let K be a field and $f_1,...,f_m$ elements of $K[x_1,...,x_n]$. There shall be developed methods for obtaining all solutions of the system S of equations

$$f_i(x_1,...,x_n) = 0 \quad (i = 1,...,m).$$

Thus, Hasse's image of algebraic knowledge is, in essence, the same as that previously seen in Weber's book, despite the substantial growth in the body of knowledge that took place during the intervening time. Hasse was able to express the classical task of algebra in more precise terms, which comprises the resolution of systems of linear equations (i.e., when the functions f_i are linear), on the one hand, and polynomial equations, on the other. At the same time, he also proposed some new generalizations of the old task. In a footnote, Hasse suggested the possibility of having a system with countably many equations, or of substituting the basis field K by an integral domain. This formulation could not have appeared in Weber's book, yet the overall conception of algebra as a discipline is not changed in Hasse's presentation of the discipline.

The two volumes of Hasse's book correspond to the two above mentioned subtasks. The first volume deals with the resolution of systems of linear equations. The second with the solvability of polynomial equations. Hasse wrote that in order to cope with this task one must consider the more general question of the structure of the field of roots of the algebraic equation, "in particular the question of its construction from the simplest possible components."[74] Here, in treating fields, Hasse followed Steinitz's presentation very closely. He did

---

73. Thus for instance on p. 13 one reads: "Bereich bedeutet zwar hiernach dasselbe wie *Ring*; jedoch ist der neutrale Ausdruck *Bereich* im angegebenen Sinne geläufiger, während man *Ring* gewöhnlich nur dort anwendet, wo wirklich kein Integritätsbereich vorliegt." (Italics in the original)

74. Hasse 1926 Vol. 2, 7: "Vom theoretischen Standpunkt erhebt sich nun aber... die hier ganz besonders interessante Frage nach der *Struktur der Wurzelkörper algebraischer Gleichungen*, insbesondere nach ihrem Aufbau aus möglichst einfachen Bestandteilen." (Italics in the original)

not, however, extend this program (as did van der Waerden) to the whole of algebra. Once again, this fact should not be attributed to Hasse's ignorance of recent developments in algebra, and the possibility of presenting other branches of algebra in a similar way, but rather to his traditional conception of the aim and scope of algebra as a separate mathematical discipline. The concept of ideal, which had become so central to algebra in the works of Emmy Noether and Emil Artin (as will be discussed in detail in Chapter 5, below), appeared in Hasse's book only in a footnote, in which Hasse remarked on its centrality for the "theory of decomposition."[75] He himself, however, made no use of that concept.

It is hard to specify how much attention Hasse's book received within the mathematical community in Germany or elsewhere, in the years following its publication. It is known, however, that in the long run *Moderne Algebra* became the standard textbook of algebra. Moreover, in the English translation of Hasse's text, based on the third German edition, one finds an interesting remark which was absent from the first edition:

> It is characteristic of the *modern* development of algebra that the tools specified above [i.e., groups and fields] have given rise to far-reaching autonomous theories which are more and more replacing the basic problem of *classical* algebra, cited above, as the center of interest. Thus in the modern interpretation algebra is no longer merely the theory of solving equations, but the *theory of formal calculating* domains, as fields, groups, etc.; and its *basic* problem has now become that of obtaining an insight into the *structure of such domains*... However in the restricted space of these volumes, it is not possible to put the more general modern viewpoint in the foreground. Therefore, we take the basic problem of classical algebra, expressed above, as a directing guide and defining frame for our presentation, though we will also be led, particularly in Vol. 2, to structural statements in the sense of modern algebra. (Hasse 1954, 11. Italics in the original)

Thus Hasse seems to acknowledge in his later editions, that the conception of the discipline as originally manifest in his textbook had become somewhat outdated with the publication of *Moderne Algebra*. Naturally, it was not possible to adapt his own book to the new dominant images of algebra without rewriting it altogether. Thus, besides this brief remark, Hasse went on to publish his book in its original format, while implicitly admitting this situation. However, the very fact that thirty years after the publication of its first edition

75. Hasse 1926 Vol. 1, 23. In the third edition Hasse added a specific reference to Emmy Noether's work on the decomposition of abstract ideals.

and about twenty five after the publication of *Moderne Algebra*, an English translation of Hasse's book was published, indicates that in spite of the great success of the new image of algebra in certain quarters, there was still room for teaching and perhaps even research following the classical image of algebra, an addressing issues such as covered by Hasse's but not by van der Waerden's textbook.

A third textbook which deserves attention here is Leonard Eugene Dickson's *Modern Algebraic Theories*, published the same year as Hasse's book. Dickson's book grew out of the courses taught by the author in the years preceding its publication[76] and it contained the essentials of several theories centering around concepts such as matrices, invariants and groups, "which are among the most important concepts in mathematics." Dickson wrote in the introduction that he envisaged four possible courses that could be taught using his book. These courses, which were seen by him as covering the entire domain of algebra, contain very little of what van der Waerden included in his own book. Thus, Dickson proposed the following four uses for his book:

1. **Higher Algebra:** Including matrices, linear transformations, invariant factors, and quadratic, bilinear and Hermitian forms. These classical issues, wrote Dickson, were presented in his book in a new and elementary fashion.

2. **Algebraic Equations:** namely, Galois theory. According to Dickson "the usual presentation of group theory makes the subject quite abstruse"; in order to overcome this problem, he introduced groups as substitution groups directly related to the solutions of the cubic and quartic equation.

3. **Icosahedron - Linear groups:** Klein's theory of the quintic equation.

4. **Algebraic Invariants:** Treated in the classical way of Paul Gordan and Max Noether.[77]

---

76. On Dickson's mathematical interests and his influence on the development of the mathematics department at the University of Chicago see Parshall and Rowe 1994, 379-381. For more detailed accounts of his work on algebra see Fenster 1997, 1998, 1999. See also Albert 1955.

77. The classical approach to invariant theory and the change initiated by Hilbert in this discipline are discussed below in § 3.1.

Not only is the idea of "algebraic structure" alien to this book, but even the abstract formulation of the central concepts, groups and fields, is almost totally absent from it. Thus, for instance, after substitution groups are defined in the discussion of algebraic equations, the definition of an abstract group is added, in small print, as an unimportant aside (p. 144). Fields of complex numbers or of rational functions are defined separately, without mentioning any connection with the idea of group (p. 150). Dickson did mention several generalizations of the concepts involved in Galois theory, but not in the abstract direction. Rather he focused on generalizations that arise directly from the kinds of problems addressed in the theory of equations: the group of a system of algebraic equations, a Galois theory for linear differential equations and for complete systems of linear partial differential equations, and so on. The theorem of Jordan-Hölder, whose development, as will be seen in the following chapters, is closely connected with the development of the structural approach to algebra, is mentioned in Dickson's book only in a footnote (p. 188).

Dickson's book covered a fairly up-to-date body of algebraic knowledge. Nevertheless, the image of algebra reflected in it has little in common with that of van der Waerden's book. Dickson's book became the most advanced algebra text available in the USA and it was not until 1941 that a new one, better adapted to recent developments of algebra, and closer to the spirit of *Moderne Algebra*, was published in the USA: *A Survey of Modern Algebra* by Garrett Birkhoff and Saunders Mac Lane.[78] The latter was bound to become the standard American text of algebra in the decades to come. It is worth mentioning that both Dickson and van der Waerden dropped the adjective "modern" from the later editions of their respective books; Dickson's book reappeared in 1959 as *Algebraic Theories*, while van der Waerden's book turned into *Algebra* in 1955.[79] This change would seem to indicate that at about that time the new, structural, image of algebra had not only been given preference over the classical one, but, in fact, that the latter had been totally superseded by the former. The image of algebra as the discipline dealing with polynomial equations was no longer the classical point of reference. The discipline dealing with alge-

---

78. The contents of the successive editions of his book are discussed in greater detail in § 9.2 below.

79. In the foreword to the fourth edition of his book (1955), van der Waerden mentioned a suggestion raised earlier by a reviewer, that the adjective "modern" be dropped from the title. The reviewer wrote that such an adjective may give rise to the suspicion that the book "is simply following a fashionable trend which yesterday was unknown and tomorrow will probably be forgotten."

braic structures ought not to be characterized anymore by the adjective "modern."

The preceding discussion can now be summarized. Weber's algebraic work provides an interesting indication of the status of the discipline by the turn of the century, and, in particular, of the interrelations between body and images of algebra. On the one hand, Weber was in the forefront of research. His 1893 article on Galois theory represents but one of his important contributions to the discipline. In writing his article, Weber clearly considered it advantageous to introduce abstractly formulated concepts and to closely examine the interplay among several of those concepts. This had been done to a certain extent also by Dedekind, with whose work Weber was well acquainted. On the other hand, however, when it came to the writing of a textbook on algebra aimed at closing the gap between the existing texts and the advances in research of three decades, the abstract approach remained, in essence, outside the book. Can an explanation be offered for this difference? A possible answer to this question is given in Melvin Kiernan's thorough account of the historical development of Galois theory, which also discusses Weber's work in detail. Kiernan claims that:

> The presentation of Galois theory in Weber's *Lehrbuch der Algebra* is, in some respects, very different from that in his 1893 article. These are, of course, written for very different audiences. The textbook presentation is much simpler and much more complete. It is not intended as a presentation of original research in a relatively untried area. (Kiernan 1971, 138)[80]

Certainly, the different expected audiences of the article and the book help account for the difference in approach. Nevertheless, it seems that there were also additional factors, and these can be accounted for by referring to Weber's images of algebra. In retrospect, the approach adopted in Weber's 1893 article is especially appealing to the modern reader, but for Weber's contemporaries it was perhaps less so. Weber himself may have considered the direction initiated in his article to be innovative and worth pursuing, but perhaps not sufficiently so as to bring about a total change in his overall conception of the aims and scope of research in algebra. By contrast, such a change came about

---

80. One finds a similar statement in Wussing 1984, 251: "...since it is intended for beginners, the first volume of the *Lehrbuch der Algebra* does not go beyond the group concept in the sense of permutation group."

in van der Waerden's textbook. The main difference between these two text-books lies not only in the bodies of knowledge they put forward, but especially in their images of knowledge. This difference is manifest, in particular, in the fact that van der Waerden turned the conceptual hierarchy of algebra upside down, by introducing groups, fields and many others concepts in all their generality from the outset. The fact that Weber did not adopt the abstract approach in his own *Lehrbuch* attests for the fact that between the dates of publication of this book and of van der Waerden's *Moderne Algebra*, the general conception of algebra had fundamentally changed. This change, however, was not brought about by the mere adoption of the abstract formulation in algebra nor by the steady growth of the body of knowledge. Rather, it was the product of a deeper, overall transformation of the aims and methods of algebra.

The three textbooks of algebra mentioned in the last section of this chapter were written by mathematicians who knew and mastered approximately the same body of algebraic knowledge that was known (and available) to van der Waerden, but whose images of algebra were closer in many respects to those manifest in Weber's textbook.[81] Van der Waerden came forward with a different presentation of the same body of knowledge, namely, with a new image of algebra. In having done so he contributed in a significant way to the development of algebra, without thereby directly adding new significant concepts, theorems, or proofs.

Chapters 2 to 5 below will provide an overview and analysis of how the change in the images of algebraic knowledge came about, thus leading to the process that culminated in the publication of van der Waerden's book. This will be done from the perspective offered by the development of ideal theory between Dedekind and Noether. In the next chapter we discuss Dedekind's work on the theory of ideals.

---

81. A further textbook which could be cited in this context is *Einführung in die Axiomatik der Algebra* (1926), by Hans Beck (1876-1942), professor at Bonn since 1921. This textbook, although adopting wholeheartedly the abstract axiomatic approach, presents a rather idiosyncratic view of the discipline, reducing it basically to the study of any domain on which four operations can somehow be defined: numbers, matrices, hypercomplex systems, etc. Groups are not discussed here in connection with solvability of polynomial equations and nothing of the sort of an algebraic structure is hinted at. But the author expresses very clearly his conception of algebra on the opening page of the book (Beck 1926, 1. Italics in the original): "Algebra *ist die Lehre von den vier Spezies,* das ist von den Verknüpfungen der Addition, Substraktion, Multiplication und Division *in endlichmaliger Wiederholung.*"

Chapter 2 **Richard Dedekind:**
Numbers and Ideals

Richard Dedekind (1831-1916) is among the nineteenth century mathematicians whose work has been most extensively studied and praised.[1] No one cared to stress the importance of his contributions, and to extol the pioneering character of his approach, more emphatically than Emmy Noether. She continually advised her students to read and re-read Dedekind's works, in which she saw an inexhaustible source of inspiration. When praised for her own innovations, she used to repeat: *"Es steht alles schon bei Dedekind."*[2] The close links between the mathematical ideas of these two masters, Dedekind and Noether, manifest themselves in their respective works on the theory of ideals more evidently than in any other context. Yet at the same time, it is precisely by examining their respective works in this field that one can best define the significant differences between their approaches and thus explain the sense in which Noether's work may be said to be more "structural" than Dedekind's. Understanding these differences enables us to identify the meaning of the change that the images of algebra, and algebraic research at large, underwent between the last third of the nineteenth century and the 1920s.

Dedekind studied in his native city of Braunschweig and in Göttingen. In 1852 he completed his doctoral dissertation, working under the supervision of Gauss. Later in 1854, he habilitated with Bernhard Riemann (1826-1866).[3] During his first years as *Privatdozent* in Göttingen he worked in close collaboration with Peter Lejeune Dirichlet (1805-1859). Thus, the decisive influence of the leading mathematicians of the early Göttingen tradition marked Dedekind's formative years. This influence was clearly manifest in all of his later

---

1. Cf. Biermann 1971; Dieudonné (ed.) 1978, Vol. 1, 200-214; Dugac 1976; Edwards 1980, 1983; Ferreirós 1993; Ferreirós 1999, 81-113 & 215-256; Fuchs 1982; Hawkins 1971, 1974; Haubrich 1988; Mehrtens 1979 (Chpt. 2), 1979a, 1982; Noether and Cavaillès (eds.) 1937; Pla i Carrera 1993; Purkert 1971, 1976; Scharlau 1981, 1982; Zincke 1916.
2. Cf. Dedekind 1964, *iv*.

work.[4] In 1858 Dedekind was appointed to the ETH in Zürich, and in 1862 he returned to Braunschweig, were he remained until his death. The notes that Dedekind prepared for his lectures, and which have been preserved in his *Nachlass*, demonstrate that he was a very dedicated and meticulous teacher.[5] Nevertheless, Dedekind neither created a circle of students around himself nor had any single important student. The influence of his ideas came mostly through his interaction and cooperation with leading contemporary mathematicians, such as Heinrich Weber[6] and Georg Ferdinand Frobenius (1849-1917),[7] and later on through David Hilbert's influential *Zahlbericht* (§ 3.2 below). Moreover, through the influence of Emmy Noether, many young students in Göttingen of the 1920s studied and worked under the spell of Dedekind's mathematics.

Dedekind's approach was idiosyncratic in many respects, and in this sense, his work provides an ideal measuring-rod for evaluating the status and the evolution of the idea of a mathematical structure by the turn of the nineteenth century. Dedekind's images of algebra were closer in spirit than those of any of his own contemporaries to both the images of Emmy Noether and the structural image of algebra manifest in van der Waerden's textbook. Where Dedekind's images of algebraic knowledge differ from those promoted in Noether's research or in van der Waerden's book, one can be certain that those of other, contemporary mathematicians differ even more.

Dedekind's overall mathematical output reflects a remarkable methodological unity of which several interconnected aspects may be mentioned. The

---

3. In his lectures on the history of nineteenth-century mathematics Felix Klein characterized Dedekind as a major representative of the Riemann tradition. Klein also reported that Dedekind was entrusted by Riemann's heirs with the edition of his *Nachlass*, shortly after the latter's death. Dedekind began working on it, and wrote illuminating comments on some of Riemann's papers, but he was not able to complete the whole undertaking. In 1871 Clebsch joined him in the task, and after the death of the latter, in 1872, the work was completed by Weber. Commenting on this, Klein added: "[Dedekind] ist wesentlich eine beschauliche Natur, der es an Tatkraft und Entschlußfähigkeit villeicht mangelt." See Klein 1926-27 Vol. 1, 274.

4. See Ferreirós 1999, 24-31 & 77-80; Scharlau 1981, 47-57. Several letters of Dedekind dating from his Göttingen period have been preserved, in which we find interesting evidence of the striking influence of his teachers upon him.

5. See for instance his lecture notes on Galois theory in Dedekind 1981. This subject is discussed in § 2.1 below.

6. The pervasiveness of Dedekind's ideas in Weber's *Lehrbuch* was mentioned in § 1.2 above. Weber's joint work with Dedekind on algebraic functions is discussed in § 2.2.4 below.

7. For the impact of Dedekind's ideas on Frobenius see Hawkins 1971; 1974; 2000, 375-378.

first, and perhaps most important of these, is a clear inclination to address mathematical problems by radically reformulating the whole, relevant conceptual setting. Dedekind ascribed to the systematic introduction of new concepts a central role in the solution of problems and in the clarification of existing mathematical knowledge. Dedekind's habilitation lecture at the University of Göttingen was exclusively devoted to developing this view.[8] Later on, throughout his long career, several of his most significant mathematical achievements were the result of a thorough and systematic application of this early formulated principle.

The kind of conceptually-oriented approach promoted by Dedekind was likewise manifest in the works of his illustrious Göttingen teachers—Gauss, Riemann, Dirichlet—all of which strongly influenced him. Thus, for instance, the latter's contribution to contemporary mathematics was summarized by Hermann Minkowski—while alluding to the well-known "Principle of Dirichlet" in analysis—by referring to "the other principle of Dirichlet", namely, the view that mathematical problems should be solved through a minimum of blind calculations and a maximum of forethought.[9] This description applies to no lesser degree to Dedekind, and, as will be seen below, to Emmy Noether as well. Yet even by the last quarter of the nineteenth-century, this approach was far from being universally taken for granted. In fact, simultaneous with Dedekind's research on the factorization properties of algebraic numbers, Leopold Kronecker (1823-1891) was working on the same problem in Berlin. Kronecker approached the whole problem from a much more computational and algorithmic perspective. His contributions in this domain, however, were to remain essentially unexplored by mathematicians of later generations, mainly because of its inherent technical difficulties, but also because of this algorithmic approach, so different from what came to dominate large portions of twentieth century mathematics.[10] In fact, among the main driving forces that led to the relative dominance of the conceptual approach over the algorithmic one during a considerable part of the twentieth century one should count Dedekind's idiosyncratic images of mathematics, which were elaborated and transmitted —as will be seen in the following chapters— through the works of Hilbert, Emmy Noether and van der Waerden.[11]

---

8. See Dedekind 1854.

9. See Minkowski 1905, 162-163.

10. Kronecker's work on algebraic number theory and its relation to Dedekind's are briefly discussed below, at the end of § 2.2.4.

Choosing adequate concepts and correctly applying them to the solution of problems was always closely related in Dedekind's mathematics with an insistence in separating notational aspects from essential traits. Attributing to mathematical concepts properties that derive from particular representation alone was a mistake that Dedekind insistently sought to avoid. He claimed to have learned this important lesson from studying the works of Gauss and of Riemann.[12] Dedekind considered that the important results obtained by Riemann in his work on the theory of functions were based on a successful application of this principle.

Dedekind's work is also characterized by a recurrent delay in publishing significant results. All his important ideas underwent a very long period of maturation before being published, and many early drafts of papers that were only published much later can be found in his *Nachlass*. This fact has been very helpful in reconstructing the development of his ideas.

Posterior elaborations of many of the fruitful concepts introduced by Dedekind in his work still influence contemporary mathematical thought and

---

11. Of course, "algorithmic" and "conceptual" cannot denote mutually exclusive approaches, but rather varying degrees of emphasis. One certainly finds algorithmic as well as conceptual elements in the works of any mathematician, and surely in the approaches of both Dedekind and Kronecker. And yet, the dominant spirit is different for each of them. Likewise, it might be misleading to associate the "conceptual" approach uniquely and exclusively with Göttingen mathematics. In a passage written in 1843 by the Berlin mathematician Ferdinand Gotthold Eisenstein (1823-1852), we read the following illuminating description of these two alternative attitudes in nineteenth-century mathematics: "In contrast to the older school, the fundamental principle of the new school, founded by Gauss, Jacobi and Dirichlet, is this. Where the old school sought to attain its ends through long and involved calculation and deductions (as is the case even in Gauss's *Disquisitiones*), the new avoids it by the use of a brilliant expedient; it comprehends a whole area in a single main idea, and in one stroke presents the final result with utmost elegance. Where the former, advancing from theorem to theorem, finally reaches fruitful ground after many steps, the latter sets down from the very beginning a formula containing in concentrated form the full circle of truths of a whole area, which need only be read off and spelled out. Following the old method, one could, in a pinch, prove the theorems; now, however, one can see the true nature of the whole theory, the essential inner machinery and wheel-work." This passage is quoted in Wussing 1984, 270 (footnote 81). Wussing refers to it as further evidence, together with Galois's papers, of "an awareness of the beginning of structural changes in mathematics" (p. 102). From the point of view of our account, however, it would seem more convenient not to use this term in this context, but rather, to emphasize the "conceptual" constituent of the approach adopted by these mathematicians, a constituent which is only one of the characteristic features of the structural approach.

12. Cf. Dedekind *Werke* Vol. 2, 54-55. On Gauss's and Riemann's emphasis on the conceptual over the representative aspects of mathematics, see Ferreirós 1999, 53-62.

practice. In particular, Dedekind introduced several of the concepts that eventually became part of what is considered the hard-core of modern algebra: ideals, fields, modules, lattices. Additional central concepts, such as group, which were not created by Dedekind himself, were applied in his works, combined with those of his own creation, in innovative contexts and with far-reaching consequences. Dedekind also introduced techniques that were later developed by Emmy Noether, and by other mathematicians, becoming central to the rise of the structural image of algebra.

All these characteristic features of Dedekind's work have contributed to his name being considered among the first—if not the very first—representative of the modern approach to algebra. It has even been suggested, for instance, that if Dedekind had not turned soon away his attention from Galois theory, on which he lectured in the early years of his career, to his later subject of algebraic number theory, he could have written the first textbook of "modern algebra."[13] Of course, such assessments can be somewhat ambiguous, if one does not specify what is actually meant by the latter term. Our interest, as was already clearly stated, focuses on the extent to which the notion of an algebraic structure appears in the images of knowledge of the various mathematicians involved in the development of algebra; this notion appears—as will be seen in the present chapter—neither implicitly nor explicitly in Dedekind's work.

In fact, a close inspection of his works on Galois theory, on algebraic number theory and the theory of ideals within it, and on other related issues, shows that Dedekind himself neither advanced a general idea of algebraic structures, nor saw those individual concepts like ideals, modules, fields, groups, etc., as particular instances of a more comprehensive, common conceptual species of "algebraic structures", giving rise to similar questions, eliciting similar answers, and deserving a unified formulation and treatment. On the contrary, as will be seen in the following analysis, these concepts enter his work on different conceptual footings and they play different roles in the different theories in which they are used. The significant change in the conceptual hierarchy that ultimately characterizes the structural image of algebra is not found in Dedekind's work.

---

13. Thus the editor of Dedekind's manuscript on Galois theory wrote (Scharlau 1981, 106): "Hätte [Dedekind] nicht in der algebraischen Zahlentheorie bald ein noch interessanteres Betätigungsfeld gefunden, so wäre es durchaus denkbar gewesen, daß aus dem Manuskript das erste Lehrbuch der 'modernen Algebra' entstanden wäre."

In order to better understand the connection between Dedekind's formulation of algebraic concepts and his methodological conception, it is first convenient to describe very briefly his contributions to the foundations of systems of real and natural numbers. Dedekind's ideas on these issues are well-known to students of mathematics. Nevertheless it is useful to examine them here briefly again, since Dedekind's basic motivations and the way he elaborates the concepts involved in these two theories resemble in many aspects those of his theory of ideals. The following account will thus help understand how Dedekind's above mentioned striving after the formulation of new, effective concepts is translated into elaborate mathematical theories in different contexts.

The first among Dedekind's foundational works we want to consider here concerns the system of the real numbers. The direct motivation for dealing with this topic was Dedekind's deep dissatisfaction with the way the differential calculus was traditionally taught in German universities. This feeling aroused as early as 1858, when he began his career as a young teacher in Zürich. In particular, he was disturbed by the accepted proof of a specific theorem, namely, the theorem stating that a monotonically increasing, bounded sequence of numbers necessarily has a limit. The validity of this particular theorem, or of any equivalent one, is enough to deduce the whole calculus satisfactorily. However, in Dedekind's view, the proof usually adduced for that theorem relied excessively on geometrical intuitions. Legitimate proofs, he thought, should be based on a coherent conceptualization of the system of real numbers, which would allow barring all geometric intuitions.[14] Dedekind's discomfort with the existing proof, coupled with his extreme pedagogical alertness and his self-imposed methodological strictures, led him through a long process of ripening whose outcome was his "theory of cuts" as presented in the now famous book *Stetigkeit und irrationale Zahlen*, which appeared in print only in 1872.[15]

Dedekind was not working in isolation when he attempted to provide a solid foundational basis for the system of real numbers. Several other mathematicians, such as Weierstrass, Cantor and Charles Mèray (1835-1911) were

---

14. Dedekind *Werke* Vol. 3, 316.
15. In 1858 Dedekind had already prepared an early version. The manuscript is reproduced as an appendix in Dugac 1976, 203-209.

attempting a similar task at roughly the same time. Still, Dedekind's own attempt clearly stands out for its characteristic abstractness, thoroughness, and conceptual clarity.[16]

To provide an adequate foundation for the calculus and for the system of real numbers meant for Dedekind to elucidate the idea of continuity, an idea that—in his opinion—had never been satisfactorily clarified in purely arithmetical terms. As a starting point, Dedekind took for granted the arithmetical properties of the system of rational numbers as well as its order, and hence those of the natural numbers. Of course, also these two systems should be themselves provided with an adequate arithmetical foundation, but this task Dedekind intended to undertake at a later opportunity, as will be seen below.

Dedekind's theory of cuts starts from an examination of the accepted correspondence between the real numbers and the points of a straight line.[17] The very existence of irrational lengths implies that while the straight line may be seen as a continuum, the system of rational numbers cannot. What should be done, then, in order to extend the system of rational numbers, using purely arithmetical methods, to constitute a continuous system? What is the essence of this continuity?

Dedekind found the answer to the above question in a seemingly trivial observation, namely, that every point of the line divides it into two separate portions, such that every point of one of these portions lies always to the left of any point of the other portion. Something similar (though not identical) happens with the system of rational numbers. In fact, given a rational number a, consider the classes A_1 of all the rationals smaller than a, and A_2 of all the rationals larger than a. If in addition we are told that a itself belongs to either A_1 or A_2, then it is clear that a is either the largest number in A_1 or the smallest number in A_2. These two classes A_1, A_2 satisfy the following three obvious properties:

1. A_1 and A_2 are disjoint.

2. Every rational number belongs to either A_1 or A_2.

3. Every number in A_1 is smaller than any number in A_2.

---

16. For a comparison among the theories of Weierstrass, Cantor and Dedekind, see Ferreirós 1999, 124-135.

17. Dedekind *Werke* Vol. 3, 319-320.

Focusing on these properties, Dedekind took a step that would become standard in all of his future works, and that led him to introduce his most important conceptual innovations: he turned this *list of properties* into a *definition*. Thus, he defined a "cut" (*Schnitt*) as a pair of classes of rational numbers A_1 and A_2, satisfying properties 1-3 above, and he denoted it as (A_1,A_2). Now any rational number defines a cut in a natural way, but Dedekind showed that there are additional cuts which are *not* defined in this was by a rational number. Thus for instance, given any positive number, whose square D is a non-square integer, if A_2 is the class of rationals having squares greater than D, and if A_1 is its complement in the rationals, then it can easily be shown that (A_1,A_2) is a cut. Nevertheless, there is no rational number which is either the greatest number in A_1 or the smallest one in A_2. It is precisely in the existence of such cuts that Dedekind saw the essence of the discontinuity of the rational numbers. On the contrary, the collection of all cuts is, like the straight line, a continuous system.

The system of real numbers can be now naturally defined as the collection of all cuts of rationals. In this way a new kind of numbers (the reals) is defined in terms of *collections* of other, previously defined, numbers (the rationals). More importantly, Dedekind was now able to prove, based exclusively on the relations of *inclusion* between those collections of rational numbers, the basic properties of the new system. Thus, he discussed the order properties of the system of all cuts of rationals and showed that this system, which is the system of real numbers, is a totally ordered one. Then, he also proved that the system of real numbers is a continuum by defining cuts of real numbers, in a similar way to the cuts of rationals. However, contrary to what is the case with the rationals, it turns out that to each cut of real numbers there corresponds one (and only one) real number. Finally, the cuts also enabled Dedekind to define the arithmetical operations with real numbers and to prove their properties, based only on inclusion relations between the collections involved. In *Stetigkeit und irrationale Zahlen* Dedekind defined actually only the addition of real numbers, but he also claimed that similar definitions might be worked out for the other operations on real numbers; not only subtraction, multiplication and division, but also radicals and even logarithms.[18]

---

18. From an early draft of the book it is known that he had also begun to develop the details of subtraction. Cf. Dugac 1976, app. XXXII, p. 208.

Dedekind closed the book by explaining how the concepts and results obtained in its first six sections could be applied to the problem that originated his whole investigation, namely, the problem of the foundation of the infinitesimal calculus. He did so in a very simple and concise way, by showing that the above mentioned theorem on the limit of a monotonic, bounded sequence of numbers is equivalent to the continuity condition just proved to hold in the real numbers.

The second work to be discussed here concerns Dedekind's elucidation of the foundations of the system of natural numbers. In this case one cannot point out a specific theorem of the calculus, or of some other mainstream mathematical discipline, as an immediate motivation for his enquiry. However, as was mentioned above, his foundational theory of the real numbers assumes that the rational numbers have been provided with their own, solid foundation. Indeed, as early as 1858, when he started developing the ideas eventually presented in *Stetigkeit*, Dedekind also began to formulate foundational theories for the rationals and the integers, defining them as paris of integer and of natural numbers, respectively. In doing this he may well have been inspired by William Rowan Hamilton's 1853 definition of complex numbers as pairs of real numbers, which he studied about that time.[19]

At any rate, Dedekind's foundation of the arithmetic of natural numbers is naturally seen as a further step in the detailed implementation of his program aimed at elucidating mathematical theories in terms of clearly defined, abstract and general concepts. If the main property to be elucidated for the real numbers was continuity, here Dedekind focused on total order as the defining property of the natural numbers, and a main question was to understand the nature and the role of mathematical induction as part of this system. His results in this field appear in what is now his second classical work: *Was sind und was sollen die Zahlen?* first published in 1888. It is in this work of Dedekind, more than in any other of his writings, that the specific concepts introduced, as well as the general approach adopted, bear the set-theoretical spirit that became characteristic of twentieth-century mathematics.[20]

---

19. See Ferreirós 1999, 218-222.

20. Pla i Carrera 1993 analyses in detail the set-theoretical content of Dedekind's works on the foundations of the systems of numbers, and explains, in present-day terms the conceptual and axiomatic implications (axiom of choice, axiom of replacement, etc.) of each step of his work. See especially his sections 2 and 4.

Without attempting to present here a detailed account of the contents of Dedekind's book, it is nevertheless relevant to stress how his theory of the natural numbers is based on a limited number of concepts involving sets and mappings. The central concept of the theory is the concept of "chain" (*Kette*): given a mapping of a system K into itself, if the image of the mapping is denoted by K', then the mapping is said to be a chain if K' is a proper subsystem of K. For every subsystem A of K, the chain of A, denoted by A_0, is defined as the intersection of all chains of K containing A.[21] This latter definition is the truly central and original one, and it allowed proving a theorem which constitutes a "set-theoretical" version of the principle of induction.[22]

Dedekind also defined infinite sets in this work. As he himself remarked, other mathematicians (particularly Georg Cantor and Bernhard Bolzano) had studied infinite sets before him, and had emphasized as their basic characteristic that they contain proper, equipotent subsets (pp. 356-357). Dedekind's innovation here was, as in other important instances, to have taken this already known *property* and transform it into a *definition*: An infinite system *is* one, that contains a proper, equipotent subsystem.[23]

The methodological approach followed by Dedekind in these two works, as well as the new concepts he introduced, might seem rather natural and standard to a modern reader. This was far from being the case for all of his contemporaries.[24] Even colleagues currently engaged in research involving incipient set-theoretical ideas (including Cantor) failed to pay due attention to them. Although some mathematical logicians (especially Ernst Schröder) did express a clear interest, other mainstream mathematicians either totally ignored or outright criticized Dedekind's books. To a large extent, Dedekind's own way of presentation was among the main reasons for these kind of reactions, but their truly innovative spirit was no less so. At any rate, it would still take many years, and the active intervention of some influential mathemati-

---

21. Dedekind *Werke* Vol.3, 353.

22. For details, see Ferreirós 1999, 222-231.

23. In his definition of finite sets Dedekind used implicitly the axiom of choice; Dedekind's use of infinite choice was criticized by several contemporary mathematicians (Cf., e.g., Hilbert 1992, 32) and the discussion that arose around this issue was central to the developments leading to Zermelo's explicit formulation of his axiom. For a detailed discussion of this point see Ferreirós 1999, 218-248. See also Moore 1982, 24-30; Peckhaus 1990, 56-57; Pla i Carrera 1993, 47-49.

24. For details see Ferreirós 1999, 248-253.

cians, before the impact of Dedekind's ideas would be properly felt. We discuss this in detail below for the specific case of algebra. But as a remarkable instance of a typical contemporary reaction to Dedekind's theory of irrational numbers, and indeed to his overall methodological approach, it is illuminating to mention here the reservations expressed by Rudolph Lipschitz (1832-1903) in the framework of the correspondence held between the two.[25]

Lipschitz was a talented mathematician from Bonn, who made important contributions to a variety of mainstream disciplines in nineteenth-century mathematics, such as real analysis, differential equations, and differential geometry.[26] In this sense, he is as good a representative as any of how contemporary mathematicians would react to Dedekind's work. In fact, from his correspondence with Dedekind in 1876 we know that he was among the first to appreciate the importance of the latter's contributions to algebraic number theory and to attempt to make them known to a broader audience (see below § 2.2.3). His overall positive predisposition to judge Dedekind's ideas cannot be doubted. Yet, concerning Dedekind's theory of irrationals Lipschitz expressed a clear dissatisfaction, and not because he considered it to be incorrect. Rather, Lipschitz thought that Dedekind's theory of cuts added nothing beyond what could already be found, in principle, in Euclid's general theory of proportions as presented in book V of the *Elements*. Specifically, he disagreed with Dedekind's claim that his book was the first to contain a legitimate proof of the validity of equalities such as $\sqrt{2} \times \sqrt{3} = \sqrt{6}$.[27]

The latter claim was indeed a deeply significant one for Dedekind and in fact, a touchstone of the importance of his contribution. Thus he answered to his colleague in great detail. He insisted that anyone claiming that the theorem had indeed been proved, the burden of proof falls fully upon him. And he added:

> Do you really think that this proof can be found in some book? I have searched for this point, of course, in a great number of works from different countries, and what have I found? Nothing but the coarsest vicious circles, such as the following: $\sqrt{a}\sqrt{b} = \sqrt{ab}$ because $(\sqrt{a} \cdot \sqrt{b})^2 = (\sqrt{a})^2 \cdot (\sqrt{b})^2 = ab$. Not even a minor explanation of the product of two irrational numbers is provided in advance, and without any kind of scruples, the theorem $(mn)^2 = m^2 n^2$, which was

---

25. See Scharlau (ed.) 1986.

26. See, for instance, Tazzioli 1994.

27. For a broader historical perspective on this theorem see Corry 1994, Fowler 1992.

proved for the rationals, is applied also to the irrationals. Is it not truly irritating that the teaching of mathematics in school is considered as an especially distinguished means for developing the understanding, while in no other discipline (grammar, for instance) would such offensive infractions against logic be tolerated even for a minute? If one does not want, or cannot for lack of time, behave scientifically, let us then at least be honest and confess it openly to the student, who is in any case inclined to *believe* a theorem just on the word of the teacher. This is better than wiping out, by means of fake demonstrations, the pure and noble sense of real proofs (*Werke* Vol. 3, 471)

Lipschitz was hardly convinced by Dedekind's vivid rhetoric. The disagreement was not just a matter of taste, but rather a fundamental one, concerning the question what is given in mathematics and what needs to be proven. For Lipschitz, just like for many of his colleagues, the general idea of continuous magnitude, underlying the specific definition of irrational number was beyond the domain of research of mathematics. It had been fully elucidated in the work of Euclid, and there was nothing to add to it. Dedekind not only thought different, but he also explained Lipschitz how he could prove that the assumption of continuity was not to be found among those underlying Euclid's system. He thus wrote:

Analyze all assumptions, both explicit and implicit, on which the edifice of Euclid's geometry is built. Grant the validity of all its theorems, and the realizability of all its constructions (an infallible method I apply for such an analysis is to replace all the technical expressions by arbitrary terms just invented (and hitherto senseless); the building should not collapse following this replacement, and I claim, for example, that my own theory of real numbers undergoes this test with success): *never*, so far as I have investigated, becomes in this way the continuity of space a condition which is indivisibly connected to Euclidean geometry. His whole system remains stable even without assuming continuity: a result that will certainly appear as astonishing to many, and which therefore I considered worthy of mention. (*Werke* Vol. 3, 479)[28]

Most probably, Dedekind was rather alone, at this stage, in demanding that this kind of proofs be provided, and in particular that they be based on an abstract, conceptual foundation such as he had presented in his theories of real and natural numbers. Both these theories were built on single abstract concepts, "cut" and "chain, respectively, which are well-defined, specific collec-

---

28. Emmy Noether liked to stress this passage as a very early manifestation of the modern axiomatic approach (see Dedekind *Werke* Vol. 3, 334).

tion of other kinds of numbers. As will be seen below, Dedekind's theory of ideals will be developed following a similar principle.

This brief account of two of Dedekind's best-known works was meant to illustrate how his methodological views were instrumental in providing the setting in which he worked out his mathematical theories. However, the earliest instance in which Dedekind's methodological guidelines were translated into elaborated mathematical work, and which is of direct interest for the account that follows, is found in the lectures on Galois theory held in Göttingen in 1856-57. The content of these lectures, and in particular Dedekind's use of the concepts of groups and fields in them, are discussed in the next section

2.1   Lectures on Galois Theory

As was already mentioned in § 1.1 above, the development of Galois theory played a central role in the overall transformation of algebra since the mid-nineteenth century. Following the publication of Galois's papers in 1846, fifteen years after his death, and the gradual adoption of the techniques and perspective promoted by his work, groups of permutations increasingly came to be considered as a main tool for dealing with the theory of solvability of polynomial equations. The development of Galois theory throughout the nineteenth century was instrumental in bringing to the fore the concepts of group and field as objects of intrinsic interest.[29]

In 1853 Kronecker published a study in which he applied to the theory of polynomial equations the kind of ideas introduced by Galois, thus being among the first mathematicians in Germany to show some interest in this theory.[30] During the years 1855-58, Dedekind lectured at Göttingen on algebra; in the winter semester of 1856-57 he taught Galois theory, thus being the first to have done so in a German university.[31] But as was later the case in their respective works on algebraic number theory, Kronecker and Dedekind approached this subject from different perspectives.[32] Kronecker concentrated

---

29. See Kiernan 1971, 103-110; 114-125.

30. See Kronecker 1853. For the early reception of Galois's work in Germany see Wussing 1984, 119 ff. See also Gray 1990, 302; Kiernan 1971, 125-132; Toti-Rigatelli 1989, 69-73.

31. Cf. Dugac 1976, 23; Toti-Rigatelli 1989, 75-84; Weber 1895, *vii*. A more general account of Dedekind's mathematical interests during his Göttingen period appears in Scharlau 1982.

on the problem of constructing all algebraic equations solvable by radicals. He thus departed to some extent from the direction stressed by Galois himself, which focused on the possibility of applying the theory of groups of permutations to study the solvability of particular equations.[33] Dedekind, by contrast, developed the view that eventually gave rise to the standard, twentieth-century treatment of Galois theory: a direct focus on the interplay between groups of permutations of the roots, on the one hand, and, on the other, the automorphisms defined on the subfields of the field of complex numbers in which the roots of a polynomial lie. This view, which was further elaborated by Weber and later by Artin, sees the solvability of particular equations as a specific application of the more general situation involved in the above mentioned interplay.[34]

Although Dedekind did not publish any work on Galois theory in the early years of his career, he studied Galois's papers thoroughly and he seems to have grasped the scope of the theory in a broader perspective than anyone else in the early stages of its development. Dedekind's Göttingen course on Galois theory was attended by only two students,[35] and thus it had no visible, direct influence. Later on, in his 1894 version of the theory of ideals, Dedekind dedicated an entire section to presenting the central ideas of Galois theory (see § 2.2.4 below). But already in his early lectures it becomes evident how Dedekind, in order to overcome the weaknesses and gaps inherent in Galois's original treatment, approached the theory by focusing on an original reformulation of the basic concepts involved as a way to gain more general, solidly built arguments. Obviously, this involved a more systematic consider-

---

32. Kiernan 1971, 126, suggests that Kronecker's acquaintance with Galois theory could date from as early as 1853, when Kronecker visited Charles Hermite (1822- 1901) in Paris. Hermite had been a student of Galois's most influential teacher, Louis Richard (1795-1849), who kept Galois's notebooks and handed them over to Hermite. An indeed, in a note to the second edition of his book on algebra (1854) Serret suggested that Hermite, not his mentor Liouville, was the great expert at the time on the theory (see Lützen 1990, 197).

33. Cf. Wussing 1984, 120 ff. According to Wussing (p. 123): "During that period [the 1850s and 1860s] Kronecker's attitude toward Galois theory was typical of the attitude of many German mathematicians of the time, who, while aiming at sharp formulations of the concepts and methods of Galois, made no decisive contribution to the development of permutation groups."

34. Cf. Kiernan 1971, 125-133.

35. See Dugac 1976, 23. One of these students was Paul Bachmann (1837-1920), later a well-known number theoretician and algebraist himself. The second one was to become a long-time friend of Dedekind's, Hans Zincke (Sommer). See Purkert 1976, 3; Toti-Rigatelli 1989, 75.

ation of ideas related with the concepts of fields and groups. It is thus pertinent to describe now briefly the contents of these early lectures on Galois theory.

Dedekind's Göttingen course on Galois theory is known to us through manuscripts found in his *Nachlass*.[36] From one of the manuscripts we learn that Dedekind divided the course into four sections. The first section, *"Elemente der Lehre von den Substitutionen"*, develops the theory of groups of permutations, or of substitutions. Dedekind defined a substitution over a finite set (*Complex*)[37] and how to multiply two substitutions of the same set. He then proved that substitutions are associative and left- and right-cancellable with respect to this product. Any finite collection of substitutions is called a group, whenever the product of any two elements of this collection belongs itself to the collection. Although throughout the article Dedekind continued to refer to substitutions and to products of substitutions, he made it very clear that he had in mind a much more general theory, with applications in many domains of mathematics. Dedekind expressed this in the following, highly significant passage:

> The following investigations are based only on the two main results proven above [associativity, right- and left-cancelability] and on the assumption that the number of substitutions is finite: The results are therefore valid for any finite domain of elements, things, concepts $\theta, \theta', \theta'', \ldots$ admitting an arbitrarily defined composition $\theta\theta'$, for any two given elements θ, θ', such that $\theta\theta'$ is itself a member of the domain, and such that this composition satisfies the laws expressed in the two main results. In many parts of mathematics, and especially in the theory of numbers and in algebra, one often finds examples of this theory; the same methods of proof are valid here as there.[38]

---

36. These were published recently (with comments) in Purkert 1976; Scharlau 1981, 59-100; and Scharlau 1982. Our account here refers to Dedekind 1981.

37. In his insightful account of the development of the concept of set in mathematics, José Ferreirós (1999, xvi) remarks that in the last third of the nineteenth century several German words were variously used to denote a set, or any collection of objects: *Mannigfaltigkeit, Klasse, Gebiet, Complex*, etc. until finally the term *Menge* became generally accepted for the concept. Ferreirós remarks: "The mere fact that one finds the word 'Menge' [set] in a text does not in itself mean that the author is employing the right notion. And the mere fact that an author uses another word does not mean that he is *not* employing the notion of set". In our account of Dedekind's work we focus mainly on the very fact that his algebraic definitions are based very often on the properties of collection, rather than on those of individual numbers, functions etc. To what extent these ideas contributed to developing the concept of set and its use in mathematics is fully discussed by Ferreirós and it falls out of the scope of our own discussion.

Dedekind would then derive all subsequent results on groups from these two assumptions alone, giving to his theory a distinct axiomatic flavor that will also be found in most of his later algebraic works. Still, this general definition of group cannot be properly called "abstract". At any rate, Dedekind also discussed other basic ideas, such as subgroups, cosets, normal subgroups, normalizer, etc., and proved many related, basic results. All knowledge about groups necessary in the ensuing discussion was already systematically introduced in this initial section, and the relevant results were thus ready to be used thereafter.

At this early stage, Dedekind also defined additional, new "algebraic" concepts, yet it is enlightening to compare his use of such concepts with his use of groups. Thus for instance, in section III, *"Über die algebraische Verwandschaft der Zahlen"*, where Dedekind strongly uses group-theoretical notions in order to present the principles introduced by Galois into the study of "higher algebra." Given a polynomial $f(x)$ of degree m, whose coefficients may be expressed as rational combinations of numbers belonging to a given domain S of complex numbers, Dedekind discussed the properties of the roots by focusing on their group of permutations. He formulated a central theorem using this group,[39] and in the proof, he referred to the domain S as follows:

> If s,s',s'' are any three roots of this equation, then one can set $s' = \theta(s)$, $s'' = \theta'(s)$, where the coefficients of the rational functions θ and θ' belong to the rational domain S.[40]

---

38. Dedekind 1981, 63: "Die nun folgenden Untersuchungen beruhen lediglich auf den beiden so eben bewiesenen Fundamentalsätzen und darauf, dass die Anzahl der Substitutionen eine endliche ist: Die Resultate derselben werden deshalb genau ebenso für ein Gebiet von einer endlichen Anzahl von Elementen, Dingen, Begriffen $\theta,\theta',\theta'',...$ gelten, die eine irgendwie definirte Composition $\theta\theta'$ aus θ,θ' zulassen, in der Weise, dass $\theta\theta'$ wieder ein Glied dieses Gebietes ist, und dass diese Art der Composition den Gesetzen gehorcht, welche in den beiden Fundamentalsätzen ausgesprochen sind. In vielen Theilen der Mathematik, namentlich aber in der Zahlentheorie und Algebra, finden sich fortwährend Beispiele zu dieser Theorie: dieselben Methoden der Beweise gelten hier wie dort."

39. Dedekind 1981, 82. Winfried Scharlau, who edited and published the manuscript, prepared an index and gave names to each subsection, whereas Dedekind had only given names to the four main sections. This subsection (§ III.3) is called by Scharlau *"Zerfällungskörper eines Polynomes. Primitives Element."* Of course, neither of the two concepts were so defined by Dedekind himself.

40. Dedekind 1981, 83: "Sind s,s',s'' irgend drei Wurzeln dieser Gleichung, so kann man $s' = \theta(s)$, $s'' = \theta'(s)$ setzen, worin die Coefficienten der rationalen Functionen θ und θ' in dem rationalen Gebiet S enthalten sind."

The term "rational domain" is used here only in passing, as short hand for a subdomain of the system of complex numbers having a specific property, namely, that of being closed under the four arithmetic operations. In the whole manuscript, this is the instance where Dedekind comes closest to identifying anything like a field. In later sections he mentions this domain again several times, and indeed its actual role within the theory becomes increasingly clear. And yet, this, or any similar kind of domain, is never systematically treated in itself.[41] Results concerning rational domains appear scattered, and they are proved whenever needed for a specific, immediate concern.

Groups, on the one hand, and specific closed collections of complex numbers, on the other, appear here as mathematical entities belonging to different conceptual spheres. The system of complex numbers and the interrelations among the "rational domains" contained in it constitute for Dedekind *the subject matter* of higher algebra; groups, on the other hand, provide a highly efficient *tool* that may be applied to tackle a problem involving polynomial equations with roots in those domains. Dedekind took this basic idea from Galois's own work, and formulated it in a very clear way, even specifying the basic assumptions needed to derive all the basic results concerning the groups. He did not do this for the "rational domains." Moreover, whereas the questions raised by Dedekind about the latter concern *properties of their elements*, his questions about groups concern *the groups themselves*, seen as autonomous mathematical entities, rather than their elements.

The different conceptual levels of reference accorded by Dedekind to groups and to "rational domains" in these early lectures will be recurrently found in later works as well. New concepts introduced in various contexts, and in particular in his theory of ideals, and which later become in *Moderne Algebra* specific instances of algebraic structures, will also appear in Dedekind's works as mathematical entities of diverse nature and playing different kinds of roles in the theories in which they are put to use. This will be seen more in detail in the next few sections, where Dedekind's theory of ideals is discussed.

---

41. Walter Purkert (1976, 10) attributes to Dedekind historical priority in developing the concept of field in these lectures. Purkert describes the conception of fields manifest in this manuscript as somewhat fuzzy: "Die hier noch etwas verschwommene Konzeption dieses Begriffs wird in den weiteren Artikeln präzisiert."

2.2 Algebraic Number Theory

Between 1871 and 1894 Dedekind published several successive versions of his theory of ideals, intended as a comprehensive treatment of the problem of unique prime factorization in fields of algebraic numbers.[42] In his work on this domain, Dedekind introduced, and illuminatingly used, several concepts nowadays identified with the modern conception of algebra: fields, modules, ideals, homomorphisms. The various versions of his theory of ideals help understanding the actual way in which Dedekind conceived, developed and applied all these concepts.

But in order to understand properly the essence of Dedekind's theory of ideals, and its historical significance, it is first necessary to describe briefly the status of the unique factorization theorem in the period of time just before Dedekind, and in particular, some of the ideas introduced by Kummer since 1847 in his works on cyclotomic fields. Kummer's theory of "ideal prime numbers" was a main starting point for Dedekind.[43]

2.2.1 Ideal Prime Numbers

The unique prime factorization theorem, also called the fundamental theorem of arithmetic, asserts that every natural number can be written in a unique way (except perhaps for the order) as a product of its prime factors. The arithmetic books of Euclid's *Elements* (Books VII-IX) already contain significant theorems dealing with the relation between natural numbers and their prime factors, and this specific property of the natural numbers was well-known to Greek mathematicians, impolitely at least. Nevertheless, the first complete formulation and proof of the theorem is usually attributed to Gauss in his *Disquisitiones Arithmeticae*.[44] To Gauss himself we also owe the first formulation and proof of the unique factorization theorem for numbers of a

---

42. As already said, Dedekind used to publish his works after long periods of maturation. Between 1871 and 1894 he worked not only on the theory of ideals, but also on many other issues. In particular he developed his foundational theories for the real and the natural numbers. Dedekind's theory of cuts was published in 1872, and there is also an early draft of it, written in 1858. Dedekind's theory of natural numbers was published in 1888, but there is an early draft dating from 1872-1878 (it was published as an appendix in Dugac 1976, 293-309). The theory of ideals was published in four versions: 1871, 1876-1877, 1879, 1894.

43. What follows is based on several existing accounts of the development of algebraic number theory from Gauss to Kummer. Especially worthy of mention are Harold Edwards's illuminating studies of Kummer's work (Edwards 1975, 1977, 1977a, 1980.)

more general kind than the usual integers. This result appeared in 1832 within his elaboration of a reciprocity law for biquadratic residues, and since then, the study of higher reciprocity laws became the main direct motivation for the study of generalized formulations of unique factorization theorems.

Gauss's interest in quadratic reciprocity arose during his years as a student. From his diary we know that in 1796 he had already proved, in two different ways, the law of quadratic reciprocity, which he published in 1801 in the *Disquisitiones*.[45] Given a prime number p, any integer q, which is relatively prime to p, is called a quadratic residue of p, if there exists an integer x, such that $x^2 \equiv q$ (mod p). Gauss's result established the conditions under which quadratic residuality is a reciprocal property, namely, under which q is a quadratic residue of p if and only if p is a quadratic residue of q. As Gauss himself remarked, this question had formerly been studied for particular cases by Lagrange and by Leonhard Euler (1707-1783), and a result similar to his own one had been formulated in 1798 by Adrien Marie Legendre (1752-1833), who however gave an invalid proof for it.[46] The law is succinctly formulated using the so-called "Legendre symbol", $\left(\dfrac{p}{q}\right)$, whose value is 1, if p is a quadratic residue mod q, and -1 otherwise. Given two odd primes, p,q, the law establishes that the following reciprocal relationships hold:

1. If p,q are not both $\equiv 3$ (mod 4), then

$$\left(\frac{p}{q}\right) = \left(\frac{q}{p}\right)$$

---

44. Gauss 1801. Cf., e.g., Bourbaki 1969, 112-113. It seems that Ian Mueller expressed an opinion shared by many when, in referring to Book VII of the *Elements*, and in particular to propositions VII 30 and VII 31, he wrote (Mueller 1981, 83): "The sensible way to describe [it] would seem to be to say that although the *Elements* contains the materials for proving the fundamental theorem, it contains neither the theorem nor the equivalent of it." For a more detailed elaboration of this view see Knorr 1976. For a somewhat unknown example of a seventeenth-century result closely related to the fundamental theorem, see Goldstein 1992.

45. See Gray 1984.

46. Gauss's own comments on the relation of his work to that of his predecessors appear in Gauss *Werke* Vol. 1, 73-74, and, in English translation, in Smith (ed.) 1929, Vol. 1, 112-113. See also Dirichlet 1894, 94 ff.; Edwards 1977, 287; Frei 1994, 69-74; Smith 1965, 56-57.

2. If $p \equiv q \equiv 3 \pmod 4$, then

$$\left(\frac{p}{q}\right) = -\left(\frac{q}{p}\right)$$

These two relations can be subsumed under the more general one:

$$\left(\frac{p}{q}\right) = (-1)^{\frac{p-1}{2} \cdot \frac{q-1}{2}} \left(\frac{q}{p}\right) .$$

The prime 2 appears as a special case:

$$\left(\frac{2}{q}\right) = (-1)^{\frac{p^2-1}{8}} .$$

Gauss considered this result to be of supreme importance for the study of binary quadratic forms. In fact, he called it the "Fundamental Theorem."[47] Over the next sixteen years Gauss published five additional proofs to his law. His continued efforts in this direction corresponded to some extent to the inherent interest and beauty of the reciprocity law. But no less than that, these efforts were motivated by Gauss's explicitly manifest ambition to find an argument for his theorem on quadratic reciprocity, that might also be extended and applied to prove analogous reciprocity theorems for congruences of degrees higher than two.[48]

In 1832 Gauss finally published the article in which the law of reciprocity for biquadratic residues appeared.[49] As it happened, however, the very formulation of the law compelled Gauss to introduce a particular kind of factors of the prime numbers of the form $4n + 1$; these factors were in fact complex numbers. These are the so-called integer complex numbers, or Gaussian integers, namely, numbers of the form $a + ib$, where a and b are integers and $i^2 = -1$. In earlier works of Euler and of Lagrange one finds complex numbers applied to the study of arithmetical properties of the natural numbers, but Gauss was the

---

47. For a discussion of Gauss's formulation, see Edwards 1977, 176-178.
48. Gauss *Werke* Vol. 2, 50. Cf. Collison 1977, 66.
49. In Gauss 1832. This was preceded by a shorter announcement in Gauss 1831.

first to systematize this direction with his work on Gaussian integers. In spite of the pioneering character of this direction of research—which developed into what became known as "higher arithmetic"—Gauss explicitly stated his opinion that this was not an arbitrary extension of the usual arithmetic, but rather one of essential importance for research in this domain.[50] Gauss studied the properties of the complex integers thoroughly and attempted to reproduce in this domain all the known arithmetical properties of the usual integers. Of particular importance was the re-elaboration of factorization properties in the domain of complex integers. Thus, Gauss defined a relation of order over this system, by introducing a norm, $N(a + ib) = a^2 + b^2$. He then defined an Euclidean algorithm for finding the *g.c.d.* of any two numbers in the domain and identified those numbers that play the role of primes in the system. Finally (and more importantly), using these prime numbers he proved the adequate version of the unique prime factorization theorem for complex integers. Gauss also used the complex integers to formulate the law of biquadratic reciprocity, but he never published the corresponding proof. Carl Gustav Jacobi found a proof soon after Gauss's publication of the law, and he communicated it to his students in Königsberg, without publishing it himself. Finally, the first to publish a proof of the law of biquadratic reciprocity was Ferdinand Gotthold Eisenstein in 1844.[51] In fact, Eisenstein published a total of five different proofs of the same law.[52] Unique factorization of the Gaussian integers was crucial to all these proofs.

The study of higher reciprocity laws led to the consideration of further extensions of the domain of usual integers, still more general than the Gaussian integers. Gauss himself had suggested in a footnote to his 1832 paper on biquadratic reciprocity, that cubic reciprocity laws should be investigated by considering numbers of the form $a + b\rho$, with $\rho^3 = 1$ ($\rho \neq 1$), but he never pub-

---

50. See Smith 1965, 71-72.

51. See Eisenstein 1844.

52. A detailed account of Eisenstein's proofs appear in Smith 1965, 81-88. See also Collison 1977, 67; Frei 1994, 85-86. In order to get a broader perspective of the kind of mathematical horizon within which this kind of work was conceived, it is worth pointing out, that two among Eisenstein's proofs applied techniques taken from the theory of elliptic functions (See Eisenstein 1845 & Eisenstein 1847, 453-465). In our discussion of Weber's textbook of algebra of 1895 (§ 1.2 above) we already mentioned that a considerable portion of the third volume of this classical book of algebra of the late nineteenth century was dedicated to discuss the connection between the theory of elliptic functions and the algebraic theory of numbers. Eisenstein's two articles constitute a classical representative of this kind of connection.

lished any results in this direction. Jacobi did elaborate this track beginning in 1827, using in this context numbers of the form $a + b\sqrt{-3}$ for the same purpose.[53] Later on, in attempting to formulate reciprocity laws for individual cases of higher degrees, Jacobi considered complex integers involving roots of unity of the fifth, eight, and twelfth degrees separately, and factorized those numbers while assuming the usual properties of the natural numbers.[54] In the wake of this line of development Ernst Edward Kummer (1810-1893) entered the picture with his theory of ideal prime numbers.[55]

The study of higher reciprocity laws was seen by Kummer as the central task and pinnacle of achievement of number-theoretical research.[56] Following in the footsteps of Jacobi, he based his own study of congruences of higher prime orders on considering numbers of the form $a_0 + a_1\theta + ... + a_n\theta^n$ ($a_i \in \mathbf{Z}$), where θ is a primitive pth root of unity (i.e., a complex number $\neq 1$, such that $\theta^p = 1$). These are nowadays called cyclotomic integers. In Kummer's time they were known simply as complex integer numbers. Kummer invested considerable efforts in working out detailed calculations with cyclotomic integers. These calculations gave few reasons to believe that unique factorization fails to hold in these domains; as a matter of fact, it does hold for all values of p, less than 23. Kummer discovered the failure of unique factorization in domains of cyclotomic integers with $p = 23$, and published his discovery in 1844,[57] while expressing deep regret for this situation.[58] Over the following years Kummer dedicated considerable efforts to developing his theory of ideal prime numbers, aimed at reversing this unfortunate state of things.

---

53. See Collison 1977, 68; Frei 1994, 82-83.

54. Jacobi 1839. See Smith 1965, 94.

55. A further mathematician involved in the early efforts to formulate and prove the law of biquadratic reciprocity was Dirichlet. On this see Rowe 1988.

56. Kummer 1850, 347: "... die Hauptaufgabe und die Spitze dieser Wissenschaft."

57. See Kummer 1844. Henry J. Smith reported in 1860 (Smith 1965, 95) that Kummer had checked the validity of unique factorization for $p = 5$ and 7, and its failure for $p = 23$ and higher primes. At the time, however, Kummer had not yet stated, according to Smith's report, whether or not it holds for $p = 11,13,17$ and 19. Harold Edwards has advanced the well-documented claim (in Edwards 1975, 233) that Eisenstein was aware of the failure of unique factorization for the case $p = 23$ some weeks before Kummer's publication. For a detailed technical discussion of the problems connected with that case, see Edwards 1977, 97-106.

58. Kummer 1844, 182.

An additional important topic where failure of unique prime factorization becomes relevant concerns contemporary attempts to prove Fermat's last theorem. Following statements by leading number-theorists such as Hilbert and Kurt Hensel,[59] it became common to assert that Kummer's discovery originated in the framework of his own attempts to prove the theorem and, consequently, this was often described as his main motivation for developing the theory of ideal numbers.[60] However, recent historical research has made absolutely clear that this was not the case.[61] Kummer applied for the first time the theory of ideal numbers to an attempted proof of the theorem only in 1847, and published his important proofs for "regular primes" only in 1858.[62] Although Kummer cared to stress after 1860 the historical importance of Fermat's conjecture, the conjecture was for him "a curiosity of number theory, rather than a major item". It is likely that he would endorse the view advanced much earlier by Gauss, who wrote in 1816:

> I must confess, however, that Fermat's theorem offers little interest for me as an isolated result, since it is easy to state many such assertions that can be neither proved nor refuted. (...) I am quite convinced that if I succeed in achieving some breakthroughs in the theory [of cyclotomic fields], Fermat's theorem will thus appear only among one of the less interesting corollaries.[63]

In order to understand the basic ideas behind Kummer's ideal number primes, it is convenient to consider a second kind of generalization of the Gaussian integers in which unique factorization also fails: quadratic number fields, i.e., collections of numbers $a + b\theta$, where a,b are any integers, and $\theta^2 = n$ for n a fixed integer. As already stated, Jacobi had suggested that they might be used in the proof of the law of cubic reciprocity. Earlier, Euler had dealt with these kinds of numbers, for instance, while studying the arithmetical properties of numbers of the form $p_2 + nq^2$. Euler implicitly assumed in his study that in the domain of numbers of the form $a + b\sqrt{-5}$, unique factoriza-

---

59. See, Hilbert 1902, 439; Hensel 1910.
60. See, e.g., Bourbaki 1969; Dickson 1920, 738.
61. See Edwards 1977a; Neumann 1981.
62. See below footnote 69.
63. In a letter to Wilhelm Olbers (1758-1840). Quoted in Neumann 1981, 186.

tion holds.[64] This assumption is however erroneous, as the following simple example shows:

$$3 \cdot 7 = (4 + \sqrt{-5})(4 - \sqrt{-5}) = 21 \qquad (*)$$

Here the number 21 is represented in *two different ways* as a product of two factors in the relevant domain.[65] Moreover, all four factors involved in this equality can be shown to be irreducible, i.e., none of them can be further decomposed in a non-trivial way into a product of numbers of the same domain. One could indeed represent the factor 3 as a product, as follows:

$$3 = (\sqrt{-2} + \sqrt{-5})(\sqrt{-2} - \sqrt{-5}) \qquad (**)$$

In this case, however, the factors clearly do not belong to the relevant domain, since a and b should both be *integers*, and clearly here $\sqrt{-2}$ is not. It so happens, that in the domain of quadratic integers $a + b\sqrt{-5}$, one can actually find eleven indecomposable factors of 21 beyond 3,7 and $4 \pm \sqrt{-5}$.[66]

What this situation implies is that, unlike the case of the usual integers, there are domains in which a number is not necessarily a prime just because it cannot be expressed as a product of simpler factors. In this example, 3 and 7 are indecomposable in the relevant domain. The numbers $4 \pm \sqrt{-5}$ are factors of neither 3 or 7 in this domain, but do divide their product 21.

The "ideal primes", by means of which Kummer was able to recover unique factorization in domains of cyclotomic integers, are factors which do not themselves belong to the domain in which the factorization is being considered. Equation (**) above illustrates the idea behind the introduction of ideal prime factors: in this case the number 3 was decomposed into two factors

---

64. Cf. Edwards 1977, 43-45.

65. Euler's assumption concerning unique factorization in this case has attracted the attention of historians, since the counterexample shown here follows directly from a result that Fermat had proved about hundred years before him and that certainly Euler himself was well-aware of, namely, that some numbers of the form $p^2 + nq^2$ have factors not of this form. See Bourbaki, 1969, 132-133; Edwards 1977, 97; Edwards 1980, 323. After Euler, at any rate, Lagrange proved again (in Lagrange 1769, 465) the existence of such factors for numbers $p^2 + nq^2$.

66. These factors are explicitly given by Dedekind who worked out the example in detail in Dedekind 1876-77, 279 ff.

which are, however, outside the domain considered. In fact, using Kummer's theory one can study factorization without having to show the factors explicitly, but rather by analyzing their *properties* as divisors of the actual numbers, as if the factors were actually given.

Kummer first announced his theory of ideal prime numbers in 1846, and published its details in 1847.[67] To any number m belonging to a given domain of cyclotomic numbers, Kummer ascribed a list of "ideal prime numbers." He then proved that this list satisfies all the division properties expected from the list of ordinary prime factors of an integer. Every "ideal prime factor" p belonging to the list of m is said to be "contained" in m. Kummer also defined for a prime ideal factor the meaning of being contained in m "with multiplicity greater than 1." In these terms, the main property expected from prime factors to satisfy in a given domain of cyclotomic numbers asserts that:

(#) m divides n, if and only if every prime factor contained in m is also contained in n with at least the same multiplicity.[68]

What this property accounts for may be illustrated by considering again the example of the numbers of the form $+ b\sqrt{-5}$. Take for instance the following decomposition of 9 in that domain:

$$9 = 3 \cdot 3 = (-2 - \sqrt{-5})(-2 + \sqrt{-5}) \qquad (***)$$

Put now $b = 3$, $b_1 = -2 - \sqrt{-5}$ and $b_2 = -2 + \sqrt{-5}$, and assume the existence of two numbers β_1, β_2 such that $\beta_i^2 = b_i$ $(i = 1,2)$ and $b = \beta_1 . \beta_2$. Here β_1, β_2 play the role of the "ideal prime factors." It is clear that every number divisible by 3 also contains these two ideal factors. Likewise, if a number is divisible by 3^2, then it is divisible by the squares of β_1 and β_2. One can thus check conditions that the β_i's satisfy as factors of 3 without having to explicitly show the factors themselves. Performing a similar procedure for all the above mentioned fifteen irreducible factors of 21 within this domain, one finds

67. See Kummer 1847; Kummer 1847a, respectively.

68. The division properties of the ideal prime numbers are discussed in Kummer 1847, 322-323. Detailed discussions of Kummer's definition appear in Edwards 1977, 76-150; Smith 1965, 107-117. See also Edwards 1980, 327-328; Ellison & Ellison 1978, 178-183.

a list of five numbers (β_1, β_2 among them)[69] playing the role of "ideal prime numbers" in this domain.

As was said above, Kummer was working in the framework of cyclotomic integers, not of quadratic integers as in our example. For domains of cyclotomic integers, Kummer was able to prove the desired divisibility properties with the help of his ideal prime numbers. Using these properties, he was able to restore unique factorization to those domains. In particular, property (#) yields the unique factorization theorem of cyclotomic integers into ideal factors, in the following sense: any two cyclotomic integers containing the same ideal prime factors with the same multiplicities, are in fact, as a consequence of property (#), one and the same, with the possible exclusion of a unit-factor. But more than restoring unique factorization Kummer's theory proved very useful indeed for his original purposes, since it allowed him to formulate and prove a very general law of reciprocity for higher congruences,[70] and, even before that, to provide a proof of Fermat's theorem valid for a very large collection of prime exponents.[71]

What happens, however, when one tries to generalize Kummer's results to more general domains? For instance, what happens when one considers exponents n (in $\theta^n = 1$) which are non-prime? Kummer himself, as part of his efforts to prove the general law of reciprocity, addressed this question and was able to generalize his theory for this case, in an article of 1856.[72] A second possible direction of generalization is that of the quadratic integers discussed

---

69. For detailed exposition of this example cf. Dedekind *Werke* Vol. 3, 281-282. A similar example is discussed in Scholz (ed.) 1990, 319.

70. Kummer formulated the law in 1848 as a conjecture (see Kummer 1850). His proof (generally valid, except for certain primes) appeared only in Kummer 1859. The first to publish a fully consistent application of ideal factors to prove reciprocity laws was Eisenstein (in Eisenstein 1850a). However, Kummer's formulation of the law was by no means the most generally possible one, and the issue continued to occupy the efforts of many mathematicians. Thus, the formulation and proof of the "most general law of reciprocity in any number field" appear as the ninth of the twenty-three famous 1900 list of problems of Hilbert (Hilbert 1902, 453). As a matter of fact, in 1899 the Göttingen Academy of Sciences established a prize-subject for 1901 for a detailed treatment of the law of reciprocity for prime, odd exponents. The prize went to Philipp Furtwängler (1869-1940) who addressed the problem in a series of articles published between 1902 and 1928. On the development of ideas connected with Furtwängler's work, see Hasse 1967, 269 ff. For an account of the attempts to solve Hilbert's ninth problem between 1900 and 1974, see Tate 1976. An alternative view, which attributes to I.R. Šafarevich the formulation of the most general law of reciprocity appears in Bashmakova et al. 1992, 106.

in the preceding paragraphs. Kummer mentioned these domains as well, in connection with Gauss's theory of quadratic forms, while suggesting that his own theory of ideal prime numbers would also work for them.[73] But neither Kummer nor anyone else seems to have elaborated further this later inquiry until Dedekind realized that it may be incorrect. For instance, in the domain G of all numbers $a + b\sqrt{-3}$, a,b integers, a direct application of Kummer's concepts turns out to be impossible.[74] Apparently, this impossibility might be bypassed by writing the numbers $a + b\sqrt{-3}$ as $a + b\sqrt{-3} = a + b + 2b\alpha$, where $\alpha = \frac{1}{2}(-1 + \sqrt{-3})$. In this case $\alpha^3 = 1$, and therefore, according to Kummer's theory, the numbers $+ b\sqrt{-3}$ have unique decompositions within the domain of cyclotomic integers defined by α. However, since α

---

71. Kummer announced his proof of Fermat's theorem for regular primes to Dirichlet in 1847 (see Kummer 1847b). The proof was published also as Kummer 1850a. When announcing his proof Kummer was certain that there are infinitely many such primes (in fact, it is not known until today whether or not this is the case). In Kummer 1857a, he proved the theorem for the only three non-regular primes up to one hundred (37, 59, 67). Kummer thus expected that the number of cases not covered by his general proof will be small enough to be treated individually (However, K.L. Jensen proved in 1915 that there are in fact infinitely many non-regular primes: see Ribenboim 1988, 257). Cauchy and Liouville raised several objections concerning Kummer's use of the ideal prime factors in his first proof. Kummer 1857 (dated 5.VI.1856) responded to all those objections, although Kummer mentioned in this article neither the objections nor the problems actually arising in his first proof. For a contemporary account (1860) of Kummer's proof see Smith 1965, 134-137. More recent accounts of these developments appear in Edwards 1975, 225-231; Edwards 1977a.

72. Kummer 1856. See Edwards 1980, 328: Edwards 1992, 6-7.

73. Kummer 1847, 209. See Edwards 1977, 152-154.

74. In fact call $a = 8$, $b = 1 + \sqrt{-3}$, $c = 2$; then $b^3 = -c^3$. Take now any exponent $k = 3j + i$ ($0 \leq i < 3$). Then $ab^k = 2^3.b^{3j+i} = 2^3.(-8)^j.b^i = \pm2^{3j+3}.b^i$, and therefore $c^k = c^{3j+i}$ divides ab^k for any exponent k.

Now c divides a, and it also divides b^3, but it does not divide b. Therefore in order for Kummer's theory to work here, c must contain an ideal factor P with multiplicity $\mu_P(c)$ greater than the multiplicity with which it is contained in b, $\mu_P(b)$.

But on the other hand, since c^k divides ab^k, then $k\mu_P(c) \leq \mu_P(a) + k\mu_P(b)$ (the multiplicities must behave here as in the ordinary case). Now obviously $\mu_P(a) \geq 0$, and therefore $\mu_P(c) \leq \mu_P(b)$, thus yielding a contradiction.

This example is fully treated in Edwards 1980, 330. Edwards 1977, Chpt. 7, studies in detail the theory of divisors for quadratic integers. The problem of extending Kummer's theory to quadratic integers with $\theta^2 = D$, $D \equiv 1 \pmod 4$, is discussed on pp. 249 ff.

itself does not belong to G, the prime factors of a given number of G may not themselves belong to G. In other words, we have the following situation: the numbers of this particular domain G cannot be uniquely factorized using Kummer's ideal prime numbers, as long as one restricts oneself to the given domain; however, by choosing a larger, suitable domain containing G one may obtain the desired factorization.[75] Thus, one learns, that when attempting to generalize Kummer's theory, it is necessary to correctly specify the domains in which one expects Kummer's procedure to be valid.

Dedekind was highly impressed by Kummer's achievements, but at the same time he was well-aware of what, from his point of view, were the short-comings of Kummer's theory. In the first place, Dedekind saw the very term "ideal prime numbers" as misleading since it mistakenly suggested unwarranted properties that the integers (but not the "ideal prime numbers") actually satisfy. Kummer had even spoken of "ideal complex numbers" without ever defining them, apparently meaning a product of ideal prime factors.[76] But Dedekind's methodological scruples regarding Kummer's work were not only a matter of personal taste; they were also reactions to gaps he had found in Kummer's proofs.[77] Such gaps were, in Dedekind's opinion, a consequence of the faulty analogy between the prime ideal numbers and the prime integers.[78] Moreover, as the example of the numbers $a + b\sqrt{-3}$ discussed above shows, particular choices of elements and notations may prove more significant than expected, when attempting to approach factorization problems using Kummer's theory. Dedekind's characteristic awareness to these kinds of difficulties throughout his work led him, in this particular case as well, to seek a more general formulation—and thus a general solution—of the problem. This he did by means of his theory of ideals, whose first version appeared in 1871.

But it took Dedekind many years of intense efforts to understand the real implications of his own ideas, and to develop them thoroughly. The evolution

---

75. In this case the contradiction pointed out in the preceding footnote disappears, since in the domain of cyclotomic integers defined by α ($\alpha^3 = 1$), one has $1 + \sqrt{-3} = 2 + 2\alpha$, and therefore $1 + \sqrt{-3}$ is divisible by 2.

76. For instance in Kummer 1851. Cf. Edwards 1977, 143-144; Edwards 1980, 328.

77. As already said, Cauchy and Liouville had also found some gaps in Kummer's proofs.

78. Cf. *Werke* Vol. 3, 251; 258.

of these ideas is manifest in his successive versions of the theory. Dedekind gradually uncovered and developed the typical proof-techniques that his theory was able to offer, as well as the potential perspectives opened by the shift of attention away from the study of operations and relations among the algebraic *numbers themselves* and increasingly towards the properties of *their collections* as such, and the interrelations among them. In the first version of the theory one can still—more or less directly—intertranslate Kummer's and Dedekind's concepts and results. It is only in the last version of Dedekind's theory that one sees an organically constructed theory in which the deductive relations among the various concepts are more clearly manifest, and in which the properties of the collections of numbers involved, rather than those of the numbers themselves, directly provide the basis for the proofs of the main theorems. The eventual elaboration of this latter feature stands at the center of our own interest in studying the rise of the structural approach to algebra, and it will be accorded much attention in this and in the following chapters.

Kummer's theory was organically rooted in the study of higher reciprocity laws. With Dedekind's theory of ideals, the problem of unique factorization in generalized extensions of the rational integers became the main focus of research in number theory. Thus, in spite of the direct connection between the two theories, the passage from the former to the latter signified in itself a major change in the conception of the whole discipline.[79] But the potential distancing from the original concerns of the discipline, to which the works of Kummer and his predecessors were bound to lead, was already felt before Dedekind's appeared on the scene. This feeling is illuminatingly expressed in a report presented as early as 1860 by the Oxford mathematician Henry J. Smith (1827-1883) to the British Association, on the current state of research in number theory. Stressing the need to remain alert to the actual motivations of this research, Smith wrote:

> The investigations relating to Laws of Reciprocity, which have so long occupied us in this report, have introduced us to considerations apparently so remote from the theory of residues of powers of integral numbers, that it requires a certain effort to bear in mind their connection with that theory... But the complex numbers of Gauss, Jacobi, and of M. Kummer force themselves upon our consideration, not because their properties are generalizations of the properties of ordinary integers, but because certain of the properties of integral numbers can only be explained by a reference to them... [W]hen we come to binomial congru-

---

79. This significant issue is discussed in (the unpublished) Haubrich 1993.

ences of higher orders, we find that the true elements of the question are no longer real primes, but certain complex factors, composed of roots of unity, which are, or may be conceived to be, contained in the real primes. (Smith 1965, 139)

The publication of the last version of Dedekind's theory of ideals in 1894 will eventually render this direct link with the original motivation of the study of generalized domains of integers almost imperceptible.

2.2.2 Theory of Ideals: The First Version (1871)

In 1855 Peter Lejeune Dirichlet came to Göttingen. Dirichlet was to become Dedekind's teacher and friend, and perhaps the mathematician who exerted the deepest influence on him.[80] In 1863 Dedekind edited and published the first edition of Dirichlet's *Vorlesungen über Zahlentheorie*, based on the latter's lectures. Dedekind also added to this edition supplements containing new material of his own creation. Dedekind's theory of ideals first appeared in 1871 as a part of his tenth supplement to the second edition of Dirichlet's *Vorlesungen*.[81] Dedekind, however, had been working on these ideas for a long time—at least since 1856—following his reading of Kummer's results.[82] Moreover, after publishing his theory of ideals for the first time, the subject continued to occupy Dedekind's thoughts, and two revised editions followed the first one.

Dedekind's theory was intended as a generalization of Kummer's results, that should also be valid for all those domains in which the latter failed. A main problem in this attempt was to give a suitable definition of the domains involved. It was thus only natural for Dedekind to organize his new theory around a few appropriately selected concepts. The first version of Dedekind's theory of factorization thus opens with the definition of its basic concepts. These concepts include fields (*Körper*), modules, congruence relative to a given module, and ideals.[83] A field of numbers is any system of numbers closed under the four operations of addition, subtraction, multiplication and

---

80. Dedekind's own appraisal of Dirichlet's influence on him appears in a letter quoted in Scharlau 1981, 35-37. On Dirichlet's influence on Dedekind, see, e.g., Edwards 1983, 10-11.

81. Dedekind 1871. For a more detailed study of this version of the theory see Haubrich 1988, 62 ff.

82. Cf. Dedekind *Werke* Vol. 1, 110.

83. Detailed studies of the basic concepts introduced here by Dedekind appear in Mehrtens 1979, Chpt. 2 and Purkert 1971, 8-11.

division (except by zero).[84] The "smallest" field of all is the field of rational numbers, while the largest possible field is the field of all complex numbers. A field A is called a divisor of another field M (and M is said to be a multiple of A) whenever all the numbers contained in A are also contained in M.[85] The "simplest" fields are those possessing only a finite number of "divisors" (i.e., subfields), and these are called by Dedekind "finite fields."[86]

The next fundamental concept is that of "algebraic integer." Consider the field of all algebraic numbers, namely, all real and complex roots of polynomials with *rational* coefficients. An algebraic integer is any algebraic number, which is a root of an irreducible *monic* polynomial of degree n with *integer* coefficients. These kinds of numbers had formerly appeared in the framework of related, number-theoretical research, but only their property of closure with respect to addition and multiplication had hitherto been highlighted.[87] To Dedekind, however, one must attribute the important insight that this is the proper domain in which factorization properties should be investigated. As was seen above, the case of domains of cyclotomic numbers covered by Kummer's theory can be generalized in various directions, suggested by different representations of complex numbers. Dedekind sought a definition that would allow him to express the *essence* of the numbers he to be investigated, rather than referring to their various possible contingent *representations*. His definition of the algebraic integers in a field of algebraic numbers was meant to encompass within a single formulation the cyclotomic integers and all their possible, relevant generalizations, while avoiding the kinds of problems mentioned above for the numbers $a + b\sqrt{-3}$.

In fact, in the example of the collection G of all numbers $a + b\sqrt{-3}$ (a,b integers), in order to obtain some version of unique factorization in terms of Kummer's theory, it was necessary to embed the given domain in a larger one. In Dedekind's new language, this situation is expressed as follows: We con-

---

84. Dedekind *Werke* Vol. 3, 224.

85. In the first version of the theory Dedekind used the terms "Multiplum" and "Divisor"; in the later version he also used the terms "Vielfacheres" and "Theiler" respectively.

86. Clearly, the modern usage of the term "finite field" is different from Dedekind's. Dedekind's finite fields contain infinitely many elements.

87. For instance in Dirichlet 1846, 640; Eisenstein 1850, 568; Hermite *Oeuvres* Vol. 1, 115. On this point see also Edwards 1980, 332 (footnote).

sider the field Ω of complex numbers $a + b\sqrt{-3}$ (a,b rational); a generalized version of the theorem of unique factorization will hold, not in the domain G (which might have been thought of as a natural candidate for it, on first sight), but rather for the domain of all algebraic integers belonging to Ω.[88] The general problem confronted by Dedekind could be restated in his own terms as follows: given a field of numbers Ω, to study the properties of the factorization into prime factors in the domain D consisting of all the algebraic integers belonging to Ω. Dedekind reformulated the problem in these terms already in the first version of the theory, yet the actual advantages of doing so would became manifest especially in the later versions.

Having defined algebraic integers, Dedekind would always use now the term "integer" to denote algebraic integers in general, whereas "rational integers" denotes for him the usual integers (p. 236).[89] Clearly, every rational integer is also an algebraic integer. But in order to study factorization properties and to identify those integers that play in this theory a role analogous to that of prime numbers in the realm of the rational integers, one first needs a concept of divisibility. Thus an integer α is said to be divisible by β, whenever there exists an integer γ such that $\gamma \cdot \beta = \alpha$.

Dedekind proceeded to prove some of the basic properties of the algebraic integers, while stressing the analogies and differences between integers in the new and in the usual sense. First, the integers are closed under addition, subtraction and multiplication. Dedekind, however, did not at this point introduce a concept similar or equivalent to an abstract ring.[90] Second, for any given algebraic number α, it is always possible to find an infinite number of integers h such that $h \cdot \alpha$ is an integer. Third, the roots of any *monic* polynomial with (algebraic) integer coefficients are also (algebraic) integers (p. 238). This result, which can be formulated in present-day terms by saying that the domain

---

88. Recall the number $\alpha = \frac{1}{2}(-1 + \sqrt{-3})$ considered above. Now one sees that α is a root of the polynomial $p(x) = x^2 + x + 1$, and hence it is an algebraic integer of the field in question, in the sense now defined by Dedekind.

89. Edwards (1980, 332) has claimed that it is not clear how Dedekind arrived at this particular definition. Haubrich (1988, esp. 47-61) contains an attempt to explain the roots of this (and other related) definitions of Dedekind.

90. Dedekind introduced in the 1879 version of the theory a formal equivalent of rings, *Ordnung*, which is discussed below.

is integrally-closed, is of central importance for Dedekind's theory, and in fact it is necessary in order to prove the main factorization theorem.[91] A more immediate consequence of this result, which is obviously false for the rational integers, is that if α is an integer, then α^r is also integer, for any rational exponent r, integer or fractional. It follows from this result that the usual definition of prime integer, as an integer having no divisors other than itself and the units[92] will not work for algebraic integers, since every integer is divisible by all its fractional powers which, as the last result implies, are also integers. Thus Dedekind introduced here "relative prime integers", one among several definitions that, according to his own judgement, accounted for the success of his investigation. Two non-zero integers are said to be "relatively prime" whenever any number divisible by both numbers is always divisible by their product (p. 240).

In the following section (§ 161) Dedekind made a digression to introduce several concepts that are needed later. He defined a module as a system of real or complex numbers closed under addition and subtraction. This is similar to the modern notion of module at the formal level; but a close inspection of the conceptual setting within which Dedekind's modules appear, indicate the significant difference, between their role at the level of the images of algebra in Dedekind's conception, on the one hand, and in the structural one, on the other. This difference will become clearer as we elaborate our account in this and in the next sections.

Whenever all the numbers of a module M also belong to another module D, M is called a multiple of D, and D is said to be a divisor of M. Dedekind was thus adopting for modules a notation completely opposed to the one introduced in earlier pages of the same text for fields. In the present context Dedekind used the notation of congruences to indicate that a certain number (and,

---

91. In present-day terms, given a ring R contained in a field Φ, an element α of Φ is called R-integral if and only if there exists a monic irreducible polynomial $f(x)$ with coefficients in R such that $f(\alpha) = 0$. The collection A of R-integral elements of Φ is called the integral closure of the ring R in Φ. Then R is said to be integrally closed in Φ whenever $A = R$. A domain which is integrally closed in its field of fractions is called, simply, integrally closed. Cf. Jacobson 1980, 605 ff. Of course, Dedekind did not use this terminology, but in the relevant places, as will be seen below, he relied on the property of integral closedness. The domain G of quadratic numbers $a + b\sqrt{-3}$ is *not* integrally closed, and therefore unique factorization does not hold in it.

92. Among all integers of a given field, units (*Einheiten*) are numbers which divide all integers (*Werke* Vol. 3, 239). In general, there are several units within a given domain of algebraic integers.

especially, that a certain product of numbers) belongs to a module (and, likewise, to an ideal). Thus $\alpha\beta \equiv 0 \pmod{P}$ means that the product on the left-hand side is contained in the module or ideal P. Likewise, if the product of two integers α, β is divisible by a third one μ this is written as $\alpha\beta \equiv 0 \pmod{\mu}$.

It is now possible to define ideals and to formulate the factorization theorems valid in the domain of algebraic integers. The basic idea is to focus on the properties of collections of numbers divided by any given factor ("actual" or "ideal"), rather than directly on the properties of divisibility by the numbers themselves, as Kummer had formerly done in his own theory. Consider the domain of integers of any given field. If one takes any integer δ in that domain, and if one denotes by $i(\delta)$ the collection of all multiples of δ (called the "principal ideal generated by δ" - *Hauptideal*), then it is clear that $i(\delta)$ satisfies the following two properties:

(a) If α and β both belong to $i(\delta)$, then both $\alpha + \beta$ and $\alpha - \beta$ must also belong to $i(\delta)$.

(b) If β belongs to $i(\delta)$ and x is any integer in the domain considered, then βx must also belong to $i(\delta)$.

Suppose now, that instead of taking an actual integer, one considers the collections of integers divided by an ideal factor of the kind defined by Kummer for domains of cyclotomic integers. One can easily see that they also satisfy properties *(a)* and *(b)*. Dedekind intended now to prove that these conditions are not only necessary, but also sufficient conditions, i.e., that the ideals so defined *fully* characterize the collection of multiples of an ideal factor. In this way his theory would allow a study of factorization properties of algebraic integers by focusing on the collections of their multiples, the ideals, while circumventing the question of the existence and nature of ideal factors.

Since in Kummer's theory every cyclotomic integer is made to correspond to a uniquely determined list of "prime ideal divisors" (each with an assigned multiplicity), then in Dedekind's theory, every ideal ought to be made to correspond to a well-defined list of "prime ideals" (each with an assigned multiplicity). Thus, whereas Kummer's theory established the unique factorization of an integer into a product of prime ideal factors with assigned multiplicities, Dedekind's theory of ideals would establish the parallel unique factorization of any ideal into a product of prime ideals with given multiplicity.

As with the other new concepts introduced in the first version of his theory, Dedekind defined ideals formally at the beginning of the relevant section, without any further motivation or explanation: an ideal is any collection of integers satisfying conditions *(a),(b)* above. As was the case for modules, and therefore contrary to the use of the same term for fields, an ideal *A* is said to be a multiple of another ideal *B* (or "divisible by" *B*) if all integers belonging to *A* belong also to B. Dedekind even says in this case that "*B* is contained in *A*."[93] For modules and ideals, then, to be a divisor or to be a multiple, expresses an *inclusion* property of collections of numbers, rather than a condition concerning a *product* in any operational sense of the word. But of course, it was necessary that Dedekind's theory of ideals could express something similar to products of ideals, since no theory of factorization can be complete if it cannot account for multiplicities. The condition of being a multiple or a divisor of an ideal, as defined up to this stage could not be directly used for that purpose, and thus Dedekind had to invest considerable effort in his first version of the theory in order to formulate the factorization theorem. He did so by defining a rather ad- hoc concept, namely, "simple ideals."

Dedekind's elaboration of the problem of multiplicities of factors is of particular importance for the present account since it sheds light on the subtle interplay between the gradual transformation of his concepts through the successive versions of the theory, on the one hand, and the role of his images of mathematics in his research, on the other hand. It is necessary, then, to discuss in some technical detail, in the first place, Dedekind's treatment of multiple factors and his proof of the main factorization theorem as they appear in the first version. In the following sections we will see how this proof evolves through later versions, thus gradually enhancing Dedekind's awareness of the structural essence of the problem. Readers who are more knowledgeable in algebra and number theory will find special interest and no difficulty following the details of the rest of this section.

Let *A* be any given ideal in a domain *D* of integers of a given field. *A* determines equivalence classes in *D* in the standard way: two numbers in *D* are said to be congruent modulo *A*, if and only if their difference belongs to *A*. Dedekind proved that the number of congruence-classes determined by any given ideal *A* in *D* is always finite, and called this number the "norm" of *A* (denoted by $N(A)$). The concept of norm in a domain of cyclotomic integers had been

---

93. Cf. *Werke* Vol. 3, 252: "[W]ir wollen sagen, *A* sei ein Multiplum von *B* oder teilbar durch *B*, *B* sei ein Teiler von *A* oder gehe in *A* auf."

formerly applied by Kummer in his work in a very significant way.[94] Dedekind introduced the norm of any integer in D as follows: Since an algebraic
integer η is a root of a polynomial of degree n with integer coefficients, let
$\eta_2, ..., \eta_n$ be the other n-1 roots of the same polynomial. The product $\eta \cdot \eta_2 ... \eta_n$
is the norm of η, $N(\eta)$. Dedekind asserted that the norm of a principal ideal
$i(\eta)$, $N(i(\eta))$, equals $\pm N(\eta)$ (p. 252).

For two ideals A,B Dedekind called the system M of all numbers belonging simultaneously to both ideals the *l.c.m.* of A and B. Clearly, every common
multiple of A and B is divisible by M. The ideal D containing all numbers
$a + b$, where a belongs to A and b belongs to B, is called the *g.c.d.* of A and B.
Every common divisor of A and B is a multiple of D. Dedekind proved that (p.
253):

$$N(M) \cdot N(D) = N(A) \cdot N(B)$$

Dedekind next defined "prime ideals" as follows: Denote by U the principal ideal generated by a unit; clearly U is a divisor of every ideal.[95] An ideal
P, different from U, is called by Dedekind prime if its only divisors are U and
P itself.[96] Dedekind then proved that the following is an equivalent definition
of prime ideals: given an ideal P, if for every product $\alpha\beta$ contained in P, either
α or β are necessarily contained in P, then P is a prime ideal (p. 253). An ideal
which is not prime is called a composite ideal (*zusammengesetzt*). According
to the second definition of prime ideals above, one can also define an ideal A
as composite if there exist two numbers α,β such that neither belongs to A but
their product does.

The basic difference between these two alternative definitions of prime
ideals can be described as follows: whereas the first is based on inclusion
properties of the ideals themselves, the second necessitates choosing particular integers and examining their behavior. From the point of view of Dede-

---

94. Cf. Edwards 1977, 83 ff.

95. In fact U equals the collection of all algebraic integers of the field considered.

96. In present-day terms, such an ideal is actually called a "maximal ideal", while an ideal is
called prime if for any product of ideals contained in it, at least one of the factors is also contained
in it. The term "Dedekind domain" has been adopted to denote those domains in which every ideal
can be uniquely written as a product of prime ideals. It can be proved that a Dedekind domain is a
Noetherian ring with unit, which is also integrally closed, and in which any non-zero prime ideal (in
the second sense of the term) is also maximal. Cf. Jacobson 1980, 600 ff.; Zariski & Samuel 1958,
270 ff. Emmy Noether stressed the centrality of this kind of rings in her axiomatic study of factorization in abstract rings. See below § 5.2.

kind's methodological preferences, then, the first definition would seem to express the essence of the concept, i.e., the one exhibiting its "intrinsic" characteristics as a collection, whereas the second appears more as a practical criterion allowing to establish whether a given ideal is or is not prime, by examining a particular choice of its elements. Thus, it should not come as a surprise that in future versions of the theory, the second definition would be reformulated in terms of the collections themselves, using the definition of a product of ideals. In the first version of the theory, however, the product of ideals is defined only in the closing sections, after having proven the main factorization theorem. This issue is not without significance for our account, and we shall return to it below.

In order to complete the proof of factorization, and in particular to be able to address the issue of multiplicities, Dedekind ought to introduce now an additional concept: "simple (*einfach*) ideals." The advantage of the latter is that—although multiplication was not defined for ideals—something similar to a power of a simple ideal can be easily defined. Their limitation, however, is that this definition—which is anyway valid only for this particular kind of ideal—relies on the choice of particular elements of the ideal in question.[97] Using the powers of simple ideals Dedekind was able to prove several results that finally led to the fundamental factorization theorem in terms of ideals. These results establish the parallels between powers of simple ideals and of prime numbers, and in fact show that Dedekind's ideals also account for division properties already covered by Kummer's ideal prime factors.[98] Thus, for any simple ideal P the following properties hold:

1. If $s \geq r$, then P^r divides P^s.

---

[97]. In order to define simple ideals Dedekind first proved the following, rather straightforward result (pp. 255-256): Let μ be any non-zero integer such that $N(\mu) \neq 1$. Then there exists number ν, such that if P is the collection of all roots π of the congruence $\nu\pi \equiv 0 \pmod{\mu}$, then P is a prime ideal. Dedekind turned this property of P into a definition: If all the numbers contained in a given prime ideal P are the roots π of a given congruence as above, then it will be called simple ideal (p. 256). "Powers" of simple ideals are defined as follows: Given a simple ideal P and an integer μ as above, then clearly also $N(\mu^r) \neq 1$. Now for every rational integer r, the roots ρ of the congruence $\pi\nu \equiv 0 \pmod{\mu^r}$ also constitute an ideal; this ideal is called the r-th power of the prime ideal P. However, in order for this definition to make sense it is necessary that the power r depend only on the ideal P, and not on the particular choice of the pair μ,ν. Dedekind proved this fact in the present version, but his proof contained a slight mistake which, as he himself mentioned some years later, could easily be corrected. Cf. *Werke* Vol. 3, 419.

2. For every integer μ, there is always a highest power r of P such that μ is contained in P^r (or "a highest power of P in μ").[99]

3. If P^r and P^s are the highest powers of P in μ,η respectively, then P^{r+s} is the highest power of P in $\mu\eta$.

4. P is the only prime ideal which is a divisor of all powers of a given simple ideal P.

Finally, divisibility of an integer by a factor is expressed in terms of powers of simple ideals as follows:

5. If all the powers of a (non-zero) prime simple ideal containing a given integer $\mu \neq 0$, also contain an integer η, then η is divisible by μ.

Since the collection of all multiples of an integer μ is the principal ideal $i(\mu)$, this last, fundamental property can be equivalently formulated in terms of the previously introduced operations on ideals as follows: every principal ideal $i(\mu)$ is the *l.c.m.* of all powers of simple ideals containing μ (p. 257). To conclude this series of results, Dedekind showed that every prime ideal is indeed a prime simple one, and thus one may finally speak only of prime ideals while forgetting about the simple ones. Further, if all the powers of the prime ideals that divide a given ideal M, divide the principal ideal $i(\delta)$ as well, then M divides the principal ideal $i(\delta)$ (p. 258). Thus in Dedekind's first version of the theory of ideals the fundamental theorem of factorization of any ideal into prime ideals could be formulated as follows:

Every ideal is the *l.c.m.* of all the powers of prime ideals that divide it.[100]

We have thus seen how to associate with any given ideal the desired, uniquely determined list of prime ideals, each with given multiplicity. Now, since to each algebraic integer contained in any extension of the field of rational numbers one can associate a unique ideal (i.e., the principal ideal generated by it), it follows that Dedekind had successfully generalized the unique factor-

---

98. Edwards 1980, 337-342, proves in detail all the results of § 163 using only ideal factors, in the spirit of Kummer, and without introducing ideals at any stage of the proofs. In this way he claims to sustain a central thesis of his article, namely, that Dedekind's main innovation consisted in having defined properly the integers of a field, as the correct generalization of a domain of cyclotomic integers, while ideals are a rather secondary, and perhaps even superfluous concept of the theory.

99. *Werke* Vol. 3, 256: "...eine höchste in μ aufgehende Potenz von P."

100. *Werke* Vol. 3, 258: "Jedes Ideal ist das kleinste gemeinschaftliche Multiplum aller in ihm aufgehenden Potenzen von Primidealen."

ization theorem into prime factors. Moreover, since the theorem holds for those domains in which Kummer's theory was previously known to work, it follows that in those domains Dedekind's ideals, defined by conditions $(a),(b)$ above, are always collections of all multiples either of an actual or of an ideal factor, as Dedekind had in fact intended them to be.

However, in spite of having successfully attained his aim by proving the factorization theorem, Dedekind did not stop there, and proved some additional, rather significant results. Up to this stage Dedekind had defined only two operations with ideals: *l.c.m.* and *g.c.d.* But in fact, the product of two ideals may also be defined without special difficulty, and this is what Dedekind did here: Given the ideals A,B their product AB is the ideal containing all the numbers of the form ab, with a belonging to A and b belonging to B, and all sums of such products (p.259). Besides the immediate properties of associativity and commutativity, the product of ideals satisfies some additional ones that turn out to be useful in dealing with problems of factorization. Thus, if P^a, P^b are the highest powers of a prime ideal P in A,B respectively then P^{a+b} is the highest power of P in AB. This property justifies the notation P^r, since clearly $P^r P^s = P^{r+s}$. Further, one can prove that $N(AB) = N(A).N(B)$. Finally, and most importantly, if an ideal A is divisible by another ideal B (in the sense that A is contained in B) then there exists a uniquely determined ideal R such that $A = BR$ (pp. 259-260).

The meaning of this result, and its location towards the end of the supplement, after the proof of the main factorization theorem, deserve further comment. According to Dedekind's definition in the initial sections of this first version, a given ideal is a multiple of a second ideal if the former is contained (as a set) in the latter. But now in the closing sections, having defined a product of ideals, one has actually defined an alternative sense of being a multiple of a given ideal. What the above-mentioned result amounts to is that both meanings are in fact equivalent; according to it, "A is a multiple of B" means both that A is contained in B and that the product of B and a third ideal C equals A.

The product of ideals did become a central concept in the future versions, since it allows a direct definition of powers, and therefore of multiplicities in the factorization. In contrast, in the first version, powers of ideals (and hence multiplicities) were defined in a rather roundabout way: first defining simple ideals, then using them in the proofs, and later overlooking their existence. But the existence of a product of ideals, defined now in the closing sections of the supplement, together with the proof of equivalence of both senses of the term "multiple of an ideal", seem on the face of it (at least in hindsight) to have allowed the possibility of formulating the factorization theorem directly in terms of products of ideals, thus circumventing perhaps the need to introduce simple ideals. Naturally the question arises, why was not the product introduced in the opening sections, and the equivalence of both meanings proven and used throughout the first version? Answering this question is necessary for understanding the evolution of Dedekind's conception of ideals.

After many years of intense efforts to work out a generalization of Kummer's results, Dedekind seems to have arrived in August 1870 at the concept of ideal as the key idea on which to base it, while preparing the second edition of Dirichlet's *Vorlesungen*. The possibility of publishing his new results as a supplement to the *Vorlesungen* offered itself to Dedekind as an ideal way to do so. Dedekind's publisher, Vieweg, on the other hand, put him under serious pressure to deliver the new edition of the *Vorlesungen* immediately together with the supplements for publication, while suggesting that further improvements in presentation could be added in future editions or in separate articles.[101] Under these circumstances, Dedekind was able to write the supplement, containing a proof of the factorization theorem in terms of ideals, but without having yet been able to develop properly the new proof techniques and, more generally, the full possibilities afforded by the concept. The contents of this supplement can thus be seen, to a considerable extent, as a more or less direct translation of Kummer's results into the language of ideals. This, however, was not accompanied by a change in the overall perspective of research.

Dedekind's definition and use of simple ideals in order to deal with multiplicity of factors in the first version epitomize, perhaps more than anything else, his approach to the theory at this stage, as well as the inherent reasons for

---

101. A very detailed account of the evolution of Dedekind's contribution to number theory, including an attentive description of his elaboration of the various versions of the theory of ideals (based on important, hitherto unpublished sources) can be found in Haubrich 1996. Events related to Dedekind's work from March to October 1870, to his discovery of ideals, and to his correspondence with Vieweg concerning publication, are described there on pp. 186-187. I am grateful to Ralf Haubrich for having enabled me to read the relevant parts of his book, while still in manuscript form, and for having mindfully commented upon earlier versions of the present chapter. The views expressed here, however, are of my own responsibility.

attempting improved versions. Simple ideals were not defined through an abstract property, as a special class of prime, or of other kind of ideals, but rather as collections of integers satisfying a specific kind of congruence. Thus the elementary building blocks of Dedekind's new factorization theorem, although couched in ideal-theoretical language, were in fact used by him within the traditional outlook. But then one can easily point out the immediate reason of Dedekind's uneasiness towards simple ideals and their subsequent omission from the theory, namely, that the definition of their powers is based on particular choices of integers. Dedekind indeed proved that the definition of power was legitimate, and in fact independent of the initial, particular choice of elements involved in it; yet *a choice must in any case be made* in order to find the power of a given simple ideal, and Dedekind would rather avoid these kinds of particular choices wherever possible. The changes introduced in the following versions regarding the treatment of multiplicity of factors, as well as concerning the definition of prime ideals, are connected with the new, more central, role accorded there to the product of ideals.

At this early stage of the theory, Dedekind has thus succeeded in finding the central concepts upon which to base his generalization of Kummer's factorization theorems, but the full possibilities afforded by them, and the change of perspective inherent in the realization of those possibilities, is yet to come. We proceed by examining how this change comes forward in the following versions of the theory.

2.2.3 Later Versions

It seems that only a few mathematicians read the first version of Dedekind's theory of ideals. From a letter sent by Dedekind to Lipschitz in 1876 one learns that, with the exception of Heinrich Weber—with whom Dedekind had worked in the edition of Riemann's works (published in 1876)—Lipschitz himself was the only other person to have explicitly expressed some interest in the theory. Lipschitz's interest stimulated Dedekind's hope that his work was not completely in vain.[102]

By mediation of Lipschitz, Dedekind was invited in 1876 to publish a French report of his new theory in the *Bulletin des sciences mathématiques*. He first sent an introduction which was immediately published.[103] After this

---

102. Cf. *Werke* Vol. 3, 464. See also Dugac 1976, 138 ff.; Edwards 1980, 349 ff.
103. Dedekind 1876-7, 278-288.

publication, Lipschitz maintained his suggestion that Dedekind introduce several changes in the presentation of his results, but Dedekind insisted on abiding by the approach that represented his twenty years of hard work.[104] Although both mathematicians agreed on the advantages of solving a problem by first elucidating and properly formulating the fundamental notions involved, rather than by analyzing specific examples, Lipschitz expressed his conviction of the need to present the results in what for him was a more didactic way, namely, by going from the known towards the unknown.[105] Therefore, he suggested that Dedekind avoid his preliminary presentation of general concepts and directly begin by describing the specific problems which he intended to solve in his research.

It is likely that Lipschitz's suggestions, together with the disappointment caused by the reduced interest aroused by the first version of his theory, led Dedekind to introduce some changes never he less. Thus, he added an introduction to his French version, in which he examined some examples in detail before developing the theory itself.[106] This addition, however, was not the only innovation of Dedekind in the 1877 version of his theory of ideals. In fact, this edition contains significant changes in the presentation, of the kind that may in retrospect be seen as illustrating Dedekind's groping towards a more "structural" formulation of the theory. These changes will now be described and analyzed, not by examining the 1877 French version, but rather the following one, published in 1879, as a supplement to the third edition of Dirichlet's *Vorlesungen*.[107] Although the later edition did not include the examples and the introductory sections, the exposition of the theory itself here was, in essence, similar to that of the French version.

The 1879 version of Dedekind's theory of factorization opens the discussion on ideals precisely where the former version had closed it, i.e., with a definition of the product of two ideals and a discussion of its immediate

---

104. One learns about this interchange from the published correspondence between Dedekind and Lipschitz. The letters appear in *Werke* Vol. 3, 468-69 (Dedekind to Lipschitz), and as Appendix XXXVI to Dugac 1976 (Lipschitz to Dedekind). Cf. also Dugac 1976, 65.

105. This very agreement is not something that should be taken for granted. It is likely that, e.g., Kummer would not have even accepted this way of presenting the problem. On the difference in styles between Kummer and Dedekind see Edwards 1992.

106. Emmy Noether commented several years later, that the many particular examples treated therein, endow this version with the "character of an elementary introduction." See her editorial remarks in *Werke* Vol. 3, 314.

107. Dedekind 1879.

properties.[108] By defining the product at the outset Dedekind could not only do away with simple ideals, but he was also able to formulate many results already known from the first version of the theory in simpler terms. In particular, several theorems formulated in the first version in terms of products of integers contained in the ideals could now be formulated and proven directly in terms of the ideals themselves and their products. Of particular interest in this regard is the characterization of prime ideals. In the first edition, as has already been said, it was proven that P is a prime ideal if and only if for every product $\alpha\beta$ contained in P, either α or β are necessarily contained in P. Having at hand a product of ideals, this characterization of prime ideals was formulated in the present version as follows (p. 301—§ 171, 9):

> If a product of ideals (or of numbers) is divisible by a prime ideal P then at least one of the factors is divisible by P.

Thus, the property of the ideal is defined in terms of operations with ideals, although at this stage Dedekind still mentions explicitly the product of the integers involved.

Similarly, the product is used to characterize a new kind of ideals introduced here by Dedekind, namely, singular (*einartige*) ideals: an ideal A is said to be singular, if it is divisible by a single prime ideal. These ideals are proven to satisfy the following important condition (p. 303):

> Every ideal that does not contain the number 1 is either a singular ideal or it may be represented in a unique way as a product of singular ideals, which are, naturally, relatively prime.[109]

Both the definition of singular ideals and the latter criterion are representative of the change in approach emerging in this version; they are not necessary for the remainder of the supplement (as a matter of fact, a stronger theorem—§ 173, 4, discussed below—is subsequently proven from which the criterion follows), yet Dedekind included them at an early stage of the discussion in order to show how far one can reach when working in a still more general setting than that of the theory of algebraic numbers.

---

108. *Werke* Vol. 3, 297-300.

109. In her editorial comments to Dedekind's works, Emmy Noether explained that in the manuscript of this version, found in his *Nachlass*, Dedekind had explicitly stressed the importance of this theorem: "... die beiden Seiten mit dem Beweis dieses Satzes die Überschrift trugen: 'Für die dritte Auflage kassiert, doch wichtig'." Noether added that this proof is to be considered the first explicit example of a factorization theorem in general ideal theory.

It was said above that one of Dedekind's important achievements in the first version of the theory was to define the domain of algebraic integers of a field as the appropriate framework in which the factorization theorems are to be proven. However, the actual presentation of the theory had not made completely transparent the crucial point where this fact was exploited. Now, in the 1879 version, this crucial point can be clearly identified. In fact Dedekind neatly separated those results which depend upon that it and those that do not. The domain of algebraic integers is seen here as a particular case of a more general kind of concept, namely *Ordnung*, for which the validity of several results are separately proved. In order to prove the unique factorization theorem it is necessary to assume, in addition to the properties of this general domain, the specific properties of the domain of algebraic integers. Of special importance among these properties is their integral closedness.

In this way, the opening sections of the 1879 version of the theory of ideals—which include the two theorems mentioned above—concern not the algebraic integers of a given field as in the first version, but rather a domain D of numbers satisfying the following three conditions:

(α) The system D is a module generated by a finite collection of n numbers. The same n numbers also generate the field of numbers which serves as reference framework.[110]

(β) The system D is closed under multiplication.

(γ) The number 1 belongs to the system D.

Any system of numbers satisfying conditions (α),(β),(γ) is called by Dedekind *Ordnung* (p. 305). From (α) and (β) it follows that all numbers of an arbitrary *Ordnung* are integers. From (γ) it follows that all the rational integers are in D. But it is still possible that an *Ordnung* D does not contain *all* the integers of the field in question. It is possible, then, to extend the concepts originally defined for modules and apply them to *Ordnungen*. One obtains, accordingly, two possible versions of the theory of ideals: a more general one in which the ideals are defined within an arbitrary *Ordnung*, and a more restricted one, when the *Ordnung* in question includes all the integers of the field. The more general theory may be of interest in itself, but a *full* translation of the factorization properties of rational integers into the realm of modules and ideals will

---

110. Here the terms "module" and "field" retain the meaning assigned to them in the first version.

only be possible in the more restricted theory. The opening sections of this version of Dedekind's theory of ideals contain some results of the general theory of *Ordnungen,* but Dedekind did not intend to develop that theory fully and, therefore, he turned to consider results concerning only the theory of ideals in the domain of all algebraic integers of a given field. Nevertheless, Dedekind did illustrate what he meant by the possibility of a more general theory with an example: given a singular ideal Q and its prime divisor P, it is not always the case, in the more general theory, that Q is a power of P (p.306). On the contrary, in the more restricted case, to be considered in the following sections of the supplement, Q is always a power of P. Although Dedekind himself did not pursue the consequences of the direction implied by the general result mentioned here, Emmy Noether did so in an interesting way in her future work on the abstract theory of ideals, but without mentioning Dedekind's original suggestion (see § 5.2. below).

The section devoted to the theory of factorization of ideals properly said opens with a theorem, in whose proof Dedekind actually uses the integral closedness of the whole domain D of algebraic integers of a given field. The theorem (§ 173, 1) reads as follows:

> If P is a prime ideal, there exists a number λ, divisible by P (i.e., a number contained in P), and a second number ρ, not divisible by P, such that $P.i(\rho)$ is the *l.c.m.* of the ideals $i(\lambda)$ and $i(\rho)$.

The proof is based on several results proved in preceding sections of the supplement, which allow to ascertain the existence of an element v in D, such that $P.i(v\rho)$ is the *l.c.m.* of $i(\lambda v)$ and $i(\rho v)$, given the above conditions,. The theorem follows from the factorization of the v in these three factors. This last step is valid only if the domain considered is integrally closed.[111]

Some additional results allow for the proof of theorem § 173, 4, which asserts that every ideal A, which does not contain the unit, is either a prime ideal or a uniquely determined product of prime ideals (p. 308). As a consequence of this result, it is possible to prove the central result that if the ideal A "is a divisor of B" (*Theiler*) in the sense that it is contained in it, then it is also a divisor in the operational sense of the word. This is formulated as a theorem as follows:

---

111. In order to prove that in an arbitrary ring, from $i(ac) = i(bc)$ there follows $i(a) = i(b)$, it is necessary to see that if $x = ra$, then there exists an element r' in the ring, such that $ra = r'ca$. This is warranted by integral closedness.

If an ideal C is divisible by an ideal A, then there exists one, and only one, ideal B such that $AB = C$. Moreover if AB divides AB' then B divides B', and from $AB = AB'$ one can deduce $B = B'$.[112]

Now the divisibility conditions may be formulated in terms of the product of ideals, but, remarkably enough, Dedekind still takes care to stress the parallel between his ideals and Kummer's ideal factors, as follows (p. 312—§ 173, 8):

An ideal A (or a number α) is divisible by an ideal D (or a number δ) if and only if all the powers of prime ideals of D (or of δ) appear also in A (or in α).

An immediate consequence of this theorem, with projections for future developments, closes the whole section and states that (p. 313—§ 173, 9):

If M is the *l.c.m.* and D the *g.c.d.* of two ideals A,B then
$$A = DA', B = DB', MD = AB$$
$$M = DA'B' = AB' = BA'$$
where A',B' are relatively prime ideals.

Dedekind's 1879 version of the theory of ideals contains a reformulation of results already known from the first version, among them of course, the unique factorization theorem of any ideal into prime ideals. Yet these results are obtained following a somewhat different approach from the original one. The difference in approach is epitomized, for instance, in Dedekind's decision to define the product of ideals at the beginning and to establish the equivalence of the two possible meanings of "being a multiple of a certain ideal."[113] This

---

112. *Werke* Vol. 3, 310 (§ 173, 6): "Ist das Ideal C teilbar durch das Ideal A, so gibt es ein und nur ein Ideal B, welches der Bedingung $AB = C$ genügt. - Ist AB teilbar durch AB', so ist B teilbar durch B', und aus $AB = AB'$ folgt $B = B'$."

113. Dedekind's own appraisal of the central role played in the improved formulation of the theory by the theorem establishing this equivalence and of the difficulties involved in its proof comes to the fore in the following passage written in 1895 (Dedekind 1895, 50-51):
"In § 172 der dritten Auflage der Zahlentheorie und ebenso in § 23 meiner Schrift Sur la théorie des nombres entiers algébriques habe ich hervorgehoben, daß die größte Schwierigkeit, welche bei der Begründung der Idealtheorie zu überwinden war, in dem Beweise des folgenden Satzes bestand: [here he quoted the equivalence theorem]... Daß dieser Satz, durch welchen der Zusammenhang zwischen der Teilbarkeit und der Multiplikation der Ideale festgestellt wird, bei der damaligen Darstellung erst nahezu am Schlusse der Theorie beweisbar wurde, machte sich in der drückendsten Weise fühlbar, besonders dadurch, daß einige der wichtigsten Sätze nur allmählich durch schrittweise Befreiung von beschränkenden Voraussetzungen zu der ihnen zukommenden Allgemeinheit erhoben werden konnten."

choice discloses Dedekind's gradual mastering of distinct ideal-theoretic proof-techniques and, more generally, his increased perception of the potential gains, in terms of new and more systematic knowledge, of shifting the focus of attention away from the study of operations and relations among the numbers themselves and towards the properties of their collections and the "structural" interrelations among the latter. Moreover the present version clarified the deductive interrelations among the central concepts of the theory itself, and in particular the reason why the main decomposition theorem is valid only when the whole collection of algebraic integers is considered.

The approach followed in the 1879 version also suggested a significant direction for future research, in which unique factorization of ideals in domains of algebraic integers would be obtained as a particular application of a more general theory. Dedekind did include in the present version one result of the kind that could be proven in such a general theory, but in future works he did not develop this direction any further. Thus, although he was apparently technically able to do this, the direct motivations of Dedekind's ideas were such, that work on a general theory of *Ordnungen* would appear as a lower priority. For Dedekind, the theory of ideals was a tool for understanding factorization properties in the most general cases of fields of algebraic numbers, rather than an end in itself. That Emmy Noether would later focus precisely on the kind of questions suggested but not pursued by Dedekind here, and fully elaborate their consequences in her own work, represent the kind of change in images of knowledge that typifies, in our account, the rise of the structural approach to algebra.

In the last version of Dedekind's theory of ideals one finds yet a further change in approach, which strengthens the trends initiated in the 1879 version, and which yields interesting new results. This last version is examined in the next section.

2.2.4　The Last Version (1894)

The last version of Dedekind's theory of ideals appeared again as a supplement to Dirichlet's *Vorlesungen*, in its fourth edition. This time the opening sections included a discussion on Galois theory, which was totally absent from earlier versions, and which sheds additional light upon the gradual process of clarifying the interplay between the various algebraic concepts and domains of inquiry that is visible in Dedekind's work throughout the years. This applies with particular interest to his elaboration of the relationship between the group of the equation, and its subgroups, on the one hand, and the fields containing the roots of the equations, on the other.[114]

First, Dedekind defined homomorphism of fields (*Permutation des Körpers*) to be any function between two given fields that preserves the four operations, and showed that they constitute a group with respect to their composition.[115] As in the previous versions, Dedekind dealt here only with fields which are subfields of the field of complex numbers. He then proved some general results concerning homomorphisms. Thus for instance, every homomorphism between two fields leaves the field of the rational numbers invariant and the identity is the only homomorphism that can be defined on the field of rational numbers. Further, the collection of numbers left invariant by an arbitrary group of permutations defined on a field is itself a subfield of the given field (p. 459). More elaborate results of Dedekind are equivalent, in modern terms, to describing any extension of a field B over A, by defining B as a vector space over A (p. 468). In fact, Dedekind even established a relation between the degree of the extension, (B,A), to the dimension of the associated vector space (p. 471). He also mentioned the fact, albeit without proving it, that given a field M and a group of homomorphisms Π defined on it, such that A is the field that remains invariant under the homomorphisms of Π, then, if Π' is any subgroup of Π, there exists a subfield A' of M, containing A, such that A' is the field that remains invariant under Π'. Moreover, $(M,A')(A',A) = (M,A)$ (p. 484).

This sharp formulation of the interplay between the two main concepts of Galois theory, and the introduction of yet a third, hitherto disconnected kind of system (essentially a vector space)[116] to this domain of discourse constitute the most salient difference between Dedekind's exposition of Galois theory in

---

114. The relation between Dedekind's ideas on Galois theory and the later versions of the theory, especially in the work of Emil Artin, is discussed in Kiernan 1971, 129-132.

115. Dedekind 1894, 457.

116. In his comprehensive account of the rise of the idea of a vectorial system Michael Crowe (1967) focuses on the line of development that starts from William Rowan Hamilton (1805-1865), on the one hand, and Hermann Grassmann (1809-1877), on the other, and from there, through the path of development connected mainly with quaternions and associated systems, to the works of Josiah Willard Gibbs (1839-1903) and Oliver Heaviside (1850-1925). Crowe's account does not analyze or comment the connection of the genesis of vectorial ideas with the rise of algebraic concepts, such as manifest in Dedekind's treatment of extensions of fields. Neither can it be fully discussed in the present account. This connection, however, deserves detailed attention on its own.

1894 and his early one, of 1856. In the earlier version, groups had provided a tool for analyzing the subject matter of the theory, i.e, the behavior of roots of equations in the field of complex numbers and in the system of its subfields. The same situation appears in the 1894 version, but the analysis is now much more structural in spirit. Nevertheless, fields and groups remain two different kinds of mathematical entities; the one numerical, the other abstract in nature. As Dedekind explicitly put it, the actual subject of contemporary algebra is the study of the interrelation among the various fields.[117] Clearly, he referred here to the field of complex numbers and the system of its subfields.

In a second preliminary section, Dedekind worked out in some detail an "algebra of modules", which had been neither elaborated nor even suggested in the previous versions, and which provided here the basis for the whole theory of factorization. Dedekind used the notation $A + B$ and $A - B$ for the g.c.d. and the l.c.m respectively and proved the following identity: if D is a divisor of M then, for any module A (p. 498):

$$M + (A - D) = (M + A) - D. \tag{σ}$$

This identity will eventually prove of great importance for Dedekind's elaboration of the concept of lattice (§ 2.3 below). Side by side with the concept of divisibility (*Theilbarkeit*) of modules, Dedekind also defined the product, while stressing that the product of A and B should not be confused with a multiple (*Vielfacher*) of either A or B (p. 500). The operation is defined here for modules in general, rather than for the particular case of ideals as in the previous version of the theory. Thus, one could now examine several properties of modules in terms of their multiplication, and it turned out that several useful properties of the algebra of rational numbers remain valid for the algebra of modules. And yet, some additional, important properties are peculiar to the algebra of modules alone. Thus, for instance, the following identity:

$$(A + B + C)(BC + CA + AB) = (B + C)(C + A)(A + B) \tag{τ}$$

The validity of this identity, called the "modular identity", may be verified simply by performing the operations on both sides and by recalling that, for every module M, one has $M + M = M$ (p. 503). Dedekind called attention to the fact that here one has two essentially different representations of a module as products of factors—the factors of the left-hand side are sums of *three* mod-

---

117. Dedekind 1894: "[In] der Verwandtschaft zwischen den verschiedenen Körpern ... besteht der eigentliche Gegenstand der heutigen Algebra."

ules while those on the right-hand are sums of *two* modules. He made important use of this identity in the present supplement as well as later, in his work on lattices.

This algebra of modules can in fact be extended to include a division operation as well, as follows. A quotient of modules $B:A$ is the set N of all numbers n such that nA is contained in B. It is easy to prove that this set is itself a module. Using the multiplication and the quotient of modules Dedekind introduced a concept that was absent from previous versions of the theory: *Ordnung*.[118] The *Ordnung* of a given ideal is the quotient $A:A$ (denoted by A^0). Given any module M contained in A^0, it is easy to prove that MA is contained in A (pp. 505-506). Reciprocally, every module A containing the rational integers and for which A^2 is contained in A (and therefore $A = A^2$) is called an *Ordnung*.[119] Negative powers of modules may also be defined as follows (p. 506):

$$A^{-n} = A^0 : A^n$$

However, not all properties of the operations on powers of rational numbers may be translated into those of modules. A full extension of the properties of the algebra of rational numbers is only possible if we restrict ourselves to modules such that $AA^{-1} = A^0$, or "proper" modules (*eigentliche Moduln*).

Finally, then, if one considers any collection of modules *having one and the same Ordnung*, then the properties of the multiplications and divisions of powers of modules are identical to those of the rational numbers. If we recall that ideals are nothing but modules whose *Ordnung* is the set of all integers of a given field of numbers, then it is clear that the former claim holds for ideals,[120] and we have thus obtained a complete set of algebraic operations on the system of ideals.

Dedekind used the term "module" in this context mainly to emphasize its direct connection with Gauss's congruences modulo a given number, introduced in the latter's work on Gaussian integers. As in the case of rational num-

---

118. As a matter of fact, these operations on modules, and the definition of *Ordnung* had already been introduced by Dedekind in a previous article of 1877, where he also explained that he had taken the term *Ordnung* from Gauss's theory of binary quadratic forms. Dedekind defined here the quotient $A:B$, of two modules A,B, as the g.c.d. of all modules C such that the product BC is divisible by A. The *Ordnung* of a given module is then the collection of all numbers α such that αA is contained in A. Dedekind also proved that the *Ordnung* of a module M coincides with the quotient $M:M$. See Dedekind 1877, 124

119. Thus Dedekind's definition of *Ordnung* here coincides with that of 1877

120. In fact, Dedekind used this property in a later section as an alternative definition of ideals.

bers, congruence modulo a given module M is preserved through addition and through multiplication by any number of the *Ordnung* M^0. In other words, if two numbers α, β belong to M, and m belongs to the *Ordnung* M^0, then both $\alpha - \beta$ and $m\alpha - m\beta$ belong to M. Further, if two numbers are congruent module M then they are congruent modulo any divisor of M. It is worth noticing, that in this version Dedekind introduced a notation which is opposed to the one presently accepted, namely: when the module M is a partial set of D, the module M is said to be a multiple of D, and this is denoted as $M > D$ or $D < M$. This notation underscores the number-theoretical motivation of Dedekind's definition of module, which happens to be opposed to the set-theoretical situation involved (i.e., that, as sets, $M \subseteq D$). Without attempting to overemphasize the importance of notational details of this kind, one should not completely overlook the significance of this divergence.[121]

If a module A contains a module B, then B determines congruence classes in A. This concept can be further generalized for the case when B is not fully contained in A. Let M be the *l.c.m.* of A and B. Two numbers α, β in A are said to be congruent modulo B, if $\alpha = \beta + \mu$, where μ is a number in M. The number of representatives of the classes of congruences determined in A by B is denoted by (A, B) and it is easily seen that (p. 510):

$$(A, B) = (A, A - B) = (A + B, B)$$

Some years later Dedekind would realize the key role played by this property in the proof of the Jordan-Hölder theorem, and he thus proposed a general formulation of that theorem based on it. In case there are an infinite number of congruence classes, then one says that $(A, B) = 0$. A further result proven by Dedekind in this context states that if C is an ideal contained in A and $(A, C) > 0$ then there is only a finite number of modules between A and C. In a footnote, Dedekind added that the reciprocal is also true. This result leads to the introduction of the concept of "proper divisor" (*echter Theiler*): A is a proper divisor of B if A is a divisor of B and if there is no other module between A and B (p. 511).

The last result of the section on modules is very significant for future developments in abstract ideal theory, and therefore, for the present discussion. It consists of the proof of the ascending chain condition (*a.c.c.*), namely, that given a chain of modules $A_1, A_2, \dots A_n, \dots$ contained in a given *finite* mod-

---

ule N (i.e., a module with a finite number of generators) such that A_i is a multiple of (i.e., A is contained as a set in) A_{i+1}, for all indexes i, then there is an index k such that $A_i = A_k$, for all $k > i$ (p. 523). The general abstract formulation of the a.c.c. and its central role in the proofs concerning factorization of abstract ideals will be one of the main contributions of Emmy Noether. Dedekind's present formulation of the a.c.c. enabled him to introduce new concepts and results in terms of inclusions of sets of numbers. Through this formulation he was able to skip both over the particular description of modules and over the need to select specific elements of them or to specify their characteristic polynomial.

When Dedekind finally arrived at the actual discussion of algebraic integers and the problem of unique factorization, then, he could rely on a wide range of concepts and results, allowing for a quite different (and certainly much more systematic) approach than that found in the previous versions of his theory. This approach may be illustrated with the following example. Theorem § 173, II states that an algebraic number is an integer if and only if it is contained in the *Ordnung* of some module A (p. 526). This is proved by considering the determinant whose rows contain the coordinates of a specific set of numbers, with respect to a given basis. Dedekind remarked that this proof is not satisfactory because it depends upon specific choices of numbers and, moreover, because the theory of determinants is "alien to the proper content of the theorem."[122] Therefore, he found it convenient to present the second half of the theorem in a different formulation, namely, claiming that the *Ordnung* of any module is composed only of integers. This is proven as a separate theorem in which Dedekind used the a.c.c. introduced in the previous section. Thus, Dedekind succeeded in exhibiting the specific inclusion property of modules, that provides the actual conceptual basis of this theorem.

Further, by considering chains of ideals, Dedekind proved two additional important theorems that have important consequences for future developments, and which we thus discuss here in some detail. The first theorem establishes the closedness of integers under addition, subtraction and multiplication. The second asserts that any root of any polynomial equation with integer (not necessarily rational integer) coefficients, is also an integer (p.

---

122. Cf. Dedekind 1894, 527: "... [die] Theorie [der Determinanten] dem eigentlichen Inhalte des Satzes gänzlich fern steht."

528). Earlier versions of these proofs were based in choosing certain polynomials and performing fatiguing manipulations on them.[123] Here, they are easy consequences of very simple inclusion relations of modules. Recall that the immediate consequence of this theorem is that if α is an integer, then $\alpha^{1/n}$ is also an integer, for all $n \in Z$. This fact was central for the elaboration of the theory since it had led Dedekind to a definition of prime numbers different from the one accepted in the rational case. The ensuing proof depended on a specific choice of equations and basis.[124] Here, on the contrary, it followed from the existence of certain *Ordnungen* and from their inclusion properties.

A further central theorem (§ 173, VI) states that:

> For every non-zero finite module M, containing only integer or fractional algebraic numbers, there exists another module N, all of whose numbers are obtained by rational operations on the numbers of M, such that the product MN contains only integers and divides the module Z of rational integers.[125]

Recall that "MN divides Z" means that, as sets, Z is contained in MN. The proof proceeds by induction on the number of generators of the module M. If M is generated by a single element α, then we choose N as the module generated by α^{-1}. If M is generated by two numbers α, β, then a specific basis for the module N is built from a particular polynomial equation of which the quotient α/β is a root. Such an equation exists because α/β is an algebraic number. Now, if $n > 2$ we assume that the theorem holds for any module generated by less than n numbers. The inductive step is based on the above mentioned modular identity:

$$(A + B + C)(BC + CA + AB) = (B + C)(C + A)(A + B) \qquad (\tau)$$

Since M is generated by at least three elements, it may be divided into three non-zero modules A, B and C, $M = A + B + C$. All the remaining modules of the equation are therefore generated by less than n numbers and this enables the completion of the proof. The module $N = (BC + CA + AB)$ is the one stipulated by the theorem.

---

123. Cf., e.g., *Werke* Vol. 3, 237-240.

124. Cf., e.g., *Werke* Vol. 3, 237.

125. Dedekind 1894, 528: "Jeder endliche, von Null verschiedene Modul M, der aus ganzen oder gebrochenen algebraischen Zahlen besteht, kann durch Multiplication mit einem Modul N, dessen Zahlen aus denen von M auf rationale Weise gebildet sind, in einen Modul MN verwandelt werden, welcher aus lauter ganzen Zahlen besteht und ein Theiler des Moduls Z ist."

This theorem is noteworthy not only because of the results that are deduced from it in the sequel, but also because, by using the modular identity in the inductive steps of the proof, Dedekind was able to avoid relying on specific choices of numbers or polynomials. Nevertheless, in the first inductive step, *a specific choice* has to be made when constructing the characteristic polynomial having α/β as a root. Obviously, given the centrality of this theorem, Dedekind would by unsatisfied by this part of the proof, and he would still keep looking for an alternative kind of argument in order to avoid such a particular choice.

The remaining results of the theory—including the main theorem asserting the unique factorization of an ideal into prime ideals—are obtained rather straightforwardly. Dedekind characterized ideals as proper modules (*eigentliche Moduln*) M whose *Ordnung* is the whole domain D of algebraic integers and which satisfy the condition $MM^{-1} = D$ (p. 553). It is here that theorem § 173, VI, mentioned above, plays a central role in the proof. In fact, the theorem asserts the existence of a second module N such that MN contains all the rational integers and is contained in D. The properties of ideals imply that $MN = D$ and that, furthermore, $M^0 = D$.[126] As Dedekind remarked, one can prove, without using theorem VI, that $M^0 = D$.

This characterization of ideals enables the use of all the algebra of modules developed in the preliminary sections. In particular, the important result proven in a previous version (i.e., that if the ideal M is divisible by the ideal A, then there exists a unique ideal B such that $AB = M$) appears here as an immediate consequence of that algebra: write $B = MA^{-1}$, and the result follows easily (p. 553—§ 177, Theorem VII).

The fourth, and last published version of Dedekind's theory of ideals, then, brings to a high point the process of translating his self-imposed methodological guidelines into well-elaborated mathematical ideas devised for solving a very specific, open problem. The theory was now built on a fully elaborated algebra of modules based on inclusions properties of distinguished sets of numbers. This approach naturally leads to focusing on the properties of

---

126. From $Z > MN > D$, multiplying by D, it follows that $DZ > MN > D^2$ (recall that $>$ means, in set theoretical terms, inclusion of the left-hand set in the right-hand one). But D is an *Ordnung* and hence $D^2 = D$, therefore $MN = D$. On the other hand, by definition $MM^0 = M$ and therefore $M^0 > D$, but also $DM > M$ and therefore $D > M^0$, so that $D = M^0$.

the sets themselves rather than on those of the individual numbers contained in them. The many new concepts introduced in developing the theory, despite their strongly abstract character, emerge organically from within the specific technical context of the problem investigated, rather than from an unmotivated drive to generalize and formulate abstract definitions as an end in itself.

Dedekind had planned, according to Emmy Noether's testimony, a fifth edition of the supplement to Dirichlet's *Vorlesungen*. This edition would contain a further elaboration of the foundations of his theory of ideals, in which the concepts of integer and fractional ideals would have been taken as starting point.[127] This plan was never realized. And yet, one can perhaps speculate about what sections in the fourth version were in need of further elaboration from the point of view of Dedekind's images of mathematics. Indeed, as was seen above, there is still a rather central point in the fourth version involving a large amount of calculations with specifically chosen polynomials and bases, namely, the first step in the inductive proof of theorem § 173, VI. This is a kind of argument that Dedekind would still probably have wanted to avoid in a future version, in which his methodological guidelines and his images of mathematics might have been fully realized.

2.2.5 Additional Contexts

An account of the successive versions of Dedekind's theory of ideals cannot be concluded without mentioning his joint research with Heinrich Weber on the algebraic theory of functions, on the one hand, and the work of Leopold Kronecker on algebraic number theory, on the other hand.

The algebraic theory of functions provides the second main context in which Dedekind's ideals, in their original conception, were successfully applied. In 1882 Dedekind and Weber published an important paper bearing the title "The theory of algebraic functions in one variable" It developed ideas originally worked out by Riemann in his innovative, but sometimes hard-to-follow, contributions to this field. In particular, it redefined Riemann surfaces and proved in an original way the Riemann-Roch theorem, and it came to play an important role in the subsequent development of the theory of functions. A brief discussion of this work is relevant to the present discussion.[128]

In the introduction to the article, the authors stated as their aim the attempt to approach the theory of algebraic functions of a single variable, as it had

---

127. Cf. the editorial comment of Noether in *Werke* Vol. 3, 314.

been conceived and developed by Riemann, but from a simpler and more general point of view than the accepted one. The proposed point of view would emphasize, in accordance with Dedekind's images of mathematics, the need to avoid, whenever possible, excessive reliance on geometrical considerations, which had not been adequately elucidated. According to the authors, earlier treatments of the theory had been impaired by superfluous, limiting presuppositions about the singularities of the functions considered, and these would be avoided in their own work. Dedekind and Weber proposed to apply to the theory of algebraic functions methods adopted from Dedekind's theory of ideals and, more specifically, to exploit the close formal analogy between the theories of algebraic numbers and that of algebraic functions.

In their article Dedekind and Weber followed very closely the formulations used by Dedekind in his 1879 versions of the theory of ideals, except for minor, necessary differences. As in algebraic number theory, a field of algebraic functions is defined as a system of functions closed under the four operations: addition, subtraction, multiplication and division.[129] Within a given field, a specific class of functions may be selected—the integral functions— whose quotients generate the whole field. Integral functions play in this realm a role similar to that of algebraic integers in their own one. The concepts and theorems of the theory of ideals are translated, using these concepts, into the theory of algebraic functions. In particular, Dedekind and Weber formulated the theorems of decomposition of polynomials into linear factors in terms of products of prime ideals. Moreover, the proofs relied exclusively on the "algebraic" properties the complex field, namely, that it is an algebraically closed field of characteristic zero (of course, these facts are not defined in these terms but only implicitly used). The "topological" properties of the field, on the contrary, are not used at all.[130]

In summarizing Dedekind's contribution to algebra in the last section of this chapter, we will comment again on this work and its significance.

---

128. For a more detailed account of Dedekind and Weber's article from the point of view of the development of the theory of functions, see Geyer 1981. A technical analysis of the contents of the article and an assessment of its central place in the historical development of algebraic geometry appears in Dieudonné 1985, 29-35. Oscar Zariski used to mention the Dedekind-Weber article as one of the main sources of his own seminal work on the foundations of algebraic geometry during the 1930s. See Parikh 1991, 68-69.

129. Dedekind & Weber 1882, 242.

130. Cf. Dieudonné 1985, 65.

Leopold Kronecker's work on algebraic number theory represented the great contemporary alternative to Dedekind's research in the same field. At roughly the same time as Dedekind, Kronecker, a student and friend of Kummer, undertook to generalize the latter's theory and to supersede the shortcomings of his teacher's achievement. Kronecker followed an algorithmic perspective, totally different from Dirichlet's and Dedekind's much more conceptual approach. The widespread adoption of Dedekind's point of view since the 1890s, especially following Hilbert's *Zahlbericht* (§ 3.1.2 below), as well as the difficulties inherent in Kronecker's writings, led to the relative oblivion of Kronecker's approach by most twentieth-century mathematicians.[131] This is not the place to describe Kronecker's theory of divisors and to compare it with Dedekind's theory of ideals.[132] What should not pass unnoticed, however, is the very existence of two theories that developed at the same time as answers to one and the same question, but which nevertheless display very different perspectives and spirit.[133] Kronecker and Dedekind were faced with the same body of mathematical knowledge, but approached it through different systems of well-defined and consistent images of knowledge; their contributions to the consequent expansion of the body of knowledge were sensibly different.[134]

---

131. In his address to the International Congress of Mathematicians in Cambridge, Mass., André Weil (1950, 90) described Kronecker's theory of divisors as follows: "In his *Grundzüge*, Kronecker did not merely intend to give his own treatment of the basic problems of ideal theory which form the main subject of Dedekind's life work. His aim was a higher one. He was, in fact, attempting to describe and to initiate a new branch of mathematics, which would contain both number-theory and algebraic geometry as special cases. This grandiose conception has been allowed to fade out of our sight, partly because of the intrinsic difficulties of carrying it out, partly owing to historical accidents and to the temporary success of the partisans of purity and of Dedekind."

Edwards (1987, 34) has remarked that "in the twentieth century, Kronecker's readers have been few, but their quality has been high. Erich Hecke, Hermann Weyl, Carl Ludwig Siegel, and André Weil are among them."

See Gray 1997, 11-30, for a brief description of Kronecker's ideas and their immediate elaboration by Jules Molk, Julius König and others.

132. See, for instance, Edwards, Neumann & Purkert 1982.

133. There is in fact a third mathematician, the Russian Egor Ivanovich Zolotarev (1847-1878) who independently addressed the same issue and came forward in 1874 with his own theory of ideals. This work however seems not to have had any direct influence on the actual development of the theory, or on the rise of the structural approach to algebra. After a first article in Russian in 1874, Zolotarev sent in 1876 a French version which was however published only four years later as Zolotarev 1880. Accounts of Zolotarev's contributions appear in Bashmakova et al. 1992, 108-115; Haubrich 1988, 27-29; Piazza 1999.

2.3 Ideals and *Dualgruppen*

Besides ideals, modules, fields and groups, a further concept investigated by Dedekind, and which was later to be considered a typical instance of an algebraic structure, is the concept of lattice. In the last decades of the nineteenth century this concept was elaborated and studied independently by Dedekind and by Ernst Schröder (1841-1902).[135] The two arrived at the concept from different directions; whereas Dedekind's concept originated in the theory of numbers, Schröder's arose within his research in the algebra of logic.[136] Both works were ignored by the mathematical community at large. It was only in the 1930s that the concept was redeveloped, especially in the works of Garrett Birkhoff and Oystein Ore, and that the theory of lattices became an independent mathematical discipline, whose results have found applications in many other fields.[137]

---

134. Harold Edwards (1987; 1988; 1990) has analyzed Kronecker's work on divisor theory, and has undertaken to bring its value back to the awareness of a wider audience of mathematicians. In a comparative analysis of Kronecker's and Dedekind's works in the field, Edwards claimed that the first version of Dedekind's theory was the best and the most effective one, and that the following versions receded from the original success, since Dedekind became "increasingly doctrinaire in his views after 1871" (Edwards 1980, 348), and that "Dedekind's theory of ideals does not appear to have opened new vistas or to have led to simplifications or improvements of the theory" (*ibid.*). Now, it is possible that from the point of view of Edward's analysis (the generalization of Kummer's ideas), the first version presents some advantages over the following ones (although perhaps this claim could also be critically appraised). However, from the point of view of Dedekind's own standards, it is clear that the first version was seen by him, from the outset, as in need of improvement; from the same point of view, Dedekind's last version was the best one, even though there was still room for further improvements. In principle, I agree with Edwards's claim that strict adherence to philosophically motivated methodological principles may hinder scientific progress. Moreover, Edwards's own contribution to pinpointing the existence of such cases in mathematics is itself important for an appraisal of his claim. However, as the above analysis shows, Dedekind did not become *increasingly doctrinaire* in the application of his principles, but only *more successful* at it. The same principles that motivated his dealing with the problem of generalizing Kummer's theory in the first place were those that motivated the successive changes introduced in the different versions of his theory.

135. Dedekind was however aware of Schröder's work and held it in great esteem. See, e.g., the preface to *Was sind und was sollen die Zahlen?* (1888), *Werke* Vol. 3, 343.

136. For an account of Schröder's work on lattices, see Mehrtens 1979, 53-56. For a general account of Schröder's work, see Dipert 1991. Peckhaus 1994 discusses Schröder's work focusing on his attempt to advance the idea of an "absolute algebra."

137. A detailed account of the development of lattice theory, in its two stages, appears in Mehrtens 1979.

Using the term *"Dualgruppen"* to denote this notion, Dedekind's published his works on lattices only after his retirement in 1894. Needless to say, Dedekind had been considering the basic ideas behind these works many years before.[138] As a matter of fact, the incipient idea of *Dualgruppe* and the importance of further developing it were mentioned by Dedekind in his 1894 version of the theory of ideals. In developing his "algebra of modules", Dedekind stressed the peculiar character of some identities satisfied by ideals, such as the "modular identity":

$$(A + B + C)(BC + CA + AB) = (B + C)(C + A)(A + B) \qquad (\tau)$$

and the identity stating that, if D is a divisor of M then, for any module A:

$$M + (A - D) = (M + A) - D. \qquad (\sigma)$$

Dedekind's proof of this last identity contains the germ of his notion of *Dualgruppen*. Indeed, from the fact that M is contained in $(M + A)$ and that $(A - D)$ is contained in D, one easily obtains the inequality:

$$M + (A - D) < (M + A) - D.$$

However, it turns out that the second inequality, namely,

$$M + (A - D) > (M + A) - D$$

cannot be derived purely in terms of the abstract properties of the operations on modules. Dedekind was thus forced to prove the second inclusion by taking specific numbers of each of the modules involved and by relying on their specific properties. But as already said, Dedekind would rather avoid such proofs whenever possible. In this case, it is not surprising that he wished to reconsider this step of the proof. Typically, he did so by formulating a new, more general concept: this was precisely his concept of *Dualgruppe*.

---

138. For instance in Dedekind 1877; 1882; 1894, 499-510. On this point, see Mehrtens 1979, 80-83. In the 1894 supplement (p. 499) Dedekind introduced the term *"Modulgruppe"* to denote a collection of modules endowed with two operations: *l.c.d.* and *g.c.d.* This willingness to use the term *"Gruppe"* in this context, should not pass unnoticed. As late as 1890, and perhaps even later, the term "group" remained often ambiguous. Whereas the abstract definition of group accepted nowadays was formulated back in the 1880s, it was still possible to find by the turn of the century working group-theorists who used the term indistinctly to mean either the general concept, or, specifically, groups of permutations (e.g., Netto 1882. Cf. Wussing 1984, 336-338). On the other hand, it was also possible to find the term used with a broader meaning, like Dedekind does here. In an early draft of his book *Was sind und was sollen die Zahlen* (1888), Dedekind had also used the term *Gruppe* for a concept he would later denote as "chains" (*Kette*). See Dugac 1976, app. LVI, 296.

A second source for Dedekind's definition of lattice arises from the phenomenon of duality, which he had also encountered in his work with modules. In fact, the operations of intersection and union of modules are dual of each other, in the sense that in any identity of modules involving these two operations one can interchange them in every instance in which they appear, thus obtaining an identity again. The same is true for the two relations of inclusion: interchanging all instances of $A < B$ and $B > A$ in any identity of modules in which they appear yields a new identity. In the 1894 supplement Dedekind called attention to this remarkable fact, and to the potentially interesting directions of research suggested by it. At that time, however, the ultimate grounds of this kind of phenomenon were unknown to him.[139]

The first publication in which Dedekind defined *Dualgruppen* explicitly and studied their basic properties appeared in 1897. In an article dealing with decompositions of integers into (not necessarily prime) factors, Dedekind defined *Dualgruppen* in a section which represented, in fact, a digression from the main subject. Given collections of three or four numbers, Dedekind calculated their *g.c.d.* and *l.c.m* and then proceeded to decompose these results into their own factors. By recombining the factors through diverse operations, Dedekind obtained certain numbers whose collection he denoted as *Kerne*. He showed that these *Kerne* play an important role in the factorization of the originally given numbers. Through the study of the properties of the *Kerne*, Dedekind introduced the concept of the *Dualgruppe*.

Given a set of n elements, take the collection of all their possible combinations and define two operations on the members of this collection: union (*Summe*), denoted by "+" and intersection (*Durchschnitt*) denoted by "-". These operations satisfy certain properties which Dedekind wrote as a list of three double assertions, and denoted as Theorem A:[140]

---

139. Cf. Dedekind 1894, 498: "...ein eigenthümlicher Dualismus, dessen *letzter* Grund schwer zu erkennen sein mag." (Italics in the original).

140. See *Werke* Vol. 2, 109: Theorem A.

$$\alpha + \beta = \beta + \alpha \tag{1'}$$
$$\alpha - \beta = \beta - \alpha \tag{1''}$$
$$(\alpha + \beta) + \gamma = \alpha + (\beta + \gamma) \tag{2'}$$
$$(\alpha - \beta) - \gamma = \alpha - (\beta - \gamma) \tag{2''}$$
$$\alpha + (\alpha - \beta) = \alpha \tag{3'}$$
$$\alpha - (\alpha + \beta) = \alpha \tag{3''}$$

The list renders manifest the dual character of the properties satisfied by these two operations. Clearly, by interchanging the two operations in one of the propositions of any pair, the second one is obtained. An immediate consequence of theorem A is that both operations are idempotent:

$$\alpha + \alpha = \alpha \tag{4'}$$
$$\alpha - \alpha = \alpha \tag{4''}$$

Dedekind considered two additional pairs of properties:

$$(\alpha - \beta) + (\alpha + \gamma) = \alpha - (\beta + \gamma) \tag{5'}$$
$$(\alpha + \beta) - (\alpha + \gamma) = \alpha + (\beta - \gamma) \tag{5''}$$

and

$$(\alpha - \beta) + (\alpha - \gamma) = \alpha - [\beta + (\alpha - \gamma)] \tag{6'}$$
$$(\alpha + \beta) - (\alpha + \gamma) = \alpha + [\beta - (\alpha + \gamma)] \tag{6''}$$

In general, these pairs of properties cannot be deduced from Theorem A alone, unless the collections involved are sets of natural numbers. Thus Dedekind raised the question of the logical interdependence among the basic properties of the operations defined on modules. He proved that although the pair (5')/(5") cannot be directly derived from theorem A, (5') can indeed be proved using the theorem and proposition (5"). Reciprocally, given theorem A and (5'), one can prove (5") (p. 110). The same holds for the relation between the theorem and the pair (6')/(6") (p. 111).

Dedekind thus defined a *Dualgruppe* as a system of elements endowed with two operations, "+" and "-", satisfying the conditions of theorem A. He stressed the fact, however, that this concept was only an auxiliary one and that its "investigation is not really necessary for our main object."[141] He also remarked that some of his results had already been given by Schröder in his *Vorlesungen über die Algebra der Logik* (1890), but pointed out the specific contribution of his own work, rooted in long years of research on systems of numbers, particularly those systems called by him modules.

Dedekind illustrated with examples the fact that the domain of ideals satisfies propositions (5). He therefore called the pair of propositions (5) *Idealgesetz*, and likewise all *Dualgruppen* satisfying (5) *Dualgruppen* of the *Idealtypus*. If the elements of the *Dualgruppe* are modules, then it satisfies propositions (6). Accordingly, Dedekind denoted propositions (6) *Modulgesetz*, and every *Dualgruppe* satisfying it, *Dualgruppe* of the *Modultypus*.[142]

Since every ideal is also a module, it is clear that every *Dualgruppe* of the ideal-type is also of the module-type. This leads to the main questions that Dedekind addressed here, namely:

* Are there *Dualgruppen* not of the module-type?

* Are there *Dualgruppen* of the module-type which are not of the ideal-type?

Dedekind answered both questions in the affirmative, by giving two specific examples, which he introduced by means of tables defining the operations. Representing operations defined on abstract elements of a finite collection by means of tables was a device already known from the theory of groups, where it had proven highly useful. However, when trying to apply it to *Dualgruppen*—even for the relatively simple example considered here— Dedekind found it to be extremely cumbersome. Dedekind thus sought an alternative definition of *Dualgruppen*.

Take an object α of a given *Dualgruppe* T and form the class α of all the objects $\alpha + \gamma$, where γ denotes any object in T. Dedekind postulated that this new object α' should be well-defined (i.e. that different objects α, β define different objects α', β'), that it contains α, and that the following properties hold:

* If α_1 is contained in α, then α_1' is also contained in α'.

* The intersection of any two systems α', β' is also a system σ'.

---

141. Dedekind *Werke* Vol. 2, 111: "... diese Untersuchung für unseren eigentlichen Gegenstand nicht erforderlich ist."

142. In present-day terms, these are known as the "distribute axiom" and the "modular axiom" respectively.

* Given two objects in T, α,β; then exists another object μ in T such that α and β' are contained in μ', and such that if α' and β' are contained in any system μ'_2, then μ' is also contained in μ'_2.

Following this alternative definition, Dedekind built new tables for the above examples. A direct inspection of the tables allow determining whether the axioms of *Dualgruppen*—as well as the distributivity and modular laws—hold (p. 119).

Dedekind also defined order within the *Dualgruppe*. He was not interested here in analyzing the axiomatic properties of order in general. Rather, he introduced order as shorthand for indicating the case—rather common in *Dualgruppen*—in which $\alpha + \alpha_1 = \alpha_1$ and $\alpha - \alpha_1 = \alpha$. In this case one writes $\alpha > \alpha_1$ or $\alpha_1 < \alpha$. This notation—it should be noticed—is opposed to the accepted one for order. Dedekind commented upon this fact:

> In the *Dualgruppe A*, containing modules, I have introduced the double notation
>
> $$\alpha > \alpha_1 \text{ or } \alpha_1 < \alpha$$
>
> which clearly might often contradict the usual sense of the signs $>,<$, if translated to other examples of *Dualgruppen*. For the general theory, however, it is rather harmless. (Dedekind *Werke* Vol. 2, 120)

Since Dedekind's ideas were taken directly from his work on modules, he preferred to remain consistent with the notation he had introduced in the supplements, in spite of the problems inherent to this notation.

After this brief discussion Dedekind returned in the remainder of the paper to deal with generalizations of the problem of factorization. In this article Dedekind defined *Dualgruppen* in general, and two special cases of the concept, with a very specific aim in mind: to establish the logical interrelations between the various kinds of properties satisfied by the operation defined on modules in the introductory chapters of Dedekind's 1894 supplement to Dirichlet's *Vorlesungen*. The discussion on *Dualgruppen* in his 1897 paper remained for Dedekind a subsidiary issue within it.

Dualgruppen are discussed again in an article published in 1900. This time, however, the study of the concept itself became the central concern, and, accordingly, the article opened with an abstract formulation of the concept and of the corresponding definition of order, together with the properties relating both definitions. Dedekind then presented a specific example of a *Dual-*

gruppe: the *Dualgruppe* generated by three modules, which is composed of 28 elements. Dedekind had already mentioned this module in a footnote added to the 1894 version of his theory of ideals.[143] But it was only here that the example was fully developed as a research subject of its own. Since it is a *Dualgruppe* generated by modules, it satisfies the modular axiom.

Dedekind built all the 28 elements of this *Dualgruppe* and defined in detail the operations among them. He then proceeded to the abstract part of the article, which is a deeper study of the properties of the modular axiom. Dedekind inspected the close relations between this axiom and a new one, the "chain axiom" (*Kettengesetz*). A chain of length n in the *Dualgruppe* is a sequence a_1, a_2, ..., a_{n+1} of its elements, such that every element of the sequence is a divisor of the former and, in addition, for every i ($1 \leq i \leq n$), there is no element b of the Dualgruppe such that $a_i < b < a_{i+1}$. Two chains are called equivalent chains if they have common extremes. Through the example of the *Dualgruppe* of 28 modules, Dedekind showed various cases in which equivalent chains of modules have the same length. The chain axiom states that any two equivalent chains within a *Dualgruppe* have equal lengths.

Dedekind showed that the modular axiom holds in a given *Dualgruppe*, if and only if that *Dualgruppe* and "all of its subgroups" (*Teilgruppen*), satisfy the chain axiom. The additional phrase "all of its subgroups" is absolutely necessary since a collection of elements may be chain within the *Dualgruppe* but may fail to be so when considered as elements of a given "subgroup" of it. Dedekind gave a proof is essentially the same that Oystein Ore gave 35 years later in his work on lattice-theoretic foundations of abstract algebra.[144]

Dedekind closed the article with some further abstract considerations on *Dualgruppen* (always exemplified by the module of 28 elements), which will not be described here. All in all, the main interest of the paper is the study of a very particular entity (the *Dualgruppe* of 28 elements), while the abstract research of the logical interrelation among the various axioms involved represents essentially a digression from the main line. Yet in spite of its marginal place in his work considered as a whole, Dedekind was conscious of the importance of this kind of analysis. As was seen above (§ 2.1), in his early lectures on Galois theory, Dedekind had mentioned the convenience of regarding groups under the perspective afforded by such an analysis. In this study we see

---

143. Dedekind 1894, 499.
144. *Werke* Vol.1, 259. See § 7.4 below

him applying the same perspective for studying the significance of chain conditions, unions and intersections.

Dedekind's ideas on *Dualgruppen* had no recognizable, immediate influence. As already stated, Dedekind had no direct students, and after his retirement—working in the rather mathematically isolated environment of Braunschweig—it was rather unlikely that he could find a receptive audience for this work. But one should also stress that the increasingly abstract character of Dedekind's later works was in itself an inherent reason for their lack of appeal to a great number of contemporary colleagues. In fact, even mathematicians whose work on algebra was significantly influenced by Dedekind, like Weber and Frobenius, seem to have overlooked Dedekind's lattices as part of his picture of algebra. To a certain extent they even seem to have estranged themselves from the direction of Dedekind's research interests during his later years. When Weber announced in a letter of 1893 to Frobenius the forthcoming publication of his *Lehrbuch*, Frobenius replied as follows:

> Your announcement of a work on algebra makes me very happy... Hopefully you will follow Dedekind's way, yet avoid the highly abstract approach that he so eagerly pursues now. His newest edition (of the *Vorlesungen*) contains so many beauty ideas... but his permutations are too flimsy, and it is indeed unnecessary to push the abstraction so far. I am therefore satisfied, that you write the Algebra and not our venerable friend and master, who had also once considered that plan.[145]

Frobenius, himself a leading group-theorist, considered Dedekind's treatment of permutations extremely abstract as to appear in a textbook of algebra. The above quotation seems even to suggest that Frobenius saw no connection at all between *Dualgruppen* and what algebra was taken to be, according to the image of knowledge embodied in Weber's *Lehrbuch*. Of course *Dualgruppen* could not appear in the first edition of the *Lehrbuch*, published before Dedekind's first articles. But neither later editions of Weber's textbook nor any

---

145. Dugac 1976, App. XLVIII, 269: "Ihre Ankündigung eines Werkes über Algebra macht mir grosse Freude... Hoffentlich gehen Sie vielfach die Wege von Dedekind, vermeiden aber die gar zu abstrakten Winkel die er jetzt so gern aufsucht. Seine neueste Auflage enthält so viel Schönheiten, der § 173 ist hochgenial [Dirichlet 1894, § 173: Ganze algebraische Zahlen], aber seine Permutationen sind zu körperlos, und es ist doch auch unnöthig, die Abstraktion so weit zu treiben. Ich bin also fast froh, dass Sie die Algebra schreiben und nicht unser verehrter Freund und Meister, der auch einmal mit solchen Gedanken sich trug."

other textbook of algebra, including van der Waerden's *Moderne Algebra*, contain any reference to lattices until the 1940s, long after lattices were reintroduced by Birkhoff and Ore.[146]

The attitude of these two mathematicians should thus help us assess the actual place that the abstract formulation of concepts was accorded in algebra by mathematicians at the turn of the century. In this regard Dedekind's approach—though moderate in itself by twentieth-century standards—was a relatively extreme one, rather than the rule at that time. Dedekind's two articles on *Dualgruppen* did not have any immediate influence on the works of his colleagues. Yet the kind of questions pursued by him, and the kind of axiomatic analysis advanced in them did strongly attract the attention of at least one person many years later: Emmy Noether.[147] She further developed these ideas in her own work.

2.4 Dedekind and the Structural Image of Algebra

Having described Dedekind's early lectures on Galois theory, the successive versions of his theory of ideals, and his definition and treatment of *Dualgruppen*, we are now in a position to evaluate Dedekind's overall conception of algebra and the place of a putative idea of algebraic structure within it.

A first, central fact to consider as part of such an evaluation is that, although Dedekind did use many of the concepts that eventually became the hard core of structural algebra, these concepts play very different roles in those of his works in which they appear. Thus, for instance, in his early lectures on Galois theory Dedekind uses both fields and groups. His fields are either the whole domain of complex numbers or certain of its subdomains (but never abstract fields) and they constitute *the subject matter* of enquiry. Groups, by contrast, provide an innovative, highly efficient *tool* that may be applied to tackle the problem of solvability of polynomial equations in those fields.

---

146. Nor there is a separate chapter on lattices in relatively late textbooks, such as Birkhoff & Mac Lane 1941 (or in its successive editions), or Chevalley 1956. Lattices appear in the exercises at the end of Chapter 3 of Bourbaki's book on set theory but not in his book on algebra. In Dubreil & Dubreil-Jacotin 1964 and Hall 1966 there are separate sections on lattices.

147. In the editorial notes at the end of the paper (*Werke* Vol. 2, 271) Emmy Noether wrote: "Und zwar zeigt sich (§ 6) die völlige Äquivalenz von Modulgesetz, Kettengesetz und 'zweitem Isomorphiesatz', letzterer ebenfalls in einer durch die Axiomatik bedingten Form: eineindeutiges Entsprechen aller Zwischengruppen durch Summen- und Durchschnittbildung."

Dedekind's fields are not abstract groups endowed with a second operation; rather, these two mathematical entities appear as constructs of a different kind. The questions raised by Dedekind about rational domains concern *properties of their elements*, while his questions about groups concern the groups themselves, considered as autonomous mathematical entities. Later on, in his 1894 version of the theory of ideals, Dedekind presented again the basics of Galois theory, insistently stressing the interplay between groups and fields, and bringing to the fore many of the structural features of this interplay. Nevertheless, the different conceptual status accorded by Dedekind to the two concepts is not altered even here. Moreover, the reader may recall (§ 1.4 above) that in Fricke's 1924 textbook of algebra—strongly influenced by Dedekind's conceptions—a similar difference in the treatment of fields and groups is still recognizable.

Fields also provide in Dedekind's work the basic conceptual framework for the study of algebraic integers. In this case, modules and ideals are the tools for investigating the properties of the factorization of those integers. Dedekind's modules and ideals are not "algebraic structures" similar to yet another "structure": fields. They are not "almost" fields, failing to satisfy one of the postulates that define the latter. It is the numbers belonging to the field considered which remain themselves the focus of Dedekind's interest, while modules and ideals provide only *tools* for that inquiry. It is not the properties of the numbers that constitute the modules which count, but rather the properties of the *collection* of numbers, *qua* collections. The study of modules and ideals depends on operations such as intersection, union and inclusion, and not on the operational relations among the *numbers* contained in the modules. In fact, the important differences among the successive versions of the theory lie precisely in the increasing elaboration of the consequences of following this perspective.

Ideals never appear in Dedekind's work as a special kind of substructure of the more general algebraic structure of a ring. *Ordnungen*—although formally equivalent to rings—do not provide the general framework for the study of ideals that rings do in the modern conception of algebra. An ideal is not a distinguished subdomain of an *Ordnung*. In fact, ideals—like other concepts included since van der Waerden under the common heading of algebraic structures—are never defined by Dedekind by endowing non-empty sets with abstractly defined operations. Thus, like groups in the case of Galois theory, modules and ideals remain in ideal theory at a conceptual level, which is different from that accorded to fields. Moreover, when dealing with modules, Dedekind did not stress the basic fact that—from an abstract perspective— they are actually groups,[148] and he certainly did not apply to the former, results already known to hold for the latter.

Our examination of the successive versions of the theory of ideals showed how Dedekind gradually mastered the structural techniques and exploited the specific insights potentially afforded by the perspective he chose for his research. Nevertheless, fields remained for him—throughout his works—an entity of a different nature, and more basic mathematical significance, than the tools he introduced for the study of the various theories (groups, ideals, modules). In fact, fields were meant to allow for a unified perspective of all those various theories. In the introduction to the first version of his theory of ideals Dedekind wrote:

> In the following paragraphs I have attempted to introduce the reader into a higher domain, in which algebra and the theory of numbers interconnect in the most intimate manner. During the lectures I gave at Göttingen [in 1856-57, 1857-58] on the division of the circle and on higher algebra, I got convinced that studying the algebraic relationship of numbers is most conveniently based on a concept that is directly connected with the simplest arithmetic principles. I had originally used the term "rational domain", which I later changed to "field."[149]

Fields, then, are meant to provide the common ground for approaching "Higher algebra" (namely, the theory of polynomial equations) and number theory. Later, in the 1894 supplement, for all the sophistication evident in the elaboration of the theory, Dedekind described the significance of the notion of fields in terms that do not disclose any clue to a direct connection between fields and the other algebraic concepts. Thus he wrote:

---

148. In the introduction to the French version of the theory of ideals (*Werke* Vol. 3, 274) did mention, vaguely, that instead of numbers like in a module, one could take arbitrary elements of a group endowed with a commutative operation. It is not even clear that the term "group" is intended here in its standard, technical sense.

149. *Werke* Vol. 3, 400: "In den nun folgenden Paragraphen habe ich versucht, den Leser in ein höheres Gebiet einzuführen, in welchem Algebra und Zahlentheorie sich auf das Innigste miteinander verbinden. Im Laufe der Vorlesungen über Kreisteilung und höhere Algebra, welche ich zu Göttingen im Winter 1856-57 (...), im Winter 1857-58 (...), gehalten habe, drängte shch mir die Überzeugung auf, daß das Studium der algebraischen Verwandschaft der Zahlen am zweckmäßigsten auf einen Begriff gegründet wird, welcher unmittelbar an die einfachsten arithmetischen Prinzipen anknüpft. Den damals vor mir benutzten Namen 'rationales Gebiet' habe ich später mit dem Worte 'Körper' vertauscht."

This term [i.e., fields] should denote here, in a similar fashion as in the natural sciences, in geometry, and in the social life of men, a system possessing a certain completeness, perfection, and comprehensiveness, by means of which it appears as a natural unity.[150]

A second, important consideration that helps us evaluating the different attitudes of Dedekind towards the various algebraic concepts introduced in his work becomes manifest in his notational choices, and in particular in the notational tensions found in the successive versions of his theory of ideals. Whereas a field A is said to be a divisor of another field M whenever all the numbers contained in A are also contained in M, a module (resp. an ideal) M is called a multiple of D whenever all the numbers in M belong also to D. One might perhaps overlook this minor notational detail, but Dedekind himself thought it worthy of comment. He thus wrote:

This notation and nomenclature may at first sight give offence, since the multiple M is in fact a *subset* of the divisor D. Nevertheless it proves later on sufficiently justified through its analogy with the divisibility of numbers.[151]

Thus, for instance, the module M containing all the multiples of 4 is, under this notation, a *multiple* of the module D containing all the multiples of 2 ($M > D$), although as sets, the former is indeed contained in the latter. In a footnote Dedekind further explained:

Even the circumstance that for *fields*, which are themselves modules, the opposite expression is used, cannot be taken here into account, since, with some degree of attention, no confusion is possible.[152]

Thus, although formally speaking a field is also a module, the two concepts correspond to different motivations. Dedekind adopted a notation that

150. Dedekind 1894, 452 (footnote): "Dieser Name soll, ähnlich wie in den Naturwissenschaften, in der Geometrie und im Leben der menschlichen Gesellschaft, auch hier ein System bezeichnen, das eine gewisse Vollständigkeit, Vollkommenheit, Abgeschlossenheit besitzt, wodurch es als ein organisches Ganzes, als eine natürliche Einheit erscheint."
151. Dedekind 1894, 495: "Diese Ausdrucks- und Bezeichnungsweise mag auf den ersten Blick Anstoss erregen, weil das Vielfache *M* in Wahrheit einen *Theil* des Theiler *D* bildet, doch wird dieselbe sich in der Folge hinreichend rechtfertigen durch die Analogie mit der Theilbarkeit der Zahlen." (Italics in the original.)
152. Dedekind 1894, 495: "Selbst der Umstand, dass bei den *Körpern*, die doch auch Moduln sind, die entgegengesetzte Ausdrucksweise gebraucht ist, kann hier nicht in Gewicht fallen, weil bei einiger Aufmerksamkeit eine Verwechselung nicht möglich ist."

stresses the particular role assigned to each concept within the theory, even if it created a certain tension, at least in appearance.[153]

As for *Dualgruppen*, we saw that Dedekind studied this concept from a rather abstract perspective. However, the concept was seen to emerge organically from within his treatment of modules of numbers and of the properties of the operations defined on them. In defining *Dualgruppen* Dedekind did not take an arbitrary set and endow it with an abstractly defined relation of order or with two abstractly defined operations, in order to check which theorems can be deduced for such a construct. Rather—motivated by the desire to improve the proofs of certain identities already known to hold for modules— he established abstract, logical interdependencies between specific properties of certain operations arising in his "algebra of modules." In his 1897 article, Dedekind defined *Dualgruppe, Idealgruppe*, and *Modulgruppe* as collections whose inclusion properties, unions and intersections satisfy certain identities (or "axioms"). He then asked under what conditions those identities are mutually equivalent. Later in his 1900 article Dedekind studied the specific example of a certain mathematical entity, which had arisen in his previous work— the 28-element lattice—and extended the previous analysis of the logical interdependence of the axioms to include now an additional one: the chain axiom.

Although Dedekind did provide examples of how *Dualgruppen* appear in the algebra of logic, fields, etc., he did not establish any direct connection between them and the other "algebraic structures." Dedekind's *Dualgruppen* do not appear as belonging to the same species of mathematical entities as fields, groups, or ideals.

A similar picture arises from the way algebraic concepts are treated in Dedekind's joint work with Weber on algebraic functions. Such concepts are not introduced there by taking a non-empty set and endowing it with one or more abstract operations whose properties are given in advance. Dedekind and Weber mentioned the fact that sums, differences and products of integral functions are again integral functions, but, like in the case of algebraic numbers, no general definition of ring appeared in their article.[154] The latter had nothing

---

153. It should be remarked, however, that in his 1894 version of the theory of ideals, Dedekind did change his initial notational choice in favor of a more unified one.

154. Thus the claim of Dugac (1976, p. 76: "L'ensemble O de toutes fonctions entières de W est un anneau") is correct only as a retrospective one, and does not correspond to Dedekind's own conception.

to contribute to the elucidation of the factorization properties of the functions and therefore, it had no place in their exposition.

This joint article is the only place in Dedekind's corpus where fields which are not subfields of the complex field were considered. The translation of concepts from the realm of algebraic numbers into that of algebraic functions, built as an almost word-by-word repetition of every step of the theory, more than evincing a putative structural conception of algebra in Dedekind's images of algebra, stresses the absence of a general concept, either formal or non-formal, of "algebraic structure."[155] Had Dedekind had a general conception of algebra in terms of structures, and given his consistent inclination to find generalizing concepts and formulations, it would be natural to expect a more general, abstract theory of ideals and modules of which numbers and functions would appear as particular cases. The redundant repetition would have thus been avoided. Dedekind, like most of his contemporary colleagues, continued to consider non-numerical realms as something essentially different from numerical systems. Dedekind and Weber's paper did explicitly show, for the first time, an important similarity connecting these two kinds of realms and, moreover, the great convenience of exploiting that similarity. But the perspective that considers those realms as being of identical nature—as the structural conception of algebra does—was still to come.[156]

In contrast to the differences found in Dedekind's approach to the various algebraic concepts appearing in his work, one can stress the interesting parallels between his definition and use of ideals, and those pertaining the main concepts of his foundational works on the systems of the real and natural numbers: cuts and chains, respectively. Rather than seeing ideals as an individual

---

155. It is noteworthy that in commenting on the historical significance of this work, Klein did not speak of any kind of unification or generalization as such, but rather of an application of number-theoretical ideas to the theory of algebraic functions (Klein 1926-27 Vol. 1, 326): "Die Einführung dieses Begriffes [Ideal] ist nur etwas Äußerliches. Viel wichtiger ist dagegen der Fortschritt, den die Verfasser machen, indem sie auch die Beweismethoden der Arithmetik auf die Behandlung des Körpers der algebraischen Funktionen übertragen."

156. Contrary to the opinion of Dieudonné (1969, 375) concerning the historical significance of Dedekind and Weber's joint paper: "On retrouve donc, ici encore, une des idées clés de la mathématique moderne, celle qui consiste à calculer sur des objects qui ne sont plus du tout des nombres ou des fonctions. De plus, le mémoire de Dedekind et Weber attirait pour la première fois l'attention sur une parenté frappante entre deux domaines mathématiques tenus jusque-là pour très éloignés, première manifestation de ce qui devait devenir un 'leitmotiv' des travaux ultérieurs: la recherche des structures communes cachées sous des apparences parfois très diverse."

manifestation of the more general species of "algebraic structures", one could more naturally associate them with a different species of mathematical idea: that of fundamental concepts on which number-systems can be built and fully elucidated. This species is not explicitly mentioned by Dedekind,[157] yet it is implicitly manifest throughout his work, playing a central role in it, much the same as "algebraic structures" constitute a tacit, leading idea of van der Waerden's book. Cuts, chains, and ideals are three concepts aiming at providing conceptual foundations for the systems of real, natural, and algebraic numbers respectively. Each of these concepts is meant to allow for the proof of some basic results, from which the most important facts concerning the domains in question may be derived. The concept of cut is meant to elucidate the idea of continuity in the system of real numbers. The concept of chain is meant to elucidate the idea of the sequence of natural numbers. In the same vein, the concept of ideal was conceived to elucidate the most important problem concerning the domain of algebraic integers, namely, the question of unique factorization.

Moreover, the preceding sections have shown that there are several additional affinities among the three concepts. First, all three concepts refer to *collections of numbers* which are defined by simple properties. Second, they are used to prove certain basic theorems in their respective theories; the proofs of those theorems are based only upon considerations of the collections of numbers involved, *qua* collections. After the collections defining the concepts are introduced, all efforts are made not to make further reference to the numbers contained in the collection, and not to base the proofs on their individual properties as elements. Rather the collections themselves, together with their inclusion relations, remain the preferred frame of reference for formulating and proving the theorems. Thus, cuts allow for a definition of continuity in the realm of the real numbers, based on questions about the relationship among certain sets (i.e.: those that constitute the cuts). Likewise, the induction principle for natural numbers is formulated, in Dedekind's theory, not in terms of "numbers" and their "successors", but rather in terms of the images of certain sets under a given mapping, and of the inclusion of these images into other sets. Finally, ideals are also collections of numbers, in this case algebraic inte-

---

157. Dedekind did mention explicitly, at least in one place, the common character of his research on the three number-systems (real, natural and algebraic), and the three fundamental concepts introduced for this purpose. Cf., e.g., *Werke* Vol. 3, 489-490.

gers. The successive versions of the theory make manifest the sustained efforts to formulate the factorization theorems in terms of the ideals themselves, while striving to reduce as much as possible the reliance on the operational properties of the individual numbers contained in them. In the last version, one sees how Dedekind was actually able to define an algebra of ideals, which he used in order to prove the theorems while relying exclusively on the properties of its operations. These are operations on *collections* of numbers, rather than on individual numbers.[158]

The present chapter has described some works of Dedekind, his images of mathematics and his overall conception of algebra. The centrality conceded by Dedekind to the adequate choice and precise formulation of concepts, the kind of axiomatic analysis incipiently contained in his work, the specific concepts that he introduced and elaborated in his theory of ideals and in other related publications, and the kind of questions and proof techniques that he gradually developed in them, were only slowly absorbed into the mainstream of algebraic knowledge. But in the long run they had an enormous influence in shaping the developments that led to the full adoption of the structural approach to algebra. Further contributions by other mathematicians were needed in order to absorb and disseminate Dedekind ideas among a much wider mathematical audience that he himself succeeded in reaching. An important role in this development was played by David Hilbert. Needless to say, Hilbert's work is not simply an elaboration or extension of Dedekind's point of view, or of his achievements. Rather, Dedekind's ideas were naturally absorbed into the far-reaching and innovative picture of mathematics advanced by Hilbert and into his substantive mathematical contributions. We thus turn in the next chapter to an analysis of Hilbert's influence on the rise of the structural approach to algebra.

---

158. Ferreirós 1999, 99-107, discusses the affinities among these three concepts from a somewhat different perspective.

Chapter 3   **David Hilbert:**
Algebra and Axiomatics

Few accounts of the development of particular mathematical disciplines around the turn of the century can be complete without analyzing Hilbert's contribution to them. Algebra, and the particular account presented here, are no exception to this rule.[1] David Hilbert (1862-1943) was the leading mathematician of his era, and the mathematical institute in Göttingen—first under the leadership of Felix Klein (1849-1925) and later on under Hilbert—became the world center of mathematics until the rise of Nazism in Germany.[2] Dedekind also spent his early career in Göttingen, many years before Hilbert's arrival there. Later on, Emmy Noether—invited to Göttingen by Hilbert in 1915—developed her own algebraic work at the same place.

Given the extent and scope of Hilbert's own contributions, his influence on modern, structural algebra should be best analyzed as part of his overall influence on twentieth-century mathematics. Obviously, such a task would be far beyond the scope of the present book. Nevertheless, one can focus on three specific components of Hilbert's work directly bearing on the development of twentieth-century algebra and—more particularly—of ideal theory between Dedekind and Noether. Those three components of Hilbert's work—invariant theory, algebraic number theory, and axiomatics—are briefly analyzed in the present chapter. Each of these contributions brings to the fore specific concepts, techniques or images of mathematics that are relevant to the present discussion. We will analyze in this chapter their direct motivations and the images of mathematics manifest in them.

---

1. General assessments of Hilbert's contribution to the rise of modern algebra appear in Hasse 1932; van der Waerden 1966, 161-163; Weyl 1944, 635.
2. Accounts of Göttingen as the world leading center of mathematics, and the roles of Klein and Hilbert in fostering this centrality appear in Reid 1970; Rowe 1989; Parshall & Rowe 1994, 150-154.

3.1   Algebraic Invariants

In 1880 Hilbert began university studies in his native city of Königsberg. Except for a semester in Heidelberg, and the customary *sojourn*—which took him after finishing his dissertation in 1885 to Klein in Leipzig, and to Hermite in Paris—all of Hilbert's early career developed in the same place. Only in 1895 Hilbert moved to Göttingen following Klein's invitation. As a student, Hilbert was strongly influenced by Heinrich Weber, whose courses he first attended by the time the latter was finishing his joint paper with Dedekind (see above § 2.2.4). The search for underlying analogies between separate mathematical realms—such as those worked out by Weber and Dedekind for fields of algebraic numbers and fields of functions—was to remain a leading motto in much of Hilbert's own research.

Beyond Weber, Hilbert's intellectual horizon in Königsberg was shaped by an extraordinary relationship with two other young mathematicians: Adolf Hurwitz (1859-1919)—first Hilbert's teacher and later on his colleague—and Hermann Minkowski (1864-1909). Before accepting in 1884 a new chair especially created for him in Königsberg, Hurwitz had studied first with Klein in Leipzig and then in Berlin, and had later habilitated in Göttingen in 1882. Hurwitz was thus well-aware of the kind of mathematical interests and techniques that dominated current research in each of these important centers.[3] Hurwitz taught eight years at Königsberg before moving to Zürich, and his influence during this period of time was decisive in shaping Hilbert's very wide spectrum of mathematical interests, both as a student and as a young researcher.

Hilbert's doctoral adviser in Königsberg was Ferdinand Lindemann (1852-1939). Lindemann—a former student of Klein's, who is remembered today mostly for his proof of the transcendence of π—was Weber's successor in Königsberg since 1883. During his Königsberg years, Hilbert lectured on many issues, covering a wide range of mathematical disciplines and distinguishing himself as a dedicated teacher. Yet the earliest stage of his research career—from 1885 to 1893—was clearly dominated by one discipline, which his dissertation had also dealt with, and which became Hilbert's first mathematical specialty: the study of algebraic invariants.

The theory of algebraic invariants was a most active field of research in the second half of the nineteenth century. It was perhaps Gauss's work on

---

3. See Freudenthal 1974.

binary quadratic forms, dating from the beginning of the century, that contained the earliest observations on algebraic invariant phenomena. But the discipline as such actually developed later in England and in Germany from two different kinds of sources. In England, Alfred Cayley and James Joseph Sylvester (1814-1897) published since 1845 the first works that contained actual contributions to the subject. They elaborated on ideas introduced earlier by Boole, when studying Lagrange's work on linear transformations of homogeneous polynomials.[4] In Germany, the second early source of the theory to algebraic invariants came from the geometrical research of Otto Hesse (1811-1874). This research triggered a line of development that led to the first articles of Siegfried Aronhold (1819-1884) on invariants, from 1849 on. Aronhold was followed in Germany by Rudolf Alfred Clebsch (1833-1872), and by Paul Gordan (1837-1912).[5] In order to understand Hilbert's contribution to the theory of invariants, a brief account of the aims and methods of the theory is in order here.

Let $f(x_1,x_2)$ be an homogeneous binary form of degree n with coefficients a_i ($1 \leq i \leq n$). Let a transformation T be defined as follows:

$$x_1 = \alpha_{11}y_1 + \alpha_{12}y_2$$

$$x_2 = \alpha_{21}y_1 + \alpha_{22}y_2$$

and let the determinant δ of the transformation be defined as

$$\delta = \alpha_{11}\alpha_{22} - \alpha_{12}\alpha_{21}.$$

This transformation implicitly defines a new form $T(f)$ of degree n on the variables y_1, y_2, $T(f) = F(y_1,y_2)$. The coefficients of this new form, b_i ($1 \leq i \leq n$), are rational functions, which are linear in the a_i's and of degree n in the α_{ij}'s. Now an "invariant" is any function I of the coefficients a_i of f, such that for some integer r

$$I(b_1,b_2,...,b_n) = \delta^r I(a_1,a_2,...,a_n)$$

This definition can be expanded to ternary forms (in three indeterminates), quaternary, and more generally to forms with n indeterminates. It can also be generalized to consider either invariants of several forms taken simulta-

---

4. Cf. Crilly 1986; Parshall 1989, 158-162; Parshall & Rowe 1994, 67-68.

5. A classical historical account of the early development of invariant theory appears in Meyer 1890. See also Fisher 1966, 141-156; Kline 1972, 924-932; Parshall 1989, 170-176; 1990, 12-13.

neously, or invariants of the coefficients and the variables taken together (in which case they are called "covariants").[6]

The work of Cayley and Sylvester concentrated on developing algorithms for finding individual invariants of particular, given systems of forms. The symbolic techniques for calculating invariants, introduced by Aronhold and adopted by his followers of the German school, were more flexible than those of their British counterparts and allowed operating at a more abstract level.[7] The most remarkable success attained through the application of these techniques was the proof of the so-called "finiteness theorem" for binary forms. This proof was published in 1868 by Paul Gordan, who was the leading authority in the discipline for many years. Gordan proved that given any system of binary forms of arbitrary degree, a finite sub-system of it can be chosen, such that any invariant of the former may be written as a rational combination of forms of the latter. Using the symbolic approach, Gordan was also able to provide, through laborious calculations, "smallest possible systems of ground-forms" for the case of binary forms of the fifth and sixth degrees.[8] In the years following its publication, various limited generalizations of Gordan's theorem were obtained, and improved proofs of his original theorem were also given.[9] At the same time, techniques for calculating invariants were significantly improved. Yet for many years the *full* generalization of Gordan's theorem remained an open problem: to prove the existence of a finite basis for any system of invariants of any degree and with an arbitrary number of indeterminates. It is here that Hilbert entered the picture.

Hilbert's doctoral dissertation, as well as his 1885 inaugural lecture and his first published articles, had been devoted to the research of invariant properties of particular algebraic forms. In 1888 he published the first of a series of proofs of the general case of the finite basis theorem.[10] However, Hilbert did not actually show how the desired basis could be constructed, as Gordan had done for the binary case: he simply proved its existence by a *reductio ad*

---

6. More detailed explanations of the basic concepts of invariant theory appear in Crilly 1986; Hilbert 1896; Weyl 1944, 618-624. A classical textbook on the issue is Study 1933. More recent expositions of the theory appear in Dieudonné & Carrell 1971; Springer 1977.

7. For a detailed account of the differences between the English and the German schools of invariant theory see Parshall 1989, 176-180.

8. Gordan 1868, 327.

9. Hilbert 1889; Mertens 1886.

10. Hilbert's proof appeared first in Hilbert 1888-9, and in an improved version in Hilbert 1890. For a reconstruction of Hilbert's proof in a more recent formulation see Springer 1977, 15-42.

absurdum argument. Many leading mathematicians of the period, and above all Gordan himself, strongly questioned the legitimacy of Hilbert's 1888 proof. Gordan's reaction has since entered the mathematical lore, as he was quoted as having claimed: "This is not mathematics, this is theology."[11]But this initially negative reaction changed soon, if only because already in 1892 Hilbert presented a constructive proof of his finiteness theorem. In fact, Gordan himself published in 1893 a simplification of Hilbert's proof.[12] It is worth noticing that Hilbert's proof elaborated upon ideas originally introduced by Kronecker in his number-theoretical research, on the one hand, and by Dedekind and Weber in their joint 1882 article, on the other. Hilbert applied a combination of these two sources of ideas for solving an open problem in an area of research originally unintended by both.[13]

In his proofs of finiteness Hilbert introduced ideas that were later to become central to Emmy Noether's elaboration of ideals in abstract rings, and therefore to the consolidation of the structural image of algebra. In his 1890 article, for instance, in order to prove the existence of the desired basis, Hilbert first proved the following theorem:

> Given an infinite sequence of forms F_1, F_2, F_3,... in n variables x_1, x_2, ...,x_n, one can always find a number m, such that any form in the sequence can be written as
>
> $$F = A_1F_1 + A_2F_2 + ... + A_mF_m$$
>
> and where the A_i's belong to the same rational domain. (Hilbert 1890, 199)

After proving this theorem by induction on the number of variables, Hilbert proceeded to prove the general theorem, independently of whether the system in question is countably infinite or not.[14] He reasoned as follows: If it is not possible to choose the desired basis, then choose from the collection a non-zero form F_1. A second form, A_2, can now be chosen such that it is not equal to a multiple of the first A_1F_1. Likewise, one can choose a third form, F_3,

---

11. Cf. Blumenthal 1935, 194; Klein 1926-27, Vol. 1, 330. Gordan conceded some years later the legitimacy of Hilbert's approach. He has then been quoted as saying (Klein 1926-27, Vol. 1, 331): "Ich habe mich überzeugt, daß auch die Theologie ihre Vorzüge hat." Gordan's reaction to Hilbert's proof is documented in the Klein-Hilbert correspondence see Frei (ed.) 1985, 61-65. See also Rowe 1989, 196-198.

12. Gordan 1893.

13. See Hilbert 1888-89, 187 & 190.

14. Hilbert 1890, 203: "Es sei nämlcih ein beliebiges System von unbegrenzt vielen Formen ... gegeben, wobei es freigestellt ist, ob diese Formen sich in eine Reihe ordnen lassen oder in nicht abzählbarer Menge vorhanden sind."

which cannot be written as $F_3 = A_1 F_1 + A_2 F_2$. In this way, one builds a sequence of forms, $F_1, F_2, F_3,...$ that contradicts the above theorem. Beyond the use of a *reductio ad absurdum* argument, Hilbert's reasoning amounts to an implicit use of the ascending chain condition, which Emmy Noether later explicitly formulated and whose specific implications she was the first to study in detail.

Using a similar approach, Hilbert obtained still more general results, pertaining to forms with an arbitrary—finite as well as infinite—number of indeterminates. They appeared in an important article published in 1893 in the *Mathematische Annalen*, meant as a synthesis of all his previous achievements in this domain, and which included also constructive techniques for describing explicitly those bases whose existence he had initially been able to prove only indirectly. Hilbert stated that the results he had achieved so far amounted to a reformulation of the whole theory of invariants. This theory, he thought, was now a part of the theory of fields of algebraic functions, much as the theory of cyclotomic fields was nothing but a noteworthy example of the theory of fields of algebraic numbers.[15]

Of particular interest for the present account is the introductory section of the article, in which Hilbert enumerated what he saw as the five most elementary properties of invariants. Understanding here by "invariants" those defined by integral linear transformations of the coefficients of the system of ground forms,[16] the first four properties mentioned by Hilbert were the following:[17]

(1) The invariants correspond to the transformations associated with a certain continuous group.

(2) The invariants satisfy certain partial linear differential equations.

(3) Every algebraic, and in particular every rational, function defined on arbitrarily many invariants, which is integral and homogeneous in the coefficients of the ground forms, is itself an invariant.

(4) If the product of two integral functions of the coefficients of the ground forms is an invariant, then also each of the factors is itself an invariant.

---

15. Hilbert 1893, 287.

16. Hilbert 1893, 288.

17. Hilbert 1893, 288. Hilbert returned to this list, with a minor change, in the lecture read before the 1893 International Congress of Mathematicians in Chicago. See Hilbert 1896, 377.

Condition (3)—Hilbert wrote—expresses the fact that the complete system of invariants determine an algebraically closed domain of functions. Condition (4) implies that the usual factorization properties are valid in the domain of invariants, i.e., that every invariant can be represented in a unique way as a product of irreducible invariants. To these four, Hilbert added a fifth condition:

(5) There exist a finite number of invariants, such that all other invariants may be expressed as rational integral combinations of them.

A system satisfying all these five conditions is called a "full system of invariants" and Hilbert raised at this point the question of the logical interdependence: which conditions are deducible from which, and which are individually valid for any system of functions?[18] The examples considered in Hilbert's former articles showed, for instance, that properties (2), (3) and (5) could be satisfied in a system for which (4) is not valid.

This list itself, and the short discussion following it, interestingly brings to mind the kind of remarks already seen in Dedekind's work concerning "axiomatic" definitions in mathematics: we saw this in his correspondence with Lipschitz, in his works on lattices and—even earlier—concerning groups, in his lectures on Galois theory (§ 2.1 above). But even more interestingly, it foreshadows the kind of analysis that Hilbert himself will wholeheartedly undertake some years later in his work on the foundations of geometry. As a matter of fact, Hilbert did not pursue in detail, neither here nor in later works, the consequences of this short discussion on the elementary mathematical facts that he identified as basic for the theory of invariants. But the very conception of the list, and its position at the beginning of this summary of the results attained in his research, betray Hilbert's awareness of the potential usefulness of such an analysis. Moreover, the nature of the five specific properties selected by Hilbert indicate the direction he would have considered as the right one to pursue in the eventuality that this kind of analysis would actually be undertaken. In Emmy Noether's work on factorization we will indeed find echoes of the direction hinted at here by Hilbert.

Besides the proof of a generalized version of Gordan's theorem, Hilbert's 1893 article also contains a further significant contribution to invariant theory: the so-called *Nullstellensatz*. This theorem, which generalized a previous one of Max Noether (1844-1921),[19] was formulated by Hilbert as follows:[20]

---

18. Hilbert 1893, 289.

Let a set of m homogenous polynomials $F_1, ..., F_m$ in n variables $x_1, ..., x_n$, and a sequence of homogeneous polynomials $F, F', F'',...$ in the same variables be given, such that each function of that sequence vanishes in all the common roots of the m given polynomials. Then there exist a natural number r, and m suitably chosen polynomials $A_1, ..., A_m$, such that any product $\Pi^{(r)}$ of r polynomials of the sequence may be represented as:

$$\Pi^{(r)} = A_1 F_1 + ... + A_m F_m.$$

Obviously, if F is any individual polynomial of the above mentioned sequence, then

$$F^r = A_1 F_1 + ... + A_m F_m.$$

Hilbert conceived the *Nullstellensatz* as a result concerning the very concrete setting of the theory of polynomials over the field of complex numbers. He saw it as a tool that might facilitate the task of actually constructing the finite bases whose existence he had already proven. Later developments in the theory of polynomials led to understand its meaning in a broader context (see § 4.7 below).[21]

Hilbert himself was the first to assess the historical significance of his own work on invariant theory.[22] In a review article read in his name at the International Congress of Mathematicians held in Chicago in 1893, Hilbert mentioned three clearly separated stages, that in his view all mathematical theories usually undergo in their development: the naive, the formal and the critical. In the case of invariant theory, Hilbert saw the works of Cayley and Sylvester as representing the naive stage and the work of Gordan and of Clebsch representing the formal stage. In Hilbert's assessment, his own work was the only rep-

---

19. See Noether 1887. Max Noether's "fundamental theorem of algebraic functions" specified the conditions under which a given polynomial in two variables can be written as a linear combination (with polynomial coefficients) of two other polynomials specified in advance. This theorem was a cornerstone of contemporary algebraic geometry and so came to be Hilbert's *Nullstellensatz*. The reader should always keep in mind that our account, focusing as it does on the rise of algebraic structures, leaves many of the main motivations of the mathematicians involved of necessity unmentioned. Such is the case in point.

20. Hilbert 1893, 294.

21. In his obituary of Hilbert, Hermann Weyl gives an account of Hilbert's work on invariants and of the *Nullstellensatz*, formulating it in terms of more up-to-date algebraic concepts, and stressing its importance for the theory of algebraic manifolds. See Weyl 1944, 617-623. For a more modern formulation of the theorem see Jacobson 1980, 424 & 564.

22. But also Minkowski and Klein stressed its importance. See Rowe 1994, 198-199.

resentative of the critical stage in the theory of invariants.[23] Moreover, in his 1893 article on invariants Hilbert explicitly claimed to have completed all the major tasks of the discipline,[24] and he in fact abandoned all research in invariant theory thereafter.[25] Hilbert's assessment has often been echoed[26] and accepted at face value.[27] Invariant theory and the algorithmic approach characteristic of its practitioners have been pronounced dead following Hilbert's achievements, thus finally clearing the way to the rise of the new structural algebra.[28]

On the other hand, however, some invariant-theorists saw Hilbert's results as having opened new avenues of research for the discipline, rather than bringing it to a dead-end.[29] Although the number of mathematicians involved in research on invariants and the quantity of published research works on the subject considerably diminished by the turn of the century, it did never completely disappear.[30] In fact, as late as 1933 a new textbook on invariant theory was published which explicitly opposed the approach introduced by Hilbert. The author of that book, Eduard Study (1862-1930), claimed that abstract general-

23. Hilbert 1896, 383.

24. Hilbert 1893, 344.

25. In 1892 Hilbert announced in a letter to Hermann Minkowski his decision in this regard. Cf. Blumenthal 1935, 395.

26. See Weyl 1939, 27-29; 1944, 627 ("Whereas Hilbert's work on invariants was an end, his work on algebraic numbers was a beginning.") Cf. also Reid 1970, 38 ("Now suddenly, in 1892, as a result of Hilbert's work, invariant theory, as it had been treated since the time of Cayley, was finished.")

27. But for a critical assessment of Hilbert's view, as it has been reiterated by many, see Parshall 1990, 11 ff. On p. 16 (ftn. 27) Parshall writes: "[Hilbert's 1893] paper is generally credited with ushering in modern algebra, although ... relative to invariant theory, these sorts of statements generally beg for deeper historical inquiry."

28. Cf., e.g., van der Waerden 1933, 401: "Aus diesem Gedankenkreis ist später in natürlicher Weise die allgemeine Theorie der abstrakten Körper, Ringe und Moduln erwachsen." Also according to Weyl (1939, 27), the impact of Hilbert's work brought about the development of "general notions and their general properties along such abstract lines as have lately come into fashion all over the whole field of algebra." Kline 1972, 931, wrote: "The subsequent history of algebraic invariant theory belongs to modern abstract algebra. The methodology of Hilbert brought to the fore the abstract theory of modules, rings, and fields." See also Dieudonné & Carrell 1971, vii.

29. Fisher 1966, 146-151, gives an account of research in invariant theory during the years following Hilbert's 1893 pronouncements. Jacobson 1946, 592, points out that classical invariant theory disappeared from textbooks of algebra only after 1930, i.e., following van der Waerden rather than following Hilbert. References to more recent mathematical works that refute the "death of invariant theory" and consider Hilbert's proof as a new starting point for further developments appear in Parshall 1989, 157-158.

ization alone, as instantiated in Hilbert's work, could be no substitute for the "real mathematics", namely the algorithmic one represented by the methods of Aronhold-Clebsch.[31] It is likewise worth noticing that in Hilbert's 1900 list of twenty three problems, the fourteenth proposes to prove the finiteness of certain kinds of systems of invariants. Relatively few efforts, however, seem to have been directed towards the solution of this particular problem until the 1950s, when a system of the kind stipulated by the problem was found, for which there is no finite basis.[32]

But we are not concerned here with the development of invariant theory as such, nor with its alleged death following Hilbert's work on this subject. Rather our concern is with Hilbert's influence on the rise of the structural approach to algebra. In connection with the theory of invariants, this influence lies—in the first place—in the stress laid by Hilbert on the conceptual analogy between the basic problems and available tools of research on number theory and on the theory of polynomials. Hilbert's contribution to algebraic number theory, as will be seen in the following section, also built on a fruitful combination of Kronecker's and Dedekind's techniques. But the setting of Hilbert's work on invariants remained the concrete one of polynomials over the field of complex numbers. More than offering completely new ideas for algebraic research, this work showed the fruitfulness of applying here techniques and concepts that had successfully been applied in other areas, thus bringing to the fore the "structural similarities" between the latter and the former.

Hilbert also isolated certain concepts and procedures which eventually became central to research in modern abstract algebra. Such is the case with

---

30. A list of selected works on invariant theory between 1890 and 1990, including publications by mathematicians such as Leonard E. Dickson, Emmy Noether, Hermann Weyl, Oscar Zariski and Claude Chevalley, among others, appears in Parshall 1990, 15.

31. See the introduction to Study 1933. Cf. Fisher 1966, 145-146.

32. The counterexample appears in Nagata 1959. Nagata opened his article by formulating Hilbert's problem in a more general setting than the original one, as follows (Nagata 1959, 766): "Let k be a field and let $x_1,...,x_n$ be algebraically independent elements over k. Let K be a subfield of $k(x_1,...,x_n)$ containing k. Is $k[x_1,...,x_n] \cap K$ finitely generated over k?" As previous works dealing with Hilbert's fourteenth problem Nagata mentioned only two earlier articles by himself, together with a 1958 article by D. Rees. In Nagata's account, the first to have considered the problem after Hilbert was Oscar Zariski, who in 1953 answered it in the affirmative, for the case $dim K \leq 2$. Zariski 1954 contains this proof, and, in fact, Zariski mentions here no other source beyond Hilbert's original formulation.

On the developments related to Hilbert's fourteenth problem see Mumford 1976. See also Springer 1977, 37.

the finite basis theorem, and with the use of chain arguments to prove it. This is also the case of the *Nullstellensatz* and of the five properties referred to by Hilbert in his 1893 paper. But also in these cases Hilbert saw their importance as part of a very specific, concrete setting, rather than as general, abstract definitions. The various systems of numbers (complex, rationals, etc.) appear in Hilbert's work on invariants as the basic mathematical entities, while systems of polynomial forms are subsidiary constructs. The properties of the latter are always deduced from those of the former. Hilbert added with his work in this domain important ingredients to the future elaboration of the structural image of algebra, but he himself applied those ingredients without essentially changing the accepted images of the discipline. The images of algebra within which Hilbert produced his work on invariant theory were similar in most respects to the classical, nineteenth-century images that dominated the works of Dedekind and of Weber.

3.2 Algebraic Number Theory

From 1893 to 1899 Hilbert's main field of research was the theory of algebraic number fields. Although Hilbert completely abandoned research on invariants after 1893, the passage from his first domain of research to the second one did not signify a complete break with the past, both because he had already lectured on number theory in Königsberg, and, more importantly, because in his research on invariants he had applied techniques originally developed for dealing with algebraic number fields. At the time, however, the techniques and concepts developed by Dedekind and Kronecker on ideal factorization were not yet commonly known and standardly used by mathematicians at large. This was partly a consequence of the novel, idiosyncratic approaches adopted by each of these mathematicians in their separate works on the subject, and partly a consequence of their intrinsic difficulty.[33] In 1893 Hilbert and Minkowski were commissioned by the German Mathematicians' Association (*DMV*) to provide a systematic and comprehensive survey of the state-of-the-art in number theory for the needs of the general mathematical audience. This commission was a clear sign of the *DMV*'s recognition of Hilbert as one of the leading authorities in the field, alongside his friend. While

---

33. Blumenthal 1935, 397, brings the following quotation from Hilbert: "Einer nahm den Kroneckerschen Beweis für die eindeutige Zerlegung in Primideale vor, der andere den Dedekindschen, und beide fanden wir scheußlich."

Minkowski was to concentrate on the theory of rational integers, Hilbert was expected to summarize the results produced by Kummer, Dedekind and Kronecker concerning fields of algebraic numbers. Minkowski, who became increasingly busy working on his own book on the geometry of numbers, decided to abandon the project and the only one to complete his part in the survey task was Hilbert.[34]

Hilbert's *Zahlbericht*, as it came to be known, was published in 1897. This was not a survey in the usual sense of the term. Hilbert did produce an impressive and exhaustive systematization of the existing results of the discipline, but he also added many important results of his own. Since Hilbert basically adopted Dedekind's approach as the leading one, and since the *Zahlbericht* became the standard reference text for mathematicians working in algebraic number theory,[35] the publication of this survey turned out to be a decisive factor for the consequent dominance of Dedekind's perspective over that of Kronecker within the discipline.

In the introductory section of the *Zahlbericht*, Hilbert gave a brief historical account of the achievements of the theory, stressing its connections with other mathematical disciplines. According to Hilbert, Gauss, Dirichlet and Jacobi had expressed their amazement in view of the close connections between number-theoretic questions and "algebraic problems" (i.e., the resolution of polynomial equations). The grounds for this connection were now clear: both sub-disciplines had their common roots in the theory of algebraic fields. The theory of number-fields had become—in Hilbert's view—the most essential component of the modern theory of numbers.[36]

Hilbert's *Zahlbericht*, together with his other contributions to algebraic number theory,[37] constituted a synthesis of the discipline, of scope and influence on future developments, comparable only to those of Gauss's *Disquisitiones* about a century before (see above § 2.2.1). Our interest in these works lies in their impact on the development of algebra, and we thus focus on Hil-

---

34. Minkowski's letters to Hilbert during these years contain many references to work on his planned section of the *Zahlbericht* and on his own book. Cf. Rüdenberg & Zassenhaus (eds.) 1973, esp. pp. 57 ff.. See also Zassenhaus 1973. Additional accounts appear in Blumenthal 1935, 396-399; Ellison & Ellison 1978, 191; Reid 1970, 42-45 & 51-53.

35. As late as 1944 Weyl wrote (p. 626) that "even today, after almost fifty years, a study of this book is indispensable for anybody who wishes to master the theory of algebraic numbers." Cf. also Hasse 1932, 529; Ellison & Ellison 1978, 191.

36. Hilbert 1897, 64.

37. Of special significance and greatest influence were Hilbert 1898; 1899a. See Hasse 1932.

bert's treatment of the various "algebraic structures" in his *Zahlbericht*, and on their respective roles within his general picture of algebra. It was already pointed out that groups and fields, although definable within a single, common formulation, played diverging roles within both Dedekind's and Weber's conceptions of algebra. For these two mathematicians, fields (not abstract fields, but rather fields of numbers) were the main subject matter of research both in Galois theory and in algebraic number theory, while groups remained a tool, effective and powerful, yet subsidiary to the main concern of that research. In the *Zahlbericht*,[38] while fields are ascribed a role similar to that conceded by Dedekind and Weber in their works, groups were barely mentioned at all, since results of group theory are much less needed here than, say, in Galois theory.[39]

Although the main concern of the *Zahlbericht* is indeed the study of fields of algebraic numbers, one also finds here the first definition of "rings" as a separate concept. At the final section of a chapter on "the theory of general fields of numbers," Hilbert provides the following definition:

> Let θ, μ, ... be any collection of algebraic numbers, whose domain of rationality is a field k of degree m, then the system of all integer functions of θ, μ, ... with rational integer coefficients will be called "ring of numbers", "ring" or "integral domain".[40]

Obviously, only limited kinds of rings—in the present sense of the term—comply to this definition, which is totally conditioned by the number-theoretical setting within which it appears. In a footnote Hilbert explained that this concept corresponds to Dedekind's *Ordnung*; thus a ring is closed under addition, subtraction and product. If $\omega_1, ..., \omega_m$ is a basis of the field k, then the ring generated by that basis is the greatest ring contained in the field. In fact, this is the ring of all the algebraic integers in that field. Moreover, every ring in the

---

38. Cf. Purkert 1971, 18: "Die Hilbertschen Arbeiten behandeln endlich-algebraische Zahlkörper. Zugrunde liegt der Begriff 'Zahlkörper', wie Dedekind ihn ursprünglich eingeführt hatte. Auf den abstrakten Körperbegriff wird—entsprechend der Zielsetzung dieser Arbeiten—nicht Bezug gennomen."

39. Thus, for instance, in Hilbert 1897, 129, where the associated group of a Galois field is defined.

40. Hilbert 1897, 121: "Sind θ, μ, ... irgend welche ganze algebraische Zahlen, deren Rationalitätsbereich der Körper k vom m-ten Grade ist, so wird das System aller ganzen Funktionen von θ, μ, ..., deren Koeffizienten ganze rationale Zahlen sind, ein *Zahlring, Ring,* oder *Integritätsbereich* genannt."

field has a basis which generates it. Thus a ring in Hilbert's sense is a system of algebraic integers of the given field, closed under the three mentioned operations.

Hilbert defined an ideal of a ring as any system of algebraic integers belonging to the ring, such that any linear combination of them (with coefficients in the ring) belongs itself to the ideal. Hilbert quoted in this context several results from Dedekind's theory of ideals. But Hilbert *did not* describe a ring as a group endowed with an additional operation, or as a field whose division fails to satisfy a certain property. Neither did he present ideals of fields as a distinguished kind of ring. Hilbert's ideals are always ideals in fields of numbers. In spite of his interest in the theory of polynomials and his acquaintance with the main problems of this discipline, Hilbert never attempted—in the *Zahlbericht* or elsewhere—to use ideals as an abstract tool allowing for a *joint* analysis of factorization in fields of numbers and in systems of polynomials. This step, crucial for the later unification of the two branches under the abstract theory of rings, will be taken only more than twenty years later by Emmy Noether. Obviously, the absence of such a step in Hilbert's work is not so much a consequence of technical capabilities, as it is one of motivations: an indication of the nature of his images of algebra, to which the idea of algebraic structures as an organizing principle was foreign.

Hilbert's conception of algebra—and his attitude towards the concepts that would eventually constitute the heart of the structural image of algebra— is similarly manifest in an article on the theory of fields of algebraic numbers, published by Hilbert in 1900 in the *Encyclopädie der Mathematischen Wissenschaften*. After defining there integral domains or rings of a field, Hilbert explained that most concepts originally introduced for the study of fields can be translated and used for the study of rings. He also defined "modules" as those systems of numbers closed under addition and subtraction: the system of integers, ideals in a field, rings in a field and ideals in a ring.[41] Like Dedekind before him, Hilbert did not equate the concept of module—a number-theoretical one in essence—to that of group, although the formal equivalence was obvious to him.

In Hilbert's conception, then—as seen in the *Zahlbericht* and in other, related works—there is a discernible separation of the roles ascribed to the various concepts involved. As in Dedekind's work, fields constitute the main

---

41. Hilbert 1900a, 687-688

subject-matter of the theory of algebraic numbers. By contrast, ideals and rings, important and interesting as they are as a subject of mathematical research, are only the main tools, rather than the subject-matter of the theory in which they appear. Groups are likewise tools, which are used to investigate the properties of fields extensions in a second domain in which fields appear as the main subject-matter: Galois theory. Groups and fields are thus for Hilbert two mathematical entities of basically different nature: the latter numerical, the former essentially non-numerical. Ideals, groups, rings and fields, although all of them appear in Hilbert's work on algebraic number theory, are neither conceived nor used as individual manifestations of the common, underlying idea of an algebraic structure.[42] Moreover, Hilbert never examined the possibility of using ideals as tools in yet a third mathematical discipline, formerly studied by him using tools taken from algebraic number theory: the theory of polynomial forms. Thus the basic mathematical idea dominating Hilbert's work on algebraic number theory is not the idea of an algebraic structure, but rather the classical nineteenth-century image of algebra: that the systems of real and complex numbers and their known properties underlie all of algebraic research. Algebraic reasoning in mathematics presupposes the properties of these systems, rather than the other way round.

An additional perspective from which to evaluate the relationship between algebraic concepts and number theory in Hilbert's images of mathematics can be found in the work of one of his doctoral students, the Swiss mathematician Karl Rudolf Fueter (1880-1950).[43] Fueter's 1903 dissertation, as well as most of his later research was done in the theory of algebraic number fields.[44] In an article of 1905, "*Die Theorie der Zahlstrahlen*"[45], he addressed, like in his dissertation, issues connected with the solution of the so-called "Kronecker's *Jugendtraum*", a reformulation of which also appeared as one of the problems in Hilbert's 1900 list of twenty-three.[46] Fueter stated that the standard text

---

42. A similar assessment was advanced by Hans Zassenhaus in his own comments on the genesis of the *Zahlbericht*. According to him, an unprejudiced reader will discover the conspicuously marginal role played by groups and other abstract concepts in this work. See Zassenhaus 1973, 19: "Dem unvoreingenomenen Leser des Zahlberichts fällt die untergeordnete Rolle der Gruppentheorie und anderer abstrakter Hilfsmittel auf."

43. For a short biographical notice on Fueter, see Frei & Stammbach 1992.

44. The dissertation's tittle was: "Der Klassenkörper der quadratischen Körper und die komplexe Multiplikation." Cf. Hilbert *GA* Vol. 3, 432.

45. Fueter 1905. The second part of the article appeared as Fueter 1907.

from which his notation, basic concepts and references were taken, was the *Zahlbericht*. He did not say, however, that his article repeated, with slight variations, much of the arithmetic part of Weber's class field theory, as developed in Vol. III of the *Lehrbuch*.[47]

Like in the *Zahlbericht*, concepts such as fields, rings, and rational domains—which are used freely in the article—denote specific collections of complex numbers satisfying certain, well-defined conditions. But Fueter also used a new concept, which was absent from Hilbert's report: the *Strahl*. A *Strahl*—a term that Fueter had originally introduced in his dissertation, and which plays a central role in the main argument of his article—is simply a collection of numbers closed under product and division. In the *Lehrbuch*, Weber had used the term *Zahlgruppe* in the same context.[48] Particular cases of *Strahlen* are fields and "regular rings." A "regular ring"—one gathers from Fueter's explanation—is a ring closed with respect to division.[49] Clearly—although Fueter did not explicitly say so—a *Strahl* is also a group. The glaring omission of any remark in this direction illuminates the essence of Fueter's most basic conceptions. Thus, on the one hand, one has a useful way of classifying collections of complex numbers according to their properties with respect to the various operations: fields, rings, rational domains, and now, *Strahlen*. On the other hand, one has an abstract tool, originally developed in algebra, that may be effectively used when analyzing the properties of those collections: groups. Concepts belonging to these two different categories are not to be mixed under this conception.

This view remained unchanged in Fueter's work throughout his career. In his textbook on number-theory—published in 1917—one finds again very idiosyncratic definitions which put forward the same separation of conceptual categories. Fueter defined a module as a domain of numbers closed under addition and subtraction, and he called a domain closed under multiplication

46. The problem in question is the twelfth: "Extension of Kronecker's Theorem on Abelian Fields to any Domain of Rationality." On Hilbert's formulation and its relation to the problem as Kronecker had originally conceived it, see Hasse 1932, 510-515; Hasse 1967, 267-268. On Kronecker's own work on the *Jugendtraum*, see Edwards 1987, 30-31. Edwards explains that Hasse recognized as wrong his own original assessment, concerning Kronecker's first statement of the theorem as incorrect. On later developments related to this problem, see Langlands 1976.

47. Schappacher 1998 describes the intricate story around the early attempts to solve Hilbert's twelfth problem and, as part of it, the import of Fueter's contribution. See also Gray 2000, 140-144.

48. Weber 1895, Vol. 3 (1908), 596.

49. Fueter 1905, 208.

and division *Strahl*. When considering congruence classes defined by an ideal in a domain of numbers, Fueter introduced the concept of abstract group, and explicitly separated the latter from all other concepts involved in the theory of numbers. Fueter wrote:

> At this stage of our discussion another mathematical discipline comes forward for the first time, the theory of groups, which is itself a sub-discipline of algebra. Since its fundamentals will not be assumed to be known here, its most essential parts must be treated, thus interrupting the course of the number-theoretical research.[50]

Thus, in a single book, Fueter used three different concepts which, abstractly seen, are identical, but which are distinctly used by him in three different contexts: groups, modules, *Strahlen*. From the point of view of later developments in algebra, this certainly looks as a somewhat ambiguous attitude towards axiomatically defined concepts, and this ambiguity underscores, by implication, what was said above about structural ideas in Hilbert's number-theoretical works. Of course, one should not forget that by the time Fueter started working on his dissertation, Hilbert's research had already moved to different concerns, and it is hard to know to what extent he was directly involved in supervising Fueter's work. Much less can we speak with certainty of a direct influence of Hilbert on Fueter in 1917. And yet, it seems extremely unlikely that if the idea of an algebraic structure underlay Hilbert's work on number theory—in a somewhat similar way, and to a somewhat similar intensity, than it later appeared with Noether and van der Waerden—it could then have been so glaringly absent from one of his pupil's work, especially in its early stages.

To summarize this section, the impact of Hilbert's work on algebraic number theory on the rise of the structural image of algebra can be described in terms very similar to those used above concerning his first domain of research. Hilbert elaborated the main concepts and contributed important results that were instrumental in building the necessary body of knowledge on which the structural approach to algebra was to be erected. Yet, the basic conceptual

---

50. Fueter 1917, 40: "An dieser Stelle unserer Betrachtungen tritt zum ersten Male eine andere Disziplin der Mathematik hinzu, die Gruppentheorie, die ihrerseits ein Teilgebiet der Algebra ist. Da wir ihre Grundsätze nicht voraussetzen werden, müssen wir kurz das Wesentlichste behandeln und solange den Gang der zahlentheoretischen Untersuchungen unterbrechen."

interrelationship dominating his work was different in nature from the hierarchical one characteristic of the structural image of algebra. The general idea of algebra as the discipline dealing with algebraic structures is still absent from Hilbert's work on algebraic number theory.

3.3   Hilbert's Axiomatic Approach

Hilbert arrived in Göttingen in 1895. The *Zahlbericht* was thus concluded at the university with which Hilbert's name has ever since come to be associated. After the *Zahlbericht*, Hilbert's next outstanding work was the *Grundlagen der Geometrie*, first published in 1899. Given the enormous impact of this work, and the interest aroused by Hilbert's metamathematical works in the early 1920s, his name has come to be associated also with research on the foundations of mathematics, and—more than anyone else's—with the widespread adoption of the modern axiomatic method. Since the latter was a necessary stage in the way to the consolidation of the structural image of algebra, it is certainly relevant to our account to examine more closely the nature of Hilbert's work on axiomatics. Of course, it is well beyond the scope of this book to analyze thoroughly this crucial feature of Hilbert's contribution.[51] We therefore limit ourselves to discuss the specific questions addressed by Hilbert in his analysis of axiomatic systems, and the role accorded by Hilbert to axiomatics within mathematics as a whole.

In the winter semester of 1898-99 Hilbert taught for the first time in Göttingen a course on the foundations of Euclidean geometry. Hilbert's interest in this mathematical domain signified, on the face of it, a sharp departure from the two fields in which he had excelled during the past years: the theory of algebraic invariants and the theory of algebraic number fields.[52] The announcement of the course came as a surprise to many in Göttingen.[53] As a

---

51. For detailed discussions see Corry 1997, 2000. See also: Bernays 1935, Mehrtens 1990, 114-142; Peckhaus 1990, Chpts. 2-3; Rowe 1993a, 20-29; Weyl 1944, 635-645. From among the works included in the bibliography, Hilbert's own published ideas on this issue appear in, e.g., Hilbert 1918, 1923, 1930.

52. For instance, in his memorial lecture on Hilbert, Hermann Weyl wrote (1944, 635): "[T]here could not have been a more complete break than the one dividing Hilbert's last paper on the theory of number fields from his classical book *Grundlagen der Geometrie*."

53. Cf. Blumenthal 1935, 402: "Das erregte bei den Studenten Verwunderung, denn auch wir älteren Teilnehmer an den 'Zahlköperspaziergängen' hatten nie gemerkt, daß Hilbert sich mit geometrische Fragen beschäftigte: er sprach uns nur von Zahlkörpern."

matter of fact, however, the issue had occupied Hilbert's thoughts long before, and he had even lectured on the issue during his Königsberg years.[54]

Modern research on the axiomatic foundations of geometry was not initiated by Hilbert. Gaps in the deductive structure of Euclid's *Elements* had been pointed out since the seventeenth century, by mathematicians like John Wallis (1616-1703).[55] Towards the end of the nineteenth century one finds several elaborate attempts to reformulate the entire body of geometry, based on a full inventory of all the presuppositions involved in it. Efforts were also directed at re-elaborating existing proofs, while avoiding all the gaps discovered over the years.

Nineteenth-century foundational investigations in geometry were strongly connected with the study of projective geometry. They culminated with the work of Moritz Pasch (1843-1930), whose book, *Vorlesungen über neuere Geometrie* presented in 1882 the first full axiomatic reconstruction of this mathematical domain.[56] Before Pasch, August Ferdinand Möbius (1790-1868), Jakob Steiner (1796-1863), and Christian von Staudt (1798-1867) had attempted in their respective works to elucidate the relationship between Euclidean and projective geometry. This undertaking involved, among others, the elaboration of the two closely related notions of cross-ratio and of general projective coordinates. Von Staudt was able to prove that these coordinates can be defined independently of any metric consideration, and that therefore such considerations are unnecessary for reconstructing the whole of projective geometry.[57] Later on, in 1872, Felix Klein—using ideas developed by Cayley[58]and aware of the independence of the parallel axiom in Euclidean geometry—reformulated the whole issue in terms of axiomatic interdependence: projective geometry should be axiomatically defined, so that the various metrical geometries can be derived by addition of new axioms.[59] Klein himself was only partially successful in pursuing this task.[60]

54. Michael M. Toepell (1986) has analyzed the development of Hilbert's ideas previous to the publication of the *Grundlagen*, and his early encounters with foundations of geometry since his Königsberg years.

55. See Torreti 1978, 44-53.

56. Pasch 1882.

57. See Klein 1926-7 Vol. 1, 132-136.

58. For an account of Cayley's contributions see Klein 1926-7 Vol. 1, 147-151.

59. In Klein 1873.

60. See Rowe 1994, 194-195; Toepell 1986, 4-6; Torreti 1978, 110-152. On von Staudt's contribution see Freudenthal 1974a.

In Pasch's axiomatic reconstruction of projective geometry, once the axioms are determined, all other results of geometry should be attained by strict logical deduction, and without any appeal to diagrams or to properties of figures involved. Yet Pasch conceived geometry as part of "natural science", whose truths can be obtained from a handful of concepts and basic laws (the axioms) directly derived from experience. For Pasch, *the meaning of the axioms themselves is purely geometrical* and it cannot be grasped without appeal to the diagrams from which they are derived.[61]

Two lines of development derived from Pasch's work: one in Italy, the other in Germany. The first of this trends is the one connected with the work of Giuseppe Peano (1858-1930). Peano was a competent mathematician, who made significant contributions in analysis and wrote important textbooks in this field.[62] But besides these standard mathematical activities, Peano invested much of his efforts in advancing the cause of the international languages and to develop a conceptual language that would allow completely formal treatments of mathematical proofs. In 1889 he succeeded in applying such a language to arithmetic, thus putting forward his famous postulates for the natural numbers. Pasch's systems of axioms for projective geometry posed a challenge to Peano's artificial language. In addressing this challenge Peano was interested in the relationship between the logical and the geometrical terms involved in the deductive structure of geometry, and in the possibility of codifying the latter in his own artificial language. This interest led Peano to introduce the idea of an independent set of axioms, namely, a set none of whose axioms is a logical consequence of the others. He applied this concept to his own system of axioms for projective geometry, which were a slight modification of Pasch's. Peano's peculiar way of dealing with systems of axioms, and the importance he attributed to the search after independent sets of postulates is similar in many respects to the perspective developed later by Hilbert; yet Peano never undertook in his works to prove the independence of *whole systems* of postulates.[63] For all of his insistence on the logical analysis of the deductive structure of mathematical theories Peano's overall view of mathematics—like Pasch's before him—was neither formalist nor logicist in the

---

61. See Contro 1976, 284-289; Nagel 1939, 193-199; Torreti 1978, 210-218.
62. A brief account of Peano's mathematical work appears in Kennedy 1981. For more elaborate accounts see Kennedy 1980; Segre 1994.
63. Cf. Torreti 1978, 221.

sense later attributed to these terms. Peano conceived mathematical ideas as being derived from our empirical experience.[64]

Several Italian mathematicians, influenced by Peano's ideas, published similar works in which the logical structure of the foundations of geometry was elaborated. Among them mention should be made of Mario Pieri (1860-1913),[65] who strongly promoted the idea of geometry as a hypothetico-deductive system, and introduced for his systems of postulates a kind of "ordinal independence", somewhat more limited than the one defined by Peano.[66] Giuseppe Veronese (1845-1917)[67] was the first to pursue in a systematic fashion the study of the possibility of a non-Archimedean geometry. He proved the independence of the Archimedean postulate from the other postulates of geometry.[68] Also Gino Fano (1871-1952) can be mentioned in this context, as having advanced an elaborated abstract view of geometry.[69]

The trend that developed in Germany following Pasch, elaborated more directly the line of thought that had in the first place led to Pasch's own achievements. It had direct mathematical, rather than logic and methodological motivations: it focused on the problem of continuity. The elucidation of the role of continuity in geometry had played a central role in Pasch's presentation, and the question remained open, whether it should be considered as given with the very idea of space, or whether it should be reduced to more elementary concepts. Klein and Wilhelm Killing (1847-1923) elaborated the first of these alternatives, while Hermann Ludwig Wiener (1857-1939) and Friedrich Schur (1856-1932) worked out the second.[70] We have also seen above, that in his correspondence with Lipschitz, Dedekind interestingly dealt with these questions as well, but his thoughts in this regard were never published.

Since his student days in Königsberg Hilbert came to know closely the German geometric tradition that led to Pasch's book and developed from it.[71] Hilbert was also aware of the achievements of the Italian school, although it is hard to say precisely which of their works he read, and how did they influence

---

64. See Kennedy 1981, 443.

65. On Pieri, see Kennedy 1981a.

66. Cf. Torreti 1978, 225-226.

67. On Veronese, see Tricomi 1981.

68. In Veronese 1891. On criticisms directed at Veronese's work by German mathematicians see Toepell 1986, 56.

69. See Freudenthal 1957, 112.

70. Cf. Contro 1976, 292-294.

71. See Toepell 1986, 9-15.

his thought.[72] In preparing his various lectures on geometrical issues in Königsberg, Hilbert studied carefully Pasch's presentation of projective geometry. He acknowledged the importance of Pasch's treatment of the system of postulates, but at the same time he came to realize some of its shortcomings, and in particular, certain redundancies that affected it. Hilbert perceived that the task of establishing the minimal set of presuppositions from which the whole of geometry could be deduced had not yet been fully accomplished.[73] Likewise, Hilbert already stated in these early lectures the need to establish the independence of the axioms of geometry.[74]

A second important source for Hilbert's axiomatic approach came from contemporary work on the foundation of physics, and in particular from the work of Heinrich Hertz (1857-1894). Without attempting to discuss this issue in detail here, it is nevertheless important to stress that, from Hilbert's point of view, an axiomatic analysis of the kind he intended to implement should in principle be applied in like manner to geometry as well as to any natural science. Indeed, Hilbert conceived geometry—at variance with other mathematical domains—to be itself like a natural science, in which sensorial intuition (*Anschauung*) played a decisive role. For him, the difference between geometry and other natural sciences would not concern essence, but rather only their current stage of development. The manuscripts of his lectures in Königsberg and in Göttingen provide plenty of evidence to this view. The following passage, to take just one illuminating example, is taken from in his 1898/99 lectures on mechanics:

> Geometry also [like mechanics] emerges from the observation of nature, from experience. To this extent, it is an *experimental science*.... But its experimental foundations are so irrefutably and so *generally acknowledged,* they have been confirmed to such a degree, that no further proof of them is deemed necessary. Moreover, all that is needed is to derive these foundations from a minimal set of *independent axioms* and thus to construct the whole edifice of geometry by *purely logical means.* In this way [i.e., by means of the axiomatic treatment] geometry is turned into a *pure mathematical* science. In mechanics it is also the case that all physicists recognize its most *basic facts*. But the *arrangement* of the basic concepts is still subject to a change in perception ... and therefore mechanics cannot yet be described today as a *pure mathematical* discipline, at least to

---

72. See Toepell 1986, 55-57.
73. See Toepell 1986, 45.
74. See Toepell 1986, 58-59.

the same extent that geometry is. We must strive that it becomes one. We must ever stretch the limits of pure mathematics wider, on behalf not only of our mathematical interest, but rather of the interest of science in general.[75]

The *Grundlagen der Geometrie* appeared in June 1899 as part of a *Festschrift* issued in Göttingen in honor of the unveiling of the Gauss-Weber monument. It consisted of an elaboration of lectures held by Hilbert in the winter semester of 1898-99. Hilbert's axioms for geometry are formulated for three systems of *undefined* objects (points, lines and planes); the axioms establish mutual relations that should be satisfied by these objects. The axioms are divided into five groups: axioms of incidence, of order, of congruence, of parallels and of continuity. These groups have no logical significance in themselves. Rather they reflect Hilbert's actual conception of the axioms as an expression of our spatial intuition: each of the groups expresses a particular way in which these intuitions manifest themselves.

Hilbert showed in the *Grundlagen* how all the known results of Euclidean, as well as of certain non-Euclidean geometries could be elaborated from scratch, depending on which groups of axioms are admitted. Thus, reconstructing the very ideas that had given rise to his own conception, Hilbert proved, for instance, two of the main theorems of projective geometry, the theorems of Desargues (Chpt. V) and of Pascal (Chpt. VI), without recourse to the axiom of continuity. It thus turned out that these particular theorems are valid in the usual, Euclidean, as well as in a non-Archimedean geometry, such as the one introduced earlier by Veronese.

Of central importance for Hilbert's undertaking was the problem of mutual independence among the individual axioms within the groups and among the various groups of axioms in the system. As was seen above, in 1893 Hilbert had mentioned his interest in a similar question in relation to his research on invariants, even though he was not yet working then within a thoroughly axiomatic conception. In the *Grundlagen* Hilbert put forward an elaborate technique, inspired perhaps on the works of Peano and Pieri, and specifically conceived to deal with the question of independence. Hilbert proved independence by constructing models of geometries which fail to satisfy a given axiom of the system but satisfy all the others. But Hilbert's attention in studying mutual independence focused *on geometry itself*; the *Grundlagen* was by no means a general study of the relations between systems

---

75. Quoted in Corry 1997, 108-109. Emphasis in the original.

of axioms and its possible models. It is for this reason that Hilbert's original system of axioms is not—from the logical point of view—the most economical possible one. In fact, several mathematicians noticed quite soon that Hilbert's system of axioms—seen as a single collection—contained a certain degree of redundancy.[76] But for Hilbert himself the actual aim was to establish the interrelations among the groups of axioms rather than among the individual axioms of different groups.

A second issue addressed by Hilbert, and perhaps the main motivation behind his whole undertaking, was the consistency of the various kinds of geometry. Hilbert established through the *Grundlagen* the relative consistency of geometry vis-à-vis arithmetic, i.e., he proved that any contradiction existing in Euclidean geometry must manifest itself in the arithmetic system of real numbers. He did this by defining a hierarchy of algebraic number fields. Hilbert first proved the existence of infinitely many incomplete models that satisfy all the axioms—except the so-called "axiom of completeness" (*Vollständigkeitsaxiom*).[77] He then proved that only *one* complete model exists that satisfies this last axiom as well, namely, the usual Cartesian geometry, obtained when the whole field of real numbers is used in the model.[78] Having done so, it became necessary to prove the consistency of arithmetic in order to establish definitely the consistency of Euclidean geometry. This was not done in the *Grundlagen* itself, but Hilbert was absolutely confident of the possibility of finding the necessary proof, and immediately assigned to it a high priority as an important open problem of mathematics. Thus, among his famous 1900 list of twenty-three problems, the second concerns the proof of the "compatibility of arithmetical axioms."[79]

---

76. Cf., for instance, Schur 1901. For a more detailed analysis of this issue see Schmidt 1933, 406-408. It is worth pointing out that Hilbert's stated aim in the first edition of the *Grundlagen* was to provide an independent system of axioms for geometry. In the second edition, however, this statement did not appear anymore, following a correction by E.H. Moore (1902) who showed that one of the axioms may derived from the others. See below § 3.5. See also Torreti 1978, 239 ff.

77. The axiom is formulated in Hilbert 1971, 26. This axiom did not appear in the first edition and, in fact, its formulation underwent several changes throughout the various later editions of the *Grundlagen*, but it remained central to this part of the argument. Cf. Peckhaus 1990, 29-35. Toepell 1986, 254-256, briefly describes the relationship between Hilbert's completeness axiom and related works of other mathematicians. The role of this particular axiom within Hilbert's axiomatics and its importance for later developments in mathematical logic is discussed in Moore 1987, 109-122.

78. For this proof see Hilbert 1971, 29-32.

These are, then, the two main questions addressed by Hilbert regarding the axiomatic system that defines geometry: independence and consistency. Now, in principle, there should be no reason why a similar analysis could not apply for any system of postulates establishing mutual abstract relations among undefined elements arbitrarily chosen in advance and having no concrete meaning. The twentieth-century profusion of mathematical theories actually constructed on the principles derived from such a practice has often been seen as evidence of the success and influence of Hilbert's own point of view and as one of his main contributions to shaping contemporary mathematical thinking. Moreover, the impact of Hilbert's axiomatic research coupled with the "formalism" associated with his name in the framework of the so-called "foundational crisis" of the 1930s have been occasionally seen as promoting the view of mathematics as an empty, formal game.[80] But as a matter of fact, Hilbert's own axiomatic research was never guided by such a view, and he often opposed it explicitly.[81] Hilbert's own conception of axiomatics did not convey or encourage the formulation of arbitrary, abstract axiomatic systems as starting points for mathematical research. Rather, his work was directly motivated by the need to better define and understand *existing* mathematical and scientific theories. In Hilbert's view, the definition of systems of abstract axioms and their analysis following the above described guidelines were meant to be conducted for established and elaborated mathematical entities. Hilbert saw natural science—and in particular geometry within it—as an organic entity, growing simultaneously in various directions.[82] In this view, the development of science involved both an expansion in scope and an ongoing clarification of

---

79. Hilbert 1901, 299-300. As it is well-known, Kurt Gödel (1906-1978) proved in 1931 that such a proof is impossible in the framework of arithmetic itself. On the subsequent evolution of this problem see Kreisel 1976.

80. A typical instance of such a view appears in Resnik 1974, 389: "[Hilbert's conception] removed the stigma of investigating axioms which do not describe any known 'reality' and opened the way to the creation of new mathematical theories by simply laying down new axioms." See also Reid 1970, 60-64.

81. For instance, in a series of lectures delivered in 1919-20 (Hilbert 1992, 14), he said: "Von Willkür ist hier kein Rede. Die Mathematik ist nicht wie ein Spiel, bei dem die Aufgaben durch willkürlich erdachte Regeln bestimmt werden, sondern ein begriffliches System von innerer Notwendigkeit, das nur so und nicht anders sein kann." For Hilbert's views on the role of *Anschauung*, as opposed to formal manipulation of empty concepts, in his system of geometry see also Toepell 1986, 258-261.

82. Hilbert's conception of mathematics as an organic unit is expressed in many places. See, for instance, his famous 1900 Paris lecture, Hilbert 1902, 479.

the logical structure of its existing parts. The axiomatic treatment of a discipline is part of this growth, but it applies properly only to well-developed theories. This view is clearly manifest in the following, typical passage, taken from a lecture of 1905:

> The edifice of science is not raised like a dwelling, in which the foundations are first firmly laid and only then one proceeds to construct and to enlarge the rooms. Science prefers to secure as soon as possible comfortable spaces to wander around and only subsequently, when signs appear here and there that the loose foundations are not able to sustain the expansion of the rooms, it sets to support and fortify them. This is not a weakness, but rather the right and healthy path of development.[83]

And indeed, Hilbert's efforts in axiomatics were actually directed only towards elaborated disciplines (geometry, mechanics, calculus of probability, etc.), rather than as starting points for new disciplines.

But for Hilbert, axiomatic research of mathematical theories not only conferred a greater degree of certainty to existing knowledge. He saw the axiomatic principles as providing mathematical concepts with justification, and indeed with their very existence. Hilbert elaborated a view of mathematical truth that equated it with logical consistency. In fact, this view endorsed *a-posteriori* the legitimacy of proofs of existence by contradiction, like the one advanced several years before by Hilbert himself for the finiteness theorem. Among the various places where Hilbert stressed the centrality of axiomatics in defining mathematical concepts and as source of mathematical truth, his 1900 list of problems is an oft-quoted one. In formulating his second problem Hilbert stated:

> When we are engaged in investigating the foundations of a science, we must set up a system of axioms which contains an exact and complete description of the relations subsisting between the elementary ideas of the science. The axioms so set up are at the same time the definitions of those elementary ideas, and no statement within the realm of the science whose foundation we are testing is held to

---

83. Quoted in Peckhaus 1990, 51: "Das Gebäude der Wissenschaft wird nicht aufgerichtet wie ein Wohnhaus, wo zuerst die Grundmauern fest fundiert werden und man dann erst zum Auf- und Ausbau der Wohnräume schreitet; die Wissenschaft zieht es vor, sich möglichst schnell wohnliche Räume zu verschaffen, in denen sie schalten kann, und erst nachträglich, wenn es sich zeigt, dass hier und da die locker gefügten Fundamente den Ausbau der Wohnräume nicht zu tragen vermögen, geht sie daran, dieselben zu stützen und zu befestigen. Das ist kein Mangel, sondern die richtige und gesunde Entwicklung."

be correct unless it can be derived from those axioms by means of a finite number of logical steps. (Hilbert 1902, 447.)

The view that equates mathematical "truth" with logical consistency was extensively developed by Hilbert in his well-known interchange with Gottlob Frege (1846-1925), following the writing of the *Grundlagen*.[84]

But in spite of the central role accorded by Hilbert to the axiomatic approach in mathematics, he did not see it as the only possible one. In a lecture delivered in 1899 in Munich before the *DMV*, and later published as "*Über den Zahlbegriff*", Hilbert discussed two different ways of dealing with concepts in mathematics: the genetic approach and the axiomatic approach.[85] The classical example of the possibility to define a mathematical entity genetically is provided by the system of real numbers. In this example one starts from the definition of the natural numbers, which arise from the basic intuition of counting. Then in order to allow for a complete definition of subtraction on the natural numbers, these must be extended to include the integers. In attempting to define division on the latter kinds of numbers one is led to define the rational numbers, and then, finally, one creates the real numbers by means of cuts of rationals. On the other hand, there is the axiomatic method, typically used in geometry. Hilbert claimed that both tendencies usually complement each other in mathematics, but he raised the question as to their relative value. Finally Hilbert stated his opinion:

> In spite of the high pedagogic value of the genetic method, the axiomatic method has the advantage of providing a conclusive exposition and full logical confidence to the contents of our knowledge. (Hilbert 1900, 184)

Thus, while axiomatics plays a central role in Hilbert's overall conception of mathematics, this conception does not involve a total conceptual break with the classical entities and problems of mathematics, but rather an improvement in their understanding.

How is Hilbert's axiomatic conception related to his images of algebra? This can be clarified by examining the manuscript of a course taught at Göttingen in 1905 on "The Logical Principles of Mathematical Thinking."[86] In

---

84. The relevant letters between Hilbert and Frege appear in Gabriel et al. (eds.) 1976, esp. pp. 65-68. For comments on this interchange see Boos 1985; Mehrtens 1990, 117 ff.; Peckhaus 1990, 40-46; Resnik 1974.

85. Hilbert 1900. For an analysis of the contents of this article, and its place in the development of Hilbert's axiomatic conceptions, see Peckhaus 1990, 29-33.

the first part of the course Hilbert put forward an axiomatic overview of arithmetic, geometry and the natural sciences. This was meant to prepare the ground for the second part of the course, dealing with the axiomatization of logic itself.[87] The section on the axiomatization of natural sciences, built mainly on materials developed by other authors, offers the first detailed account of what Hilbert understood by the "axiomatization of physics", an issue which he had included as the sixth problem of his 1900 list of problems.[88] What is of particular interest for the present account is Hilbert's axiomatic treatment of the addition of vectors.

The axiomatization of physics and of natural science, according to Hilbert, was a task whose complete realization was still very far away. Yet, one particular issue for which it had been almost attained (and only very recently, for that matter) was the "law of the parallelogram" or—what amounts to the same—the laws of vector-addition. In the lectures, Hilbert based his own axiomatic presentation of this domain on works of Gaston Darboux (1842-1917), of Georg Hamel (1877-1954), and of one of his own doctoral students, Rudolf Schimmack (1881-1912).[89]

Hilbert's axiomatic treatment starts by defining a force as a three-component vector. No additional assumptions are made on the nature of the vectors themselves, but it is implicitly clear that Hilbert means the collection of all ordered triples of real numbers. Hilbert was not referring here to an arbitrary collection of abstract objects, satisfying certain well-defined properties, but rather to a very concrete mathematical entity that had been increasingly adopted over the past decades in the treatment of physical theories.[90] In fact, in Schimmack's article of 1903—based on his doctoral dissertation—a vector was explicitly defined as a directed, real segment of line in the Euclidean space. Moreover, two vectors were said to be equal if their lengths as well as their directions coincide.[91] As with other works of Hilbert's students, we have no reason to assume that this differed with the master's view.

---

86. Hilbert 1905. There is a second manuscript of the same course in Hilbert's *Nachlass*, at the Niedersächsische Staats- und Universitätsbibliothek Göttingen, (Cod. Ms. D. Hilbert 558a) annotated by Max Born.

87. See Corry 1997, for a detailed account of the first part of this course., and Peckhaus 1990, 61-75, for the second.

88. See Hilbert 1902, 452.

89. The works referred to by Hilbert are Darboux 1875, Hamel 1905, Schimmack 1903. For additional details on these works, see Moore 1995, 273-275. An additional related work, also mentioned by Hilbert in the manuscript, is Schur 1903.

The axioms are then meant to define the addition of two such given vectors. This addition—said Hilbert—is usually defined by means of the vector whose components are the sums of the components of the given vector. On first thought, this could be taken as the single axiom needed to define the sum. But the task of the axiomatic analysis is precisely to separate this single idea into a system of several, mutually independent, simpler notions that express the *basic intuitions* involved in it.[92] Six axioms are then put forward, that define the vector addition: the first three assert the existence of a well-defined sum for any two given vectors (without stating what its value is), and the commutativity and associativity of this operation. The fourth axiom connects addition to the direction of the vectors as follows:

Let αA denote the vector $(\alpha A_x, \alpha A_y, \alpha A_z)$, having the same direction as A. Then every real number α defines the sum:

$$A + \alpha A = (1 + \alpha)A.$$

i.e., the addition of two vectors having the same direction is defined as the algebraic addition of the longitudes along the straight line on which both vectors lie.[93]

The fifth one connects addition and rotation:

If D denotes a rotation of space around the common origin of two forces A and B, then the rotation of the sum of the vectors equals the sum of the two rotated vectors:

$$D(A + B) = DA + DB$$

---

90. The mathematical roots of the idea of a vector space are discussed in Dorier 1995. The contributions of Oliver Heaviside, Josiah Willard Gibbs and their successors to the development of the concept of a vector space, in close connection with physical theories from 1890 on is described in Crowe 1967, 150 ff. See also above Chapter 2, note 105. A significant clue to the actual context within which the axiomatization of vector addition was seen by Hilbert, and by the other authors quoted by him in his lectures, comes from the fact, that the articles by Hamel and by Schur were published in the *Zeitschrift für Mathematik und Physik*—a journal that bore the explicit sub-title: "*Organ für angewandte Mathematik.*" This journal was founded by Oscar Xaver Schlömilch (1823-1901). By the turn of the century the editor of the *Zeitschrift* was Carl Runge (1856-1927), a leading Göttingen applied mathematician (see Forman 1974).

91. Schimmack 1903, 318.

92. Hilbert 1905, 123.

93. Hilbert 1905, 123: "Addition zweier Vektoren derselben Richtung geschieht durch algebraische Addition der Strecken auf der gemeinsame Geraden."

i.e. the relative position of sum and components is invariant with respect to rotation.[94]

The sixth axiom concerns continuity:

> Addition is a continuous operation, i.e., given a sufficiently small domain G around the end-point of $A + B$ one can always find domains G_1 and G_2, around A and B *respectively*, such that the end-point of the sum of any two vectors belonging to each of these domains will always fall inside G.[95]

These are all simple axioms—continued Hilbert—and if we think of the vectors as representing forces, they also seem rather plausible. The axioms thus account for the basic known facts of experience, i.e., that the action of two forces on a point may always be substituted by a single one, that the order and the way in which they are added does not change the result, that two forces having one and the same direction can be substituted by a force having the same direction, and that the relative position of the components and the resultant is independent from rotations of the coordinates. Finally, the demand for continuity in this system is similar to that of geometry, and it is thus similarly formulated as in that case.[96]

That these six axioms are in fact necessary to define the law of the parallelogram was first claimed by Darboux, and later proved by Hamel. The main difficulties for this proof arose from the sixth axiom. In his 1903 article, Schimmack proved the independence of the six axioms (in a somewhat different formulation from that adopted by Hilbert in his 1905 lectures), using the usual technique of models that satisfy all but one of the axioms.

---

94. Hilbert 1905, 124: "Nimmt man eine Drehung D des Zahlenraumes um den gemeinsamen Anfangspunkt vor, so entsteht aus $A + B$ die Summe der aus A und aus B einzeln durch D entstehenden Vektoren:

$$D(A + B) = DA + DB.$$

d.h. die relative Lage von Summe und Komponenten ist gegenüber allen Drehungen invariant."

95. Hilbert 1905, 124: "Zu einem genügend kleinem Gebiete G um den Endpunkt von $A+B$ kann man stets um die Endpunkte von A und B solche Gebiete G_1 und G_2 abgrenzen, daß der Endpunkt der Summe jedes im G_1 u. G_2 endigenden Vectorpaares nach G fällt."

96. This axiom of continuity is central to Hilbert's overall conception of the axiomatization of physical science. It would be beyond the scope of the present book to discuss it in detail here. However, it is worth pointing out that, according to Hilbert, the adoption of such an axiom is forced upon us by *experiment*. (Cf. Hilbert 1905, 125). From a strictly mathematical point of view, says Hilbert here, it would be possible to conceive interesting systems of physical axioms that do without this continuity, that is, it would be possible to formulate systems of axioms that define a kind of "non-Archimedean physics." For additional details, see Corry 1997, 131-136.

So much for Hilbert's discussion of the axioms of vector addition in his lectures. What is remarkable from the point of view of the present account is that Hilbert indicated no connection whatsoever between the above axioms and the axioms for abstract groups, which were by then already well-established. Nor did he mention any connection with the field properties of the real numbers, and much less with abstract fields. The picture is thus very clear: the concept of vector had come to be a very basic one in various domains of physics; vectors are concrete mathematical entities, like the real numbers and like Euclidean geometry, whose essential properties can be best understood if analyzed with tools provided by the axiomatic method. This analysis shows that the axioms defining addition of vectors are the six ones presented above, and these axioms manifest our basic intuitions of the ideas involved. The apparent similarities of this system of axioms with the ones defining groups or fields is not directly relevant to the motivations and the actual details of this analysis. Hilbert used in this case the axiomatic approach to analyze a concept that was later recognized as a particular instance of algebraic structure, but his own conception of the former arose and developed without any direct connection with the latter.

Hilbert's contributions to the foundations of geometry turned out to be of momentous importance for this particular discipline, as well as for mathematics at large, and it further enhanced Hilbert's already well-established prestige. But new challenges and important contributions still stood ahead of him. Around 1901 Hilbert began to develop his new field of interest, integral equations. This domain would occupy his efforts until about 1912, when he began to publish on the foundations of physics. Around 1920 Hilbert renewed his interest in the foundations of mathematics, starting to develop his metamathematical researches. The results of this later period of mathematical activity, combined with his former successes in the foundations of geometry, have remained associated with Hilbert's name as the champion of the modern axiomatic method. But as our account above shows, it was not part of Hilbert's axiomatic conception—at least in its early stages[97]—to encourage a new conception of algebra, in which concepts defined by abstract systems of postulates would assume the central role, and to which the systems of real and complex

---

97. On the relation between Hilbert's earlier and later views on axiomatics see Rowe 1994, 199-201.

numbers would be conceptually subordinate. In the following section we summarize Hilbert's contribution to algebra and its repercussions on future developments within the discipline. In § 3.5 we discuss how the immediate elaboration of the methods applied by Hilbert in the study of the foundations of geometry led to new kinds of inquiries not originally contemplated by Hilbert himself.

3.4   Hilbert and the Structural Image of Algebra

In summarizing Hilbert's contribution to the rise of the structural image of algebra, we can paraphrase a formulation that has been used in other contexts, as follows:[98] although Hilbert's works contain most of the materials needed for elaborating the structural image of algebra, Hilbert himself neither put forward this image in its completed form in his works, nor suggested that it, or something similar to it, should be adopted in algebra.

In his work on invariant theory Hilbert provided a viable alternative to the existing algorithmic approach and showed the fruitfulness of introducing into the realm of polynomial research techniques previously developed in the algebraic theory of numbers. This transposition of methods brought to the fore the similarity of roles played by fields of numbers and fields of functions in number theory and in function theory respectively, thus stressing the convenience of focusing on structural parallels between separate theories. More specifically, Hilbert's implicit use of the ascending chain condition for the proof of the finite basis theorem, his proof of the *Nullstellensatz* with its stress on the power of the polynomial rather than on the polynomial itself, and the particular five properties of invariants chosen to stand as basic principles of the theory, foreshadow many of the central points that the structural re-elaboration of the theory, in terms of abstract rings, would eventually adopt.

Hilbert's work on the theory of algebraic number fields, on the other hand, helped making the techniques and concepts created by Kronecker and Dedekind, and especially the approach advanced by the latter, widely known and accepted by the practitioners of the discipline. It refined existing concepts, introduced new ones, and systematized with their help the study of extensions of fields.

---

98. Cf., e.g., the footnote at the beginning of § 2.2.1 above.

Finally Hilbert's axiomatic elaboration of the foundations of geometry—with its concomitant axiomatic conception of mathematics—was instrumental in encouraging the definition of axiomatic systems for elaborate mathematical theories as well as for their metatheoretical analysis. This was a necessary condition for the trend that would directly lead, as will be seen in the next two chapters, to the definition of abstract rings. The latter provided in turn the conceptual framework for Emmy Noether's structural theory of rings.

But although all these features could have at this stage been combined into a new image of algebra, Hilbert's motivations were very firmly grounded in nineteenth-century's conceptions. Rather than promoting a new view of the aims and scope of research in algebra, all the above achievements were put in Hilbert's works—with unprecedented success—to the service of the general concerns (as well as of the more specific questions) of that classical mathematical tradition of which Hilbert was part and parcel.

Hilbert's treatment of fields of algebraic numbers followed Dedekind's approach. Like the latter, it preserved the separate conceptual statuses of the various concepts involved. Hilbert's fields are always sub-fields of the field of complex numbers, while his groups are non-numerical entities used as tools to elucidate the properties of the former. Likewise, Hilbert's polynomials are always defined over the field of complex numbers: all the properties of the former—including the finite basis theorem, the implicit chain conditions and the *Nullstellensatz*—are derived from the known properties of the latter. Finally, Hilbert's own use of the axiomatic method involved, by definition, an acknowledgment of the conceptual priority of the concrete entities of classical mathematics, and a desire to improve our understanding of them, rather than a drive to encourage the study of mathematical entities defined by abstract axioms devoid of immediate, intuitive significance. Thus in Hilbert's work, one finds no idea of algebraic structure as an organizing principle of algebraic knowledge. All the concepts introduced in algebra derive their meaning, their justification and their properties from those of the systems of complex and real numbers, rather than the other way round.

In fact, the proximity between Hilbert's and Dedekind's ideas goes much farther. Although none of these two mathematicians based the study of algebra on abstractly formulated axioms, in the present, formalistic, sense of the term, both did introduce a kind of axiomatic analysis when dealing with their algebraic entities, especially groups and polynomials. Thus in his lectures on Galois theory and in his works on *Dualgruppen* Dedekind addressed the question of what are the basic statements from which the whole theories considered could be deduced. As we saw, Hilbert's five basic properties for invariants play a similar role. Hilbert's axiomatic method is an elaboration of this view and an application of it to the most central entities of mathematics, rather than to subsidiary ones. But on the other hand, the algebraic works of both Dedekind and Hilbert displayed and promoted all those structural features mentioned above, *independently of any adoption of the modern axiomatic approach*, and in fact, quite before this method became an important concern for Hilbert himself, and for his successors.

This assessment of Hilbert's contribution to the rise of the structural image of algebra is further confirmed when examining the works of his many students and of the many mathematicians that worked under his influence. Hilbert supervised no less than sixty-eight doctoral dissertations, sixty of which were written between 1898 and 1915.[99] Only four among these sixty dealt with issues directly or indirectly related to Hilbert's first domain of research: invariant theory. Not one dissertation deals with problems connected with the theory of factorization of polynomials, although at that same time important works were being published by other mathematicians—such as Emanuel Lasker and Francis S. Macaulay—which elaborated on Hilbert's own ideas (see below § 4.7). Nor is there any dissertation dealing with topics that later came to be connected with modern algebra—such as abstract fields, or the theory of groups in any of its manifestations—and that at the time attracted intense activity throughout the mathematical world. Another remarkable fact is that, although five among the twenty-three problems that Hilbert included in his 1900 list can be considered in some sense as belonging to algebra in the nineteenth-century sense of the word, none of them deals with problems connected with more modern algebraic concerns, and in particular not with the theory of groups.

In 1922, the journal *Die Naturwissenschaften* dedicated one of its issues to Hilbert's sixtieth birthday. Several of his former students wrote summary articles on Hilbert's achievements in his various fields of activity. Otto Toeplitz (1881-1940) described Hilbert's contribution as algebraist.[100] Toeplitz explained in considerable detail Hilbert's contribution to the theory of algebraic invariants, stressing the significance of his achievements as com-

---

99. The list of Hilbert's doctoral students appears in Hilbert *GA* Vol. 3, 431-432.
100. See Toeplitz 1922. For Toeplitz's biography see Behnke & Köhte 1963.

pared to the earlier works of Gordan and of Max Noether. As a second contribution of Hilbert to algebra Toeplitz mentioned the algebraic aspects of
Hilbert's research on integral equations and, in particular, those contributions
connected with Hilbert's theory of infinite determinants.[101] In his account,
Toeplitz even spoke about Hilbert's foundational work, in view of his current
debate with the intuitionists—especially Luitzen E. Brouwer and Hermann
Weyl. Toeplitz praised Hilbert's position and his use of the axiomatic
approach, and justified his willingness to accept non-constructive existence
proofs. Yet in no way did Toeplitz connect these issues with the approach that
by then was already being increasingly adopted in algebra, and that accorded
a central place to concepts defined by means of systems of abstract postulates.

   In view of the account offered in the present chapter, and given that Hilbert's work already contained most of the features that would later characterize the structural image of algebra, one may wonder what was Hilbert's actual
attitude towards this image of algebra as it consolidated around 1930, after his
retirement. In particular—one may ask—what was Hilbert's attitude towards
the treatment accorded to the discipline in van der Waerden's *Moderne Algebra*? What did Hilbert think, for instance, of van der Waerden's definition of
the fields of rational and real numbers—those mathematical entities laying at
the heart of Hilbert's approach to invariants, to algebraic number theory and
to geometry—as particular cases of more general, abstractly defined algebraic
constructs? Could in his view the conceptual order be turned around so that the
system of real numbers be dependent on the results of algebra rather than
being the basis for it? We have no direct evidence to answer these questions,
yet one can conjecture that such answers would not be straightforward but
rather ambiguous. Support for this conjecture can be gathered from Hermann
Weyl's remarks in his obituary to Hilbert. Regarding the role of axiomatics in
modern algebra he stated:

---

101. It is worth stressing that in describing the manifold connections of Hilbert's theory of integral equations with other parts of mathematics, Toeplitz elaborated on Hilbert's favorite metaphor
for mathematical knowledge, the "building" metaphor. Toeplitz wrote (1922, 76-77): "So hat
Hilbert in seiner Theorie der unendlichen Variablen ein großes Haus errichtet, in dem viele einzelne
Untersuchungen geräumige Wohnungen gefunden haben; bequeme Treppen führen von einem
Stockwerk zum anderen; Insassen, die vorher einander nicht gekannt hatten, verkehren hier
freundschaftlich miteinander. Manche von ihnen auch sind enttäuscht; sie hatten sich ihre Zukunft
nicht so vorgestellt, manch imponderabler Traum ihrer Jugend war hier nicht erfüllt. Aber es ist ein
freies Haus, in dem sie wohnen, keinem ist es benommen, den alten Traum zur Erfüllung zu bringen;
und es ist festgefügt, daß niemand es mehr einzureißen vermag."

Hilbert is the champion of axiomatics. The axiomatic attitude seemed to him one of universal significance, not only for mathematics, but for all sciences. His investigations in the field of physics are conceived in the axiomatic spirit. In his lectures he liked to illustrate the method by examples taken from biology, economics, and so on. The modern epistemological interpretation of science has been profoundly influenced by him. Sometimes when he praised the axiomatic method he seemed to imply that it was destined to obliterate completely the constructive or genetic method. I am certain that, at least in later life, this was not his true opinion. For whereas he deals with the primary mathematical objects by means of the axioms of his symbolic system, the formulas are constructed in the most explicit and finite manner. In recent times the axiomatic method has spread from the roots to all branches of the mathematical tree. Algebra, for one, is permeated from top to bottom by the axiomatic spirit. One may describe the role of axioms here as the subservient one of fixing the range of variables entering into the explicit constructions. But it would not be too difficult to retouch the picture so as to make the axioms appear as the masters. An impartial attitude will do justice to both sides; not a little of the attractiveness of modern mathematical research is due to the happy blending of axiomatic and genetic procedures. (Weyl 1944, 645)[102]

It seems likely—and it is at least arguable—that Hilbert could have fully endorsed this quotation. And yet, irrespective of Hilbert's own views, it is clear that some trends already present in his work were instrumental in giving rise to the structural approach in algebra. The elaboration of such trends started immediately after the publication of the *Grundlagen*. Without attempting to present a comprehensive account of such developments, the next chapter will discuss some of them. But first, in the closing section of this chapter we discuss one of the most immediate reactions to Hilbert's work on axiomatics, with later influences on the development of structural algebra.

3.5   Postulational Analysis in the USA

The *Grundlagen* influenced many branches of mathematics, not only algebra, and sometimes in ways not initially envisaged by Hilbert himself. If we explore the deep transformations undergone by the whole discipline of topology, for instance, at the turn of the century, we will must likely find clear evidence of such an influence. An important instance of this, that must be briefly

---

102. The accepted identification of Hilbert's work with axiomatic formalism is interestingly manifest in the fact that this well-known passage of Weyl is usually quoted half-way, while its second part ("I am certain that ...") is ignored. Cf., e.g., Mehra 1974, 66 (ftn. 87a).

mentioned here, appears in the topological sections of Felix Hausdorff's *Grundzüge der Mengenlehre*.[103] This is perhaps the earliest, significant manifestation of the trends that would dominate the discipline over the new century, and that may perhaps justifiedly be dubbed "structural" in the same sense intended here for algebra. The *Grundlagen* had an enormous impact on Hausdorff from very early on, especially because of the new view of geometry that, in Hausdorff's opinion, it embodied: geometry as a fully autonomous discipline, independent of any kind of *Anschauung* or empirical basis. In fact, Hausdorff was among the first to consistently develop "formalist" views under the influence of the *Grundlagen*, and his work on topological issues was significantly guided by such a conception.[104]

Anothe specific context in which the impact of Hilbert's work, and especially of his work on axiomatics, was strongly felt was in the American mathematical research community, which at the time had only recently been established. This is particularly the case among mathematicians associated with the University of Chicago under the leadership of E.H. Moore. A singular kind of work produced under this impact—which has sometimes been known as "postulational analysis"—illustrates the direction in which Hilbert's original views could be further elaborated. This kind of work is also of particular interest for understanding the developments discussed below in Chapter 4, and therefore it deserves a more detailed consideration here.

The University of Chicago, and its mathematics department, initiated activities in 1892. Although it was not the first American higher-education institution committed to outstanding original research as a top priority, it was to be the first one to succeed in establishing a long-lasting tradition of significant mathematical research and training of future generations of scientists.[105]

---

103. Hausdorff 1914.

104. Purkert 2002 contains unpublished sources dating from the early 1900's were such "formalist" views are clearly and interestignly stated. I am indebted to Walter Purkert for providing me a draft of his forthcoming text.

105. For a detailed account of the rise and consolidation of the American mathematical research community between 1876 and 1900, see Parshall & Rowe 1994. The scientific and institutional contributions of three mathematicians stand at the center of this account: James Joseph Sylvester, and the creation of the department of mathematics at Johns Hopkins University (1876-1883); Felix Klein and the mathematical education of a whole generation of young American mathematicians in Germany (1884-1893); and Eliakim Hasting Moore and the creation of the department of mathematics at the University of Chicago (1892-1900). On the latter, see especially Parshall and Rowe 1994, Chpt. 9.

Eliakim Hastings Moore (1862-1932)[106] was the man chosen to build the mathematics department and to transform it into a leading scientific center.

In the decades preceding the creation of the Chicago department, the standard track opened to Americans seeking a higher degree in mathematics took them to German universities and, above all, to Göttingen. The success of the Chicago department was to be measured, to a large extent, by its ability to offer an appealing alternative to this established practice. Moore, working together with two other faculty members, Oskar Bolza (1857-1942) and Heinrich Maschke (1853-1908), soon created a dynamic department that was indeed able to attract many young and gifted graduate students. By the turn of the century its reputation was more than firmly established.

Moore's research interests in the earlier phase of his career concentrated on algebraic geometry and group theory.[107] But in 1901, under the impact of Hilbert's *Grundlagen*, a completely new area of interest opened for him, as well as for many of his students and collaborators. This interest led to the development of postulational analysis in the USA.

In the fall of 1901 Moore conducted a seminar on Hilbert's *Grundlagen*. Special attention was accorded to the possibility of revising the proofs of independence between the axioms of Hilbert's various groups. One step in this direction had already been taken by Friedrich Schur in an article published in 1901. Schur claimed that the axioms of connection and of order taken together contained a logical redundancy that Hilbert had not pointed out. In pursuing this kind of question Schur was simply going one step further beyond Hilbert's own analysis of the logical structure of geometry, yet in a direction and to an extent not formerly contemplated by Hilbert himself. As already said, Hilbert was interested in the interrelation among the *groups* of axioms, in which he saw the isolable facts of our spatial intuition—rather than on the *individual* axioms. In his seminar Moore followed the direction indicated by Schur, and discussed the latter's article with his students. Moore realized that, although Schur's criticism was essentially correct, the latter's more specific identification of the redundancy presumably contained in Hilbert's system of axioms was not accurate. Moore proved that this redundancy involved, in fact, one of the axioms of connection and one of order. Moore's results were published in

---

106. Details on Moore's work and of his role in the rise of the mathematical community in America appear in Bliss 1933; Parshall 1984, esp. 321-328; Parshall & Rowe 1994, 279-286 & 363-392; Siegmund-Schultze 1998a.

107. For details on Moore's works see Parshall & Rowe 1994, 372-378.

the 1902 volume of the *Transactions of the American Mathematical Society*.[108] Moore's article betrayed a certain degree of incertitude concerning the real scope and meaning of the mathematical activity involved in it. This incertitude creates the feeling of a noteworthy, latent tension aroused by the interplay between the various traditions manifest in the article. Thus Moore wrote:

> Clearly the body of axioms of a system depends essentially upon the choice of the basal notions of the system. In this connection a remark is pertinent with respect to one's attitude concerning the foundations of geometry. I suppose that if geometry [footnote: for $n = 3$. In case $n > 3$ the geometry is perhaps essentially abstract] is taken to be a natural science—the science or a science of space in which we live—it would, as contended by Pasch and Peano, be undesirable to introduce the line as the basal notion. But we may indiscriminate between that part of geometry which establishes a body of postulates based as directly as may be on spatial experience or intuition, and that part of geometry which consists in the organization of science on the basis of the accepted body of axioms; and so we understand that it may in the development of the theory be convenient to replace the body of primary notions and relations by another body of notions or relations, less fundamental, but, with respect to the deductive geometry, more convenient. (Moore 1902, 144)

The perspective adopted by Moore—following Schur and in opposition to Hilbert—would stress the elaboration of the body of postulates so as to formulate them more conveniently from the "deductive" point of view, while abandoning spatial intuition as the primary criterion for their choice and treatment. This perspective implied a subtle shifting of the focus of interest away from geometry as the science of space and towards the study of the system of axioms as an issue of inherent interest.

Pressing this perspective just a bit further, the Harvard mathematician Edward Huntington (1847-1952)[109] took the next step, that in retrospect would seem to have been unavoidable. As suggested in the former section, the kind of analysis applied by Hilbert to the axioms of geometry could in principle have been applied as well to any other collection of abstract postulates, although Hilbert's actual motivations did not contemplate the realization of that possibility. In an article published in the *Bulletin of the American Mathematical Society*, Huntington analyzed two axiomatic definitions of abstract groups: the one advanced by Weber in his 1893 article (§ 1.2 above), and a

---

108. Moore 1902. For more details on this, see Parshall & Rowe 1994, 382-387.
109. For a short biography of Huntington see Boas 1974.

second one proposed by William Burnside (1852-1927) in his 1897 book on finite groups.[110] As Huntington remarked, no such analysis had been previously done, and the accepted systems of postulates defining groups contained many redundancies.[111] To his analysis Huntington added two proposals of his own for defining abstract groups, using systems of postulates presumably irredundant. E.H. Moore soon published his own study of the two systems originally analyzed by Huntington, as well as of the new two systems proposed by the latter.[112]

The systems of postulates for groups and for geometry were theretofore seen—at least tacitly—as belonging to different conceptual categories. The latter was meant to provide a solid basis to the science that elaborates our intuitions of space. The former simply defined an abstract concept with useful applications in various mathematical domains. The articles of Moore and of Huntington manifested a willingness to analyze them from a common perspective, and therefore to ascribe them identical conceptual status. The noteworthy significance of this elusive change was underscored by the uncertainty expressed in Moore's passage quoted above. Hilbert's original motivations in writing the *Grundlagen* not only did not contemplate such a view: as far as we know, Hilbert did never express a direct interest in the perspective opened by the application a similar analysis to arbitrary systems of abstract postulates. But be Hilbert's putative position on this question what it may, Moore's and Huntington's articles clearly opened the way for new, similar analyses of systems of axioms for additional mathematical domains.

Several other American mathematicians soon followed suit. E.H. Moore's first doctoral student and later colleague at Chicago, Leonard Eugene Dickson (1874-1954), himself a distinguished group-theorist, published his own contributions on the postulates defining fields, linear associative algebras, and groups.[113] Oswald Veblen (1880-1960) was foremost among Moore's students to pursue this trend of research wholeheartedly. In fact, Veblen's inter-

---

110. Burnside 1897. On the significance of this book for the development of group theory, see Wussing 1984, 243 & 251-252.

111. Huntington 1902, 296. Huntington 1902 analyzes Weber's definition while Huntington 1902a analyzes Burnside's. It should be pointed out that Klein—although far from having developed the kind of axiomatic approach elaborated by Hilbert and implemented here for the first time for groups—recognized the merits of the axiomatic method precisely in cases such as that of group theory, in which he saw the axioms as having emerged organically from within an elaborate mathematical theory. On Klein's attitude toward axiomatics see Rowe 1994, 192-195.

112. In Moore 1902a.

ests in logic and foundational research, stemming from his early work on postulational analysis, became some thirty years later one of the moving forces behind the rise of Princeton as a leading center for research in logic.[114] In 1903 Veblen completed his dissertation in Chicago. He presented in it a new system of axioms for geometry, using as basic notions point and order, rather than point and line. Veblen proved that his axioms were complete and independent. His results were also published in the *Transactions* in 1904.[115] He was joined in the same direction by another distinguished doctoral student of E.H. Moore's, Robert Lee Moore (1882-1974).[116]

Works on postulational analysis continued to appear in the leading American journals, mostly in the *Transactions* but also in the *Bulletin of the American Mathematical Society*. According to the statistics of E.T. Bell,[117] 8.46% of the mathematical papers written in the USA between 1888 and 1938 were devoted to postulational analysis. In the index to the first ten volumes of the *Transactions* (1900-1909), works on postulational analysis appear under the classificatory heading of "Logical Analysis of Mathematical Disciplines." The disciplines whose systems of postulates were analyzed include "real and complex algebra", groups, fields, algebra of logic, and geometry (especially projective geometry). The same heading appears in the index to the next ten volumes of the *Transactions* although the number of works included under it is considerably lower. In those years (1910-1919) the main issues of research were the postulate systems for Boolean algebra and for geometry (which included projective geometry and "analysis situs"). This change of emphasis

---

113. Dickson 1903, 1903a, 1905, respectively. On Dickson and his mathematical work, see Albert 1955.

114. For Veblen's contribution to the development of the Princeton department see Aspray 1991. In particular, Veblen's contribution to postulate analysis is considered there in pp. 55-56. A comparison of Hilbert and Veblen regarding their respective approaches to axiomatics appears in Scanlan 1991, 989-997. See also Montgomery 1963; Moore 1991, 7-8; Parshall & Rowe 1994, 438-439.

115. Veblen 1904. Veblen published additional works in this trend: see Veblen 1905; 1906; Veblen & Wedderburn 1907. According to Hourya Sinaceur's account (Sinaceur 1991, 228) Veblen 1905 advanced many of the central ideas that were later developed in Artin & Schreier's characterizations of real fields. Artin and Schreier, as well as van der Waerden in 1930, were probably unaware of this article of Veblen's.

116. On R.L. Moore's life see Wilder 1976. The first work of Moore on postulational analysis appeared under the name of his teacher at Austin, Texas, George Bruce Halsted, (Halsted 1902). More significant, however, was his contribution in Moore 1908.

117. Bell 1938, 6.

in the disciplines investigated indicates that standard postulate systems for algebraic disciplines had already been adopted in algebraic domains in the first decade of the century as part of this trend.

The analysis of these basic notions—sometimes called postulates, sometimes axioms—is similar in all the articles on postulational analysis. All postulate systems are required to comply with the guidelines established by Hilbert in the *Grundlagen*. These requirements are explicitly stated again in each new paper. Thus, it is required that the postulates be independent and consistent.[118] If there exists more than one possible system for a given mathematical branch, equivalence among different systems should be proved. Some articles also introduced new concepts and ideas into postulational analysis, that appeared neither in the *Grundlagen* nor in previous works of the same kind. Some of these ideas were to become significant for future developments of foundational research, others led nowhere. To the first kind belongs the concept of "categoricity", first introduced by Veblen in 1904. Veblen called a system of postulates categorical if the addition of some new axiom necessarily renders the system redundant; otherwise the system is called disjunctive.[119] Veblen apparently believed (wrongly so) that Hilbert had investigated the categoricity of the system in his analysis of the postulates of geometry.[120] An additional requirement which was absent from the *Grundlagen* and was first introduced in this trend—but which contrary to categoricity was not to remain central to the study of postulational systems—is that the postulates be "simple." This requirement, however, was seen as somewhat problematic from the outset, since "the idea of a simple statement is a very elusive one which has not been satisfactorily defined, much less attained."[121]

Research on postulational analysis evolved into a specialized discipline, particularly in the USA. All articles of this trend include constant references to previous, related works. They also included brief remarks explaining the

---

118. But Veblen 1904, for instance, does not discuss consistency of the system of postulates.

119. Veblen 1904, 346. Veblen added in a footnote that the terms had been suggested to him by the philosopher John Dewey (1859-1952). A similar idea had already been discussed in Huntington 1902b.

120. Cf. Scanlan 1991, 994.

121. Huntington 1904, 290. See also Moore 1905, 179. As was seen above, in his 1905 lectures Hilbert himself will also introduce the requirement of formulating as simple axioms as possible. This requirement was explicitly stated for the first time by a member of the Hilbert school (as far as I know) in Schimmack 1903, 317: "Ein Axiom soll in mehrere zerlegt werden, wenn sich aus ihm allein schon ein wesentlicher Schluß ziehen läßt."

relative advantages and merits of the particular new system proposed in them, as compared to the preceding ones. Thus they constitute a long series of interconnected articles with small variations from one to the next. The bulk of the work is usually invested in defining models, specifically conceived to prove the independence of the individual postulates comprising the systems. Often, several abstract concepts are treated in a single paper; most of the papers on fields, for example, open with a discussion on groups.

The motivations behind these works on postulational analysis involved a twofold interest: on the one hand, the most fitting postulate systems were sought, which could be used as a starting point for coherent research of a particular mathematical discipline. In such cases, the discipline in question (group theory, projective geometry, etc.) was at the center of interest, while the system of postulates was only a subordinate tool, meant to improve research in the former. Hilbert's own foundational research, as was seen in the preceding section, is the foremost example of this tendency. But on the other hand, a new, autonomous mathematical activity was developing, which focused on the postulate systems themselves as an object of inquiry. The postulate systems in question were taken from different branches of mathematics, but now the latter provided only an excuse for dealing with the former. Postulational analysis was not concerned with the specific problematic of the system of complex numbers, of the continuum, or of the abstract theory of groups, but rather with the various postulate systems proposed as alternative definitions for them.

The accumulated experience of research on postulational analysis brought about an increased understanding of the essence of postulational systems as an object of intrinsic mathematical interest. In the long run, it had a great influence on the development of mathematical logic in America, since it led to the creation of model theory.[122] But as a more direct by-product of its very activity, postulational analysis also provided a collection of standard axiomatic systems that were to become universally adopted in each of the disciplines considered. By affording standards for axiomatic *definitions* of various mathematical branches, it provided the natural framework within which abstract, structural *research* on those issues was later to proceed. This was indeed the case, as will be seen in the next chapter, with the definition and early structural research of rings. These arose in Germany in direct connection with the elab-

---

122. See Scanlan 1991, 996-999.

oration of techniques introduced in postulate analysis, as practiced in the USA in the first two decades of the century.

These works on postulational analysis had noticeable connections with the images of algebra as well. On the one hand, they meaningfully affected those images in the long run, by laying increased stress on the similarly formulated abstract definitions of what came to be seen as the various algebraic structures. This point is discussed in greater detail in the next chapter, with particular attention to fields and abstract rings. On the other hand, publications related to postulational analysis offer interesting glimpses into contemporary images of algebra. This is specifically true concerning the articles of Edward Huntington, who was probably the mathematician with the largest number of articles published on postulational analysis in the USA. Beyond the above mentioned analysis of the axioms of a group, Huntington also published later additional work on the postulates for the algebra of logic, the systems of real and complex numbers, the continuum, and several others.[123] Huntington had a respectable position among American mathematicians, but at the same time, he is perhaps—from among those involved in the trend—the one whose overall mathematical contribution can be considered the least impressive and the least original. One can thus assume that the images of algebra manifest in his articles are representative, at least to a considerable extent, of those shared by the majority of his contemporaries, rather than idiosyncratic.

Especially relevant for the present discussion is the way in which the "concrete", classical entities of mathematics appear intermingled in Huntington's articles, in the analysis of the various systems of postulates. Algebra appears invariably to be based on the study of the real and complex numbers, while the aim of the analysis of its postulate systems is to provide a solid foundation for those two number systems. Although interest in groups appears sometimes as an intrinsic one, usually it does so as subsidiary to interest in the system of real numbers. Thus for instance, in the following passage:

> The well-known algebra which forms one of the main branches of elementary mathematics, is a body of propositions expressible in terms of five fundamental concepts—(the class of complex numbers, with the operations of addition and of multiplication, and the subclass of real numbers, with the relation of order)—and deducible from a small number of fundamental propositions, or hypotheses. (Huntington 1905b, 209)

---

123. Huntington published articles in postulational analysis until as late as 1938. See the list of his contributions in the bibliography at the end of this book.

The striving after a clear separation of the different abstract features (order, algebraic properties, density, etc.), characteristic of the "classical mathematical entities", like the real numbers, became a central concern of modern, structural algebra, especially as embodied in van der Waerden's textbook (§ 1.3 above). Hilbert's introduction of the five groups of axioms reflects a similar attitude concerning geometry. In Huntington's article, in spite of the familiarity with which he handles the technical aspects of postulational analysis, such a separation for the case of algebra, as the above quotation shows, is still absent. As a matter of fact, the separation gradually emerges as postulational systems develop. One of the improvements which several authors explicitly mention as innovative in their successive analyses of particular systems is precisely a new separation attained since the preceding article on the same issue. Although the concept of "simple" postulate is not a definite one, the search for "simpler" postulates appears as a central motivation leading to the separation of abstract characteristics of mathematical entities. The most important case of this is the separation of the axioms of arithmetic and the axioms of order in the system of real numbers.[124]

Moreover, a certain ambiguity is manifest here regarding the question whether the real and complex numbers are to be considered through the "genetic" point of view or through the "abstract" point of view. In other words, the question arises whether or not the existence of any system of numbers is to be assumed as a starting point for algebra. In the works on postulational analysis considered here, the issue is not always explicitly addressed and there are, of course, no definite answers to it. But the minor changes appearing from one paper to the next, and the ambiguous use of some basic concepts in its treatment, bear witness to a definite interest in these kinds of question and to a search for a more clearly articulate stand on them.

In addition, foundational concerns are also manifest in these works, as doubts are stated concerning the legitimacy of the basic notions, on which the whole axiomatization project is built. Echoes to such concerns we find in the following quotation:

> It should be said, in conclusion, that no attempt is here made to give a metaphysical analysis of the concepts *class*, *operation*, and *relation*, on which algebra is based, or of the *laws of deductive logic* by which its propositions are deduced—

---

124. Cf. e.g. Huntington 1905, 18. Similar considerations are mentioned in Blumberg 1913. This particular separation had already been introduced by Hilbert in the *Grundlagen*, chapter 3.

the discussion of these more fundamental notions, here assumed as familiar, being matter for the trained student of philosophy. I hope, however, that a paper like the present may be of indirectly service on the philosophical side of the subject, by enabling one to formulate very precisely the problem involved in the question: *What is algebra?* (Huntington 1905b, 211. Italics in the original).

Thus, these articles help understanding the status of axiomatic formulations and the place of abstractly defined concepts in algebra. In spite of the fact that Huntington's articles deal with the very definition of concepts like groups, fields, and the algebra of logic, in abstract terms, they imply no overall change concerning the basic assumptions and the scope and aims of the discipline of algebra. In particular, nothing similar to an idea of algebraic structure is implied or immediately derived from the abstract treatment of the concepts, and the classical dependence of algebra on the systems of real and complex numbers—considered as given beforehand—seems even to be reinforced.

To summarize, one can say that postulational analysis was a direct offshoot of the kind of research initiated by Hilbert in his *Grundlagen*, but an offshoot that followed a direction not originally intended by him and which went much farther than Hilbert would have thought mathematically worthwhile. Gradually, the main interest of research in postulational analysis shifted towards the formulation of ever improved postulates systems for individual mathematical domains and away from the implications of this analysis on the study of the theories defined by these systems. The intrinsic interest offered by the study of postulational systems was thus underscored. At the same time, the ongoing confrontation with the technical details involved in the treatment of those systems refined existing perceptions of geometry.[125] It also helped understanding the mathematical significance of the particular abstract concepts analyzed, their relation to the real and complex numbers, and their role in algebra in general. This process was essential for the final shaping of the structural conception of algebra. But as the above discussion shows, by the turn of the century such a structural conception was yet to crystallize.

---

125. The actual contribution of the American postulational analysts to the study of the foundations of geometry is discussed in Torreti 1978, 237-246.

Chapter 4   **Concrete and Abstract: Numbers, Polynomials, Rings**

One obvious, main difference between Dedekind's and Noether's theories of ideals lies in the fact that, whereas the former studies properties of numbers belonging to sub-fields of the field of complex numbers, the latter is framed on an abstract, axiomatic theory built around the concept of rings. Many of the concepts introduced by Dedekind in his work on ideals, such as fields, modules, *Ordnung*, etc., are so closely related—from a formal point of view—to the concept of an abstract ring, that in retrospect one could easily be led to think of the process leading from research using the former concepts to the introduction of the latter as a very straightforward, almost natural step. But when one examines the actual historical process that brought about the formulation and early research on abstract rings, a completely different picture emerges. One sees that this process was rather slow and convoluted. The present chapter discusses Abraham Fraenkel's definition of and early research of abstract rings, and the complex development of ideas immediately preceding it. It also considers some parallel advances in the theory of polynomial forms that completed the necessary background for Emmy Noether's research on abstract rings. As will be seen, the development of the main algebraic disciplines between Dedekind and Noether involved and necessitated not only the addition of new concepts, theorems and proofs: it also implied significant qualitative changes, which modified the images of knowledge. As already stated, a central figure in this process of modification of the images of knowledge—in mathematics in general, and in algebra in particular—at the turn of the century was David Hilbert. His contribution to these developments was discussed in some detail in the preceding chapter. The present chapter opens with a discussion of Kurt Hensel's theory of p-adic numbers and then turns to examine the way in which the theory was reformulated by his student, Abraham Fraenkel. Fraenkel tended to adopt many of the newly introduced images of algebra, which were absent from the work of his teacher.

Of central importance for our account in the present chapter is the publication in 1910 of Ernst Steinitz's abstract theory of fields, which constituted a decisive step on the way to the consolidation and widespread adoption of the structural approach in algebra. Steinitz worked out the abstract formulation of the concept of field, that had been first introduced by Weber. Yet, at variance with the latter, Steinitz not only *formulated* the concept abstractly, he also *investigated* it abstractly.

It will be seen that one of the main sources of inspiration for Steinitz's innovative work came from Hensel's ideas. In turn, Steinitz's work had a definite influence on Fraenkel's images of algebra. But interestingly enough, Hensel himself expressed little interest in the new approach to algebra implied by the work of his followers, such as Steinitz and Fraenkel. Hensel's work was itself more directly connected to Kronecker's tradition of nineteenth century research on algebraic number theory, than it was to what came to be considered as "modern algebra." Thus, Hensel's work has a transitional character which bestows a special interest upon the study of its historical significance; naturally, for lack of space, we will limit ourselves here to a very brief description of it, mentioning only those aspects directly connected to the developments which constitute the main issue of the present book.

As for Fraenkel's works considered here, it should be stated from the outset that—although important in several matters of detail—they cannot be considered among the main shaping forces behind the rise and widespread adoption of modern abstract algebra in general. Moreover, their main contribution to the development of the theory of abstract rings is connected only to the fact that Emmy Noether adopted Fraenkel's definitions in her early works on this subject. Nevertheless, these works provide an illuminating example of the way in which structural concerns subtly penetrated existing mathematical traditions and gradually led to the systematic adoption of the structural image of algebra. These are the reasons for including in the following sections a somewhat detailed study of the works of Hensel, Steinitz and Fraenkel.

4.1   Kurt Hensel: Theory of *p*-adic Numbers

Kurt Hensel (1861-1941) was born in Königsberg and studied in Bonn and Berlin. Among his teachers, those who exerted the deepest influence upon him were Lipschitz, Weierstrass and Kronecker, but he also studied thoroughly the works of Dedekind. In 1901 Hensel accepted a position in Marburg, where over many years he was intensively active in both research and teaching. He also served as editor of the prestigious *Journal für die reine und angewandte Mathematik*.[1]

In 1902 Hensel co-authored with Georg Landsberg (1865-1912) a book on the theory of algebraic functions which became a classic in its field.[2] This book, which the authors dedicated to Dedekind, attempted to summarize the vast amount of research conducted in the field during the preceding four decades, and especially the contributions of Weierstrass, Kronecker, Dedekind and Weber. In particular the influence of Dedekind and Weber's 1882 joint article (see § 2.2.4 above) is pervasive throughout the book; Hensel and Landsberg explicitly pointed out the close connections with the theory of algebraic numbers, that had characterized the latest development of the field presented in their book. And yet, the book presented a very-well balanced combination of Dedekind's and of Kronecker's respective approaches, in the relevant sections. The approach developed by Hensel and Landsberg displays very clearly the close attachment of Dedekind's images of mathematical knowledge to those characteristic of nineteenth-century algebra. It showed the extent to which the ideas contained in Dedekind's work could be further developed, without necessarily leading to the structural approach that later developed in algebra.

But Hensel's name has remained associated—more than with anything else—with his most personal invention: the theory of *p*-adic numbers. Hensel first published his ideas on *p*-adic numbers in an article of 1899: "*Über eine neue Begründung der Theorie der algebraischen Zahlen.*" Over the following years, and especially since his arrival at Marburg in 1901, he continued to develop these ideas systematically. Hensel published his work on *p*-adic numbers in two textbooks: *Theorie der algebraischen Zahlen* in 1908 and *Zahlentheorie* in 1913. Although these two books presented basically the same body of knowledge, several important differences in style and approach between them clearly indicate how structural considerations were increasingly permeating Hensel's mathematical thinking, perhaps inadvertently. The transitional character of Hensel's work on *p*-adic numbers is epitomized in the changes manifest in the successive publication of his two expository textbooks, which we now proceed to discuss.

---

1. For further biographical details on Hensel see Hasse 1949.
2. Hensel & Landsberg 1902.

Dedekind and Weber had applied to the study of algebraic functions ideas originally developed in order to deal with problems of algebraic number theory. In writing his book on algebraic functions in cooperation with Landsberg, Hensel had come to be very well acquainted with Dedekind and Weber's work. In developing his own theory of p-adic numbers, Hensel also combined ideas from these two closely connected disciplines—the theory of algebraic functions and algebraic number theory—but he went in a direction contrary to that followed by Dedekind and Weber. Influenced by Weierstrass's lectures on the representation of complex functions as infinite power series, lectures which he had attended as a student in Berlin,[3] Hensel applied similar ideas to the research of fields of algebraic numbers. In the domain of rational complex functions, linear factors of the form $(z - \alpha)$ possess both a zero (at $z = \alpha$) and a pole (at $z = \infty$). In this sense, the linear factors behave similarly to the quotients $(z - \alpha)/(z - \beta)$. This poses a difficulty when trying to develop a theory of factorization in this domain, since these factors cannot play here the same role of basic building blocks that they play in the theory of factorization of polynomials. From Weierstrass's work, however, Hensel learnt that such factors may be used in a different sense as basic elements of the theory of complex functions. In fact, in the surroundings of a given point, every algebraic or rational complex function can be represented as an infinite series of integer and rational powers of linear factors $(z - \alpha)$. These representations had proven extremely useful in Weierstrass's research, and Hensel expected to exploit their usefulness in order to attain a deeper understanding of the domain of algebraic numbers. According to Hensel, many of the limitations encountered in the study of specific domains of numbers were due to the fact that numbers had traditionally been represented in a single way: through the decimal representation. Any function, on the other hand, can be variously represented as a power series by choosing different points around which to develop it. If functions could only be represented only either around zero, or around an infinitely distant point—claimed Hensel—then one would find in the theory of functions the same limitations hitherto encountered in the study of fields of algebraic numbers. Thus Hensel intended to enlarge the scope of the existing theory, by inquiring into the possibility of providing various alternative representations of algebraic numbers, through suitable "changes of basis."

---

3. On Weierstrass's lectures see Dugac 1973.

A full account of the theory of p-adic numbers would certainly be beyond the scope of the present book. It is nonetheless pertinent to present here some of its basic concepts, in order to understand the historical significance of Hensel's work.[4]

A rational g-adic number is a series

$$A = a_r + a_{r+1}g^{r+1} + a_{r+2}g^{r+2} + ...,$$

where r is any integer and g is a positive integer, and where the a_i's are rational numbers whose denominator (after eliminating common factors with the numerator) has no common factors with g. Addition and subtraction of two rational g-adic numbers are defined in the obvious way. The product AB of the g-adic rational A and a second one, $B = b_r + b_{r+1}g^{r+1} + b_{r+2}g^{r+2} + ...,$ is defined by:

$$AB = a_r b_r g^{2r} + (a_r b_{r+1} + a_{r+1} b_r)g^{2r+1} + ...$$

Division is in general not defined for any two numbers in this domain.

If the a_i's are restricted to be only positive integers between 0 and g - 1, then A is called a *reduced* g-adic number. It can be shown that, for any given g, every rational number can be uniquely represented as a reduced g-adic number.[5] This was the starting point for Hensel's theory, since it afforded—as he wished—the possibility to represent a rational number in several different ways by taking different bases.

If one takes a prime number p as basis, instead of an arbitrary rational g, it is possible to define a division between p-adic numbers. In his first textbook, Hensel began by considering that case.[6] He noticed that the product of two p-adic numbers is zero if and only if one of the factors is zero. By contrast, this is not the case if one chooses a non-prime basis g.[7]

Hensel addressed two subjects in his 1908 textbook. First, the issue of factorization in its many manifestations. The second subject concerned what came to be known later as valuation theory, when developed by Hensel's students Jószef Kürschák (1864-1933) and Alexander Ostrowski (1893-1986).[8] This was essentially a book on number theory which introduced many concep-

---

4. For an account of the theory, in terms similar to those developed by Hensel himself, see Dickson et al. 1923, 59-75. For a more modern approach to the theory, see Koblitz 1979, esp. Chp. 1.

5. See Dickson et al. 1923, 60.

6. Cf. Hensel 1908, 34 ff.

7. Hensel 1908, 28.

tual innovations, but which, at the same time, retained the classical nineteenth-century image of the discipline. Concepts that are seen today as the main core of algebra hardly appear in it.

The concept of group, for instance, is barely mentioned, in the framework of a discussion on Galois theory.[9] Hensel explained that the permutations of the roots of a given equation constitute a group, because the combination of any two of them yields a third permutation. Hensel did not mention any other properties of groups, such as distributivity, the existence of a zero element, or the existence of inverses. Beyond this brief allusion, the concept of group appears again in the last section of the book, but only in the same mathematically superfluous fashion as before.[10]

Similar is the case with ideals, which appear first in Chapter 9 of the book. That chapter and the one preceding it deal with the factorization properties of the algebraic integers. Hensel followed here an approach akin to that introduced by Kronecker in his early work on algebraic integers, an approach based on the concept of "divisors." At a certain stage Hensel considered the collection of all the multiples of a given divisor δ, denoted by $J(\delta)$. Hensel claimed that if one wishes to call that collection by a name, then one can use the name ideal, a name which had been introduced by Dedekind.[11]

A general idea of an algebraic structure is totally absent from Hensel's 1908 book. Even the abstract formulation of concepts is not exploited here in order to attain generality of any kind. This situation changed slightly in Hensel's book of 1913. Hensel's 1908 textbook had been originally conceived as the first of two volumes in which the author would exhaustively present his novel approach to the theory of numbers. However, instead of the second volume, Hensel preferred to write a completely new book in which the theory would be presented in a wider, didactically clearer, and more systematic fashion. In writing this new book, Hensel received significant help from his student Abraham Halevy Fraenkel (1891-1965).

---

8. See Kürschák 1913; Ostrowski 1917. A comprehensive treatment of valuation theory, in modern terms, appears in Jacobson 1980, chapter 9.

9. Hensel 1908, 338-339.

10. Groups appear in a similar way in Hensel & Landsberg 1902. On pp. 106 ff., groups of permutations are defined as any collection of substitutions closed under their product. Later on (p. 499), a group of geometrical transformations is defined in similar terms. Hensel and Landsberg added that in group theory, one distinguishes between finite and infinite groups.

11. Hensel 1908, 237. Also here Hensel was following an approach similar to that of Hensel & Landsberg 1902.

Fraenkel had come to Marburg in 1910 in order to complete his doctoral dissertation under Hensel.[12] Already in 1912 he published his first paper presenting an axiomatic foundation for Hensel's system of p-adic numbers. This paper can be more naturally associated with the American trend of postulational analysis than with Hensel's own line of interests. Coming from a different perspective enabled Fraenkel, as will be seen below, to consider as equally worthy of interest the g-adic numbers (with g an arbitrary integer), previously left aside by Hensel. Fraenkel's work during these years seems to have convinced Hensel of his student's expository and didactic capabilities. Fraenkel actually wrote up almost all chapters of Hensel's second book as they finally appeared in print in 1913.[13] Fraenkel's influence is clearly felt in several changes of emphasis and overall approach manifest in the new exposition of Hensel's theory.

The first chapter of Hensel's 1908 book had discussed the properties of the rational integers in order to provide the conceptual basis for the theory of p-adics. His 1913 book starts from a different point: modern algebraic concepts, such as fields, rings and groups are defined there and some of its immediate properties are deduced. In Weber's *Lehrbuch* the whole of algebra had appeared as conceptually dependent on the properties of the real number system; he therefore opened his book with a detailed account of their properties (§ 1.2 above). By contrast, in van der Waerden's *Moderne Algebra* the image of the discipline will be dominated by the notion of algebraic structure. The latter appears as prior and more fundamental than any other mathematical entity, and thus the system of real numbers is presented at a rather late stage, described in terms of suitable algebraic structures. Thus, the differences between Hensel's two books would seem to encapsulate, on a small scale, the overall passage from the old to the new images of algebra. This parallel, however, is more apparent than real. The changes introduced in Hensel's 1913 book remain on the surface and they influence neither the actual content of the book nor its presentation of the conceptual relationship between systems of numbers and "algebraic structures" of any kind. The algebraic concepts defined in the first chapter of Hensel's 1913 book seem to have been introduced by Fraenkel more as an aid for clear exposition than as an object of intrinsic interest. The book reflects a kind of compromise between Hensel and

---

12. The dissertation was published as Fraenkel 1914.
13. Cf. Fraenkel 1967, 110-111.

Fraenkel, as a result of which the modern concepts were used as a shorthand allowing a handy reference to certain properties of the systems considered in the discussion.

The algebraic concepts presented in Hensel's 1913 book follow closely Hilbert's definitions and his number-theoretic orientation (§ 3.2 above). Thus, the arithmetic of rational numbers is introduced by formulating seven basic properties, from which all the others may be logically derived. These seven properties are, in fact, axioms satisfied by the field of rational numbers. A field is thus defined as a domain on which two operations are abstractly postulated, satisfying these seven properties. The concept of field, it is stated, will dominate the forthcoming discussion on the theory of numbers; this is one among several concepts introduced as of late into arithmetic and—according to the opinion stated in the book—the most important among them. Additional algebraic concepts, which are defined in the sequel are presented as subsidiary to the concept of field.

A module is defined as a domain endowed with an operation of addition, which satisfies three conditions: associativity, commutativity, and the existence—for any given pair of elements a,b in the module—of a unique element x, such that $a + x = b$. Some basic results are derived from those assumptions, and several additional basic concepts are also introduced, such as basis and module generated by several given elements.

Hensel now defined groups in a rather noteworthy fashion. He translated word-by-word his own definition of module, changing only "addition" and replacing it with "multiplication." Naturally the above mentioned basic results and additional concepts introduced for modules are obtained *mutatis mutandis* for groups. What is then the difference between the two concepts, module and group? From the context it becomes clear that a field is considered a module with respect to the first of its operations (or addition) and a group with respect to the second (or multiplication). Thus, Hensel explained that if a group contains a non-zero element, then all of its elements are also non-zero (p. 11). Obviously this mentioning of a zero element (the neutral element with respect to multiplication being called 1) makes no sense if the group is abstractly considered. If the elements of the group, on the contrary, are actually members of a field, then there is a zero in the field which must be left out when considering the multiplicative group of that field.

There is a further detail in the way groups are defined, that helps us understand Hensel's own conception of those abstract algebraic concepts which appear in his treatment of number theory. The book mentions the existence of two alternative denominations for the concept in question: groups, following Weber, and *Strahlen* following Fueter. We have already commented on Fueter's number-theoretical work and the images of mathematics manifest in it (§ 3.2. above). After taking his doctorate in Göttingen, Fueter habilitated in 1905 in Marburg with Hensel.[14] Fueter had provided advice to Hensel on his previous book, and had even helped with the proof-reading. Fueter's nomenclature, and his distinction between groups, modules and *Strahlen*, were well-suited to Hensel's own conception of the relationship between the various algebraic concepts and the systems of real and complex numbers. Hensel and Fueter worked within a frame of ideas provided by the classical images of mathematics implied in number-theoretical research of the late nineteenth century, as epitomized in Hilbert's *Zahlbericht*.

The first chapter of Hensel's 1913 book closes with the definition of rings. The term is reportedly taken from Hilbert and it denotes a field in which division is not always warranted (p. 13). Some basic facts about rings are proven and the procedure for building a field of quotients out of two rings is also introduced. In fact, Hensel's rings are integral domains, namely, domains in which non-zero elements cannot have zero product.

Hensel's 1913 book includes many of the central concepts of modern algebra, which were absent from his first one. But the role accorded to them as part of the general conception of the discipline, and the interrelations among these concepts, are rather ambiguous. Hensel himself did not advance a consolidated idea of algebraic structure, nor did he seem to have been interested in anything similar to it. Yet his own work and the specific ideas he developed within the body of knowledge were instrumental in advancing still further the trends that eventually led to the full-fledged development of the structural image of algebra. Two works, that were written under the explicit influence of Hensel's ideas, represent crucial crossroads on the way to consolidating the idea of algebra as a science of structures: Steinitz's theory of abstract fields and Fraenkel's theory of abstract rings. We now proceed to analyze them in some detail.

---

14. See Frei & Stammbach 1992.

4.2   Ernst Steinitz: *Algebraische Theorie der Körper*

A major milestone in the way to the overall transformation of the images
of algebra at the beginning of the twentieth century was the publication in
1910 of Ernst Steinitz's work on the theory of abstract fields: *Algebraische
Theorie der Körper*. In this work, Steinitz presented an exhaustive account of
the results of the theory to that date and, at the same time, opened new avenues
for research on any abstractly formulated algebraic concept. The innovative
character of this work has been repeatedly stressed and its deep influence on
the future development of algebra has been remarked upon insistently.[15] In
fact, it has also been considered by many as the "cradle of modern algebra."[16]
Naturally, one should be careful when evaluating these kinds of appraisals,
since this distinguished title has been attributed with similar enthusiasm to
many other important papers and books, from Galois's pioneering work, on
the one hand, to *Moderne Algebra*, on the other.

We have seen that the concept of field had already been formulated in a
completely abstract fashion several years before Steinitz's work, by Weber and
by others. Likewise, many results regarding fields were well-known among
Steinitz's colleagues, and the abstract theory of groups had been established as
an independent research discipline. Thus, the very fact of considering an
abstract concept as worthy of inquiry was not really new. What was, then, so
innovative in Steinitz's paper?

The importance of Steinitz's paper lies in the new image of knowledge it
suggested for the domain of its inquiry, which was later adopted for algebra at
large. Steinitz's contribution cannot be summarized in a single definition or
result introduced by him, or even in a collection of definitions and results, but
rather in the program of research that his article exemplified and encouraged.
It is true that fields had been *defined* in abstract terms before Steinitz, but
Steinitz was the first to present an *abstract investigation* of fields. Moreover,
Steinitz specified what kind of questions are to be pursued in order to do this
kind of research and what kind of legitimate and interesting answers are to be
expected.

---

15. See, e.g., Mehrtens 1979; 145; Purkert 1971, 20; Sinaceur 1991, 200-206.
16. Cf. Artin & Schreier 1926, 85; Bourbaki 1969, 109; Scholz (ed.) 1990, 405; van der Waerden 1972, 244; Wussing 1989, 281. Cf. also the preface of H. Hasse and R. Baer to the 1930 re-edition of Steinitz 1910.

Ernst Steinitz (1871-1928) completed his doctoral dissertation in 1894 in Breslau and habilitated in 1897 at the Charlotenburg technological institute in Berlin. He was professor since 1910 in Breslau, and from 1920 until his death in Kiel. The direct source of inspiration for Steinitz's elaboration of abstract fields was Hensel's work on the theory of numbers. In the introduction to his 1910 article Steinitz wrote:

> I was led into this general research especially by Hensel's *Theory of Algebraic Numbers*, whose starting point is the field of p-adic numbers, a field which counts neither as the field of functions nor as the field of numbers in the usual sense of the word.[17]

This testimony stresses the significant fact that, in spite of the existence of abstract formulations of the concept, the only fields considered by algebraists before Steinitz were particular fields of numbers or fields of functions. Steinitz was certainly well-acquainted with Dedekind's and Hilbert's works, but he did not learn from them that there is a special mathematical interest in the abstract treatments of fields; this he learnt only from Hensel's work. Steinitz was a personal friend of Hensel and the two had close professional contacts. Steinitz had the opportunity to discuss at length Hensel's theory of p-adic numbers directly with its creator. This work confronted Steinitz with a completely new instance of field, one which was neither the typical field of numbers nor the typical field of functions. This was the ultimate example that led him to reflect upon the value of an abstract, structural investigation of this kind of mathematical entity.

Steinitz opened his article by explicitly stating his methodological outlook: he announced what might be called today a "structural research program", albeit only for a particular algebraic domain, i.e., for the study of fields. His subject matter would be abstract fields as these were defined by Weber in his article of 1893. However, his research would diverge from that of Weber in an important sense. In Steinitz's own words:

> Whereas Weber's aim was a general treatment of Galois theory, independent of the numerical meaning of the elements, for us it is the concept of field which represents the focus of interest.[18]

---

17. Steinitz 1910, 5: "Zu diesen allgemeinen Untersuchungen wurde ich besonders durch Hensels Theorie der algebraischen Zahlen (Leipzig, 1908) angeregt, in welcher der Körper der p-adischen Zahlen den Ausgangspunkt bildet, ein Körper, der weder den Funktionen- noch den Zahlenkörper im gewöhnlichen Sinne des Wortes beizuzählen ist."

Considering an *abstract concept* as the focus of the inquiry is a central element of Steinitz's approach. This in itself, however, would not have signified a real innovation, since it was already being done for groups—to a considerable degree at the least—starting three decades earlier. Steinitz's innovation in this sense was to state exactly which questions are the important ones to be asked in the framework of such an abstract inquiry. He put this in the following words:

> The aim of the present work is to advance an overview of all the possible types of fields and to establish the basic elements of their interrelations. [19]

Steinitz also explained in detail the steps to be followed in order to attain this aim. First, it is necessary to consider the simplest possible fields. Then, one must study the methods by which, from a given field, new ones can be obtained by extension. One must then find out which properties are preserved when passing from the simpler fields to their extensions. Here we find an illuminating example of the kinds of procedures that would later come to be central for any structural research in mathematics, explicitly formulated and effectively realized for the first time.

A central tool for the study of the simplest fields is the characteristic, whose importance Steinitz claimed to have realized while studying Kurt Hensel's work on p-adic numbers. Weber, it should be recalled, had not envisaged in his work the possibility of fields of characteristic other than zero. Steinitz showed that any given field contains a "prime field" which, according to the characteristic of the given field, is isomorphic either to the field of rational numbers or to the quotient field of the integers modulo p (p being a prime number). Then, after thoroughly studying the properties of these prime fields, Steinitz proceeded to classify all possible extensions of a given field and to analyze which properties are transferred from any field to its various possible extensions. Since every field contains a prime field, by studying prime fields, and the way properties are passed over to extensions, Steinitz would attain a full picture of the structure of all possible fields.

---

18. Steinitz 1910, 5: "Während aber bei Weber das Ziel eine allgemeine, von der Zahlenbedeutung der Elemente unabhängige, Behandlung der Galoisschen Theorie ist, steht für uns der Körperbegriff selbst im Mittelpunkt des Interesses."

19. "Eine *Übersicht über alle möglichen Körpertypen zu gewinnen und ihre Beziehungen untereinander in ihren Grundzügen festzustellen, kann als Programm dieser Arbeit gelten.*" (*loc. cit.*, Italics in the original)

An illustrative example of the kind of results proven by Steinitz in this context is provided by his theorem on "algebraically closed" extensions. An extension K' is said to be algebraically closed if every polynomial with coefficients in K' factorizes into linear factors. Steinitz proved that for every field K there exists an algebraically closed extension A (called the "algebraic closure" of K): A is uniquely determined up to isomorphism, and it is not properly contained in a larger algebraically closed field. Since every field contains a prime field, and since all the prime fields have been completely classified, one has much information about all those algebraically closed fields which are algebraic closures of the different prime fields. Having proved this theorem, one can attach a more concrete meaning to the claim that one has "elucidated the structure" of the algebraic extensions of fields.

Another remarkable feature of Steinitz's work is the peculiar way in which set-theoretical considerations were introduced into the arguments of the proofs. By the time of publication of Steinitz's article, Dedekind's and Cantor's early works on set theory were already well-known, and so was Zermelo's more recently published work on the axiom of choice. Steinitz was aware of the problematic status of the axiom and he thus took pains to separate all those proofs whose arguments depended on the axiom. Thus the axiom of choice appears only towards the end of the work; there the well-ordering of certain sets is necessary for proving the above mentioned theorem on the existence of the algebraic closure. A clear understanding of the specific role of set-theoretical considerations was a necessary step on the way to the new image of algebra, and Steinitz was indeed among the first to make use of them in a conscious and systematic fashion in the framework of algebraic research.[20]

Now, it has already been said that groups had been abstractly investigated before Steinitz presented his own structural program for the research of abstract fields. Is there, nevertheless, any specific innovation in Steinitz's work, compared to what had been attained in group theory, which may justify the claim that this was the "cradle of modern algebra"? Without attempting to summarize in a short paragraph a comparative account of the rise of the theories of groups and fields, we can refer here to our account up to this stage in order to answer this question. Whereas the abstract theory of groups had been motivated, at least to a great extent, by the study of systems which were from

---

20. For a more detailed discussion of Steinitz's attitude towards the axiom of choice, see G.H. Moore 1982, 171-174.

the outset *non-numerical domains* (in particular permutations and continuous transformations),[21] the study of fields arose almost exclusively from the study of numerical systems. Thus, besides providing a second, meaningful instance of abstract research of a mathematical entity of this kind, there was a fundamental feature of the structural program which did not became evident from the achievements of group theory, but did arise from the abstract study of fields. The study of fields suggested that it is not necessary to start from the various systems of numbers with their known properties in order to secure the conceptual foundations of algebra. Rather, one can go in the opposite direction: as far as the study of polynomial forms is concerned—in particular the question of the solvability of polynomial equations and the relationship between the roots and the coefficients of a polynomial—the number systems may be seen as specific manifestations of a more basic kind of mathematical entities. This was in essence the implication of the new perspective suggested by Steinitz's work.

But although Steinitz imprinted a peculiar, structural spirit on his work on abstract fields, a full-fledged articulation of the structural image of *algebra at large* was still to come. It was still necessary, in order to produce a definite change in the images of algebra, that the approach followed here in studying this particular issue, would be extended to all of algebra. Such a step, which might seem in retrospect as obvious, was in fact taken only in 1930 in van der Waerden's book. The passage from Steinitz to van der Waerden was is no way a trivial or an immediate one. Two significant, intermediate stages between these two works were taken with Emmy Noether's general theory of factorization in abstract rings, and, prior to it, with Fraenkel's early definition and research of this latter entity, which was directly modelled after Steinitz's. Before finally describing Fraenkel own contribution, we will briefly examine a last work which preceded and prepared the way for it.

4.3   Alfred Loewy: *Lehrbuch der Algebra*

Fraenkel helped Hensel in writing his 1913 book and it is likely that he was responsible for the approach adopted in it. Fraenkel also collaborated in the writing of yet another textbook, this one by his uncle, Alfred Loewy (1873-

---

21. Wussing 1984 bases his authoritative account of the rise of the abstract concept of group on developments arising from *three* main sources: permutations, continuous transformations and the theory of numbers.

1935). Loewy studied in Breslau, Berlin, München and Göttingen between 1891 and 1895, receiving his doctorate in München in 1894. He taught in Freiburg, where he was *Privatdozent* since 1897 and *ausserordentlicher Professor* since 1902. Loewy was a prolific researcher who published in fields as diverse as algebra, group theory, differential equations and actuarial mathematics. However, given his Jewish descent (he was in fact a professed, Orthodox Jew), it was not before 1919 that he became a full professor. In 1933 he was forced into retirement by the Civil Servant Law introduced by the Nazi government.[22]

Loewy exerted a decisive influence in shaping Fraenkel's early academic career; for one, it was Loewy who induced Fraenkel to travel to Marburg to study under Hensel. Moreover, Fraenkel derived his early interest in the study of axiomatic systems from his uncle. Loewy's *Lehrbuch der Algebra* did not reach a wide audience, but it was among the first to introduce in Germany the methodology, the terminology and the achievements of postulational analysis as practiced in the USA. Loewy—who had published several of his articles in the *Transactions of the American Mathematical Society* during the first decade of the century—was well-acquainted with the aims and methods of this trend. Fraenkel published his first article on the axiomatic foundations of the *p*-adic numbers whilst collaborating in the preparation of Loewy's book.[23] A brief account of the contents of Loewy's book further clarifies the state of the abstract axiomatic approach to algebra in Germany at the time of Fraenkel's first publications.

The first volume of Loewy's textbook appeared in 1915 under the title: *Grundlagen der Arithmetik*. A projected second volume, dealing with the theory of algebraic equations, never appeared in print.[24] By "*Grundlagen der Arithmetik*" Loewy meant an overview of the foundations of the system of real numbers, as well as of the basic operations that can be defined on them. Thus in the first three chapters, he discussed two approaches to the definition of the real numbers, while in the fourth and fifth chapters he dealt with topics such as powers and logarithms, limits, infinite series, and infinite products.

Following an approach similar to Kronecker's, Loewy took the series of natural numbers as the basis for the whole of arithmetic. The ontological justification of this series he saw as the business of philosophers, rather than of

---

22. For more details on Loewy's biography, see Fraenkel 1938, Remmert 1995.
23. Cf. Fraenkel 1912, 45 (footnote).
24. Cf. Fraenkel 1938, 18.

mathematicians. Once the series is given, however, one is lead to realize it through a precise conceptual definition (*Gedankenbildung*). This definition he characterized by means of Peano's postulates, defined for a system N of abstract elements. However, instead of the usual formulation of the postulate of induction, Loewy introduced the following, equivalent one:

P_5.The system N includes the number 1, 1^+ (the successor of 1), 1^{++} (the successor of 1^+), etc., and *only* the elements of this sequence. (Loewy 1915, 2)

Using the methods of postulational analysis, Loewy proved that his postulates are independent. Further, he claimed, relative consistency follows from the fact that the system of natural numbers satisfies the postulates, while, in addition, the series of natural numbers had been considered a-priori to be logically possible. Relative consistency is the most one can prove at this stage; a method for proving absolute consistency—asserted Loewy—is not likely to be found by mathematicians.[25] Thus, for Loewy, it was the acceptance of the natural numbers which justified the acceptance of Peano's postulates, rather than the other way round. The duality earlier discussed by Hilbert between the genetic and the axiomatic approach is thus manifest in Loewy's work, and moreover, in this case the second approach has not in any way overshadowed the first. He then proved the principle of induction as a theorem deducible from the above five postulates; since the system of natural numbers satisfies the postulates, it also satisfies the theorem of induction.

Loewy proceeded to define the integers. Although he had declared the series of natural numbers to be the only concept that must be accepted as a basis for arithmetic, Loewy actually also took the existence of the series of the integers for granted, and admitted—implicitly and a-priori—its logical possibility. In fact, he formulated five postulates defined on an abstract system N' as follows:

P'_1) N' contains at least one element.
P'_2) Every element x of N' has a successor, x^+, (a predecessor of x is called x^-.)
P'_3) Every element is a successor of one and only one element.
P'_4) An element x is always different from $x^+, x^{++},...$
P'_5) For each element x of N', the series

---

25. Loewy 1915, 3: "Eine allgemeine Methode aufzufinden, um die absolute Widerspruchslosigkeit eines Postulatensystems zu beweisen, dürfte den Bemühungen der Logiker und Mathematiker nicht gelingen."

$$\ldots, x^{--}, x^{-}, x, x^{+}, x^{++}, \ldots$$

contains all the elements of N'.

After proving the independence of the postulates, Loewy asserted their relative consistency; like in the case of the natural numbers, a proof of absolute consistency is unlikely to come forward, and thus one can only prove relative consistence. Thus Loewy wrote:

> Assume that the existence of a system N' of elements satisfying the above postulates P'_1 - P'_5 is logically possible. Then the system N' completely generates the series of natural numbers.[26]

It is not clear on what basis Loewy ascertained the relative logical possibility of this system of postulates. From the text, it would seem that he took the series of the integer numbers for granted; in such case, however, his argument would turn circular, since he went on to characterize the system of the integers by establishing a correspondence between a specified element x of N' and 0, between x^+ and 1, x^- and -1, etc. Moreover, Loewy also defined, based on this system of postulates, the arithmetic operations on the integers and their basic properties.

Starting with the postulates, Loewy defined the addition and multiplication of *positive* integers, in the usual, inductive way. Before extending these definitions to the negative integers, he introduced several abstractly defined binary relation, equivalence and order relations. Then he proceeded to work with an abstract set on which a composition law has been defined (*Verknüpfbares System*). An equality (*Gleichheit*) is an equivalence relation which is compatible with the law of composition defined on the set.

Next Loewy provided postulate systems defining groups and fields. These systems were taken from the works of Dickson and of Huntington.[27] Both systems account equally for the finite and the infinite cases; Loewy used the above-introduced concept of "equality" in order to differentiate between the two cases. Groups and fields are always defined on a set endowed with an equivalence relation, which is required to be compatible with the operations. If the equality relation divides the set into a finite number of equivalence

---

26. Loewy 1915, 7: "Setzen wir die Existenz eines Systems N' von Elementen, das den vorauf-gehenden fünf Postulaten P'_1) bis P'_5) genügt, als logisch möglich voraus, so führt das System N' zu der vollständigen Zahlenreihe."

27. More specifically: Dickson 1905; Huntington 1905a.

classes, then one has a finite group; otherwise one has an infinite one (p. 26). This kind of definition appears again in Fraenkel's work on rings.

Another specific point that allows seeing the direct influence of Loewy on Fraenkel concerns the way divisors of zero are treated in his book. Loewy introduced the rules for the product of negative integers, and proposed to see in them, quoting Hankel, arbitrary conventions, which however "favor the preservation" of the existing formalism of the operations.[28] Under these definitions, the system of integers satisfies all the properties of a field, except the existence of inverses for the division. Loewy did not introduce here a special concept of ring to describe this situation. Rather he stated that the system of integers is not a field, but nevertheless, some of the theorems that had been proved for fields hold for this system as well. Thus for instance, the product of any two integers is zero, if and only if one of them is zero. As a consequence of this theorem, one can also prove the cancelability with respect to the product. Now, as will be seen below, Fraenkel's direct motivation for defining abstract rings was the study of systems with zero-divisors; the integers, as Loewy showed here, are not among those systems, and thus the system of postulates proposed by Fraenkel in 1914 would not contemplate the integers as a ring.

Loewy completed his construction of the reals by defining, first, the field of rational numbers as classes of equivalences of pairs (fractions) of integers, and then, the field of reals by equivalence classes of infinite series of rationals.

In chapter three Loewy developed the abstract theory of the real numbers. This implies a further elaboration of the basic concepts of the theories of groups and of fields. Loewy was thus able to prove theorems such as (p. 173):

> The smallest possible field "of characteristic zero" is isomorphic (in a unique way), to the field of rational numbers.

He also showed that there exists a one-to-one correspondence between the system of real numbers and the points of the straight line. This is the kind of theorems that Loewy, following the ideas of the American school of postulational analysis and certainly also following Hilbert, saw as part of the axiomatic approach to the definition of the field of rational and real numbers.

---

28. Cf. Loewy 1915, 43: "Die Gleichungen, durch welche wir die Multiplikation der ganzen negativen Zahlen definierten, sind 'arbiträre Konventionen zugunsten der Erhaltung des Formalismus im Kalkül' (H. Hankel, Theorie der komplexen Zahlensysteme. Leipzig 1867, S. 41)."

Loewy's book, published as late as 1915, provides thus a further example of how the use of abstract methods does not necessitate or imply structural conceptions of the concepts considered. Loewy used the methodology of postulational analysis as originally conceived by Hilbert, namely, as a means of analyzing *known* mathematical entities in the first place. He made a consistent use of concepts such as fields and groups, and yet the aims of his inquiry had nothing to do with structural questions. Loewy was interested in clarifying the nature and the properties of the classical systems of numbers; he asked classical nineteenth-century-type questions about those systems and his answers echoed the same spirit. Postulational analysis, together with some newly introduced algebraic concepts appeared in his works as new tools to pursue the same classical aims. At the same time, however, several new specific questions arose within his discussion, in particular, questions concerning the potential interest offered by the study of number systems containing zero-divisors. These questions were to be addressed by Fraenkel, who took Loewy's and Hensel's trains of ideas into a new direction, leading to the definition and early research of abstract rings.

4.4   Abraham Fraenkel: Axioms for *p*-adic Systems

Several months after his arrival at Marburg, Abraham Fraenkel wrote the first article which brought him some recognition: an axiomatic foundation for Hensel's system of *p*-adic numbers.[29] Following the conception behind Hilbert's *Grundlagen der Geometrie* and behind those works of the American postulationalists which were intended to provide minimal systems of independent postulates for the *known*, concrete mathematical entities—such as the real and natural numbers, the continuum, etc.—Fraenkel took another known entity which had not been considered thus far—the system of *p*-adic numbers—and provided a suitable system of independent postulates for it. The influence of this article on the further development of the ideas initially introduced by Hensel was rather marginal. Later in his life, Fraenkel himself would claim that the importance then accorded to this work was far greater than it actually deserved.[30] However, Fraenkel's axiomatic treatment of the *g*-adic system led directly to his later definition of abstract rings.

---

29. Fraenkel 1912.
30. Cf. Fraenkel 1967, 111.

Fraenkel's 1912 system of postulates was meant to characterize a "concrete" mathematical entity, rather than an abstract one. In order to attain this task Fraenkel formulated three separate sub-systems of postulates which are necessary to account for all the features of that entity. The first system of postulates defines the order-type (*Ordnungstypus*) of the p-adic numbers, the second one defines their arithmetic properties as a field, and the third system of postulates is needed for rendering the system of p-adic numbers a categorical one.

Consider a system S of abstract elements, each of which is contained in one out of a collection of pairwise disjoint classes C_i. An order-type is defined on this system by means of two binary relations: for any two elements of the system α, β one can say either that α is "smaller" than β (denoted $\alpha < \beta$) or that α is "lower" than β (denoted $\alpha \prec \beta$). These two relations are defined through the postulate systems Γ and Λ respectively.

Before introducing the postulate systems themselves, Fraenkel explained the meaning accorded to the term "equality" (*Gleichheit*) in the article: since a p-adic number admits several different representations (though a unique *reduced* one), and in order to avoid unwieldy formulations, the sign "=" is to be understood exclusively as identity (*Identität*). Two different representations of one and the same element are therefore to be considered "equal."[31] Loewy's book of 1915 would define *Gleichheit* in more precise, and more general, terms using equivalence classes; in his 1914 definition of abstract rings Fraenkel would also discuss again this concept and stress its importance for the theory. Although this is a rather marginal issue in Fraenkel's work, it provides further indication of the ambiguous attitude of Fraenkel, as well as of other mathematicians, towards the abstract formulation of algebraic concepts. The abstract concepts are defined once as operations on congruence classes, then as arbitrary sets endowed with an abstractly defined operation, and later on as a set of numbers in which equality is given a new interpretation.

The system Γ, of seven independent axioms postulates that the collection of classes C_i is a totally ordered one, and that it contains a single smallest class C_∞. Moreover, every class C in the collection (except for C_∞) has an immediate predecessor, denoted by $/C$, and also an immediate successor, $C/$. In order to ensure that the first six axioms of the system completely characterize the col-

---

31. Fraenkel 1912, 48.

lection of classes C Fraenkel introduced a kind of generalized induction principle, as follows:

Γ_7. If a system S of classes as above satisfies the following two conditions:
    I. It contains an additional class besides C_∞,
    II. If it contains the class C then it also contains the classes $/C$ and $C/$,
then the system in question contains all the classes of S.

As was seen in the preceding section, a similar version of this postulate would also appear later on in Loewy's axiomatic characterization of the integers. Fraenkel remarked in his 1912 paper, that the system defined by these postulates was a natural extension of the system of natural numbers as defined by the axioms of Peano. He also indicated how the operations of addition and product could be defined in this system, based on postulate Γ_7.

Fraenkel summarized the discussion on the system Γ by claiming that, if one takes ">" as the basic order relation, the system defines an order of the type $^*\omega + \omega + 1$, which is equivalent to the order-type of the rational integers, together with an additional element placed at the end. This justifies denoting the classes with indexes C_i, and asserting that $C_i > C_j$, whenever $i > j$ with respect to the usual order of the rational integers.

The second binary relation, "$\prec$", is meant to characterize the inner order within each of the classes of S taken separately. It is defined by nine independent postulates of the system Λ. This system establishes, first, that each class C is totally ordered by $\prec$, while the class C_∞ contains a single element (postulates Λ_1-Λ_4). Further, within each class C there exists a special subset N, which is countably infinite and everywhere dense in C with respect to $\prec$, and each of whose elements (except probably the lowest one) has an immediate predecessor belonging to C-N (Λ_5-Λ_6, p. 53). Postulate Λ_7 establishes that each class C is continuous in the sense of Dedekind; this may be formulated as follows:

Λ_7. Let C_1, C_2 be a partition of C, such that for any two elements, c_1 of C_1, and c_2 of C_2, always $c_1 < c_2$ (i.e., let (C_1, C_2) be a cut). Then there exists either a "highest" element of C_1 or a "lowest" element C_2.

Finally, Λ_8-Λ_9 establish the existence of both a (unique) lowest and a (unique) highest element in each class of the system (except for C_∞).

Thus the relation "<" orders the classes of S with respect to each other, while the relation "$\prec$" orders the elements within each given class. In order to complete the definition of the order-type of S it is also necessary to consider two elements α and β belonging to two different classes C_m and C_n. In such case one says that $\alpha < \beta$ if and only if $C_m < C_n$. If α and β belong to the same class C, one says that α is equivalent to β ($\alpha \sim \beta$); this is clearly an equivalence relation. The two order relations, $<$ and $\prec$, can be now combined into a single one, and the whole system becomes what Fraenkel called a "*Zweifach geordnete Menge*".

Now, among several examples of systems satisfying the above two collections of postulates, Fraenkel discussed in some detail the domain of (reduced) rational p-adic numbers, with p a fixed prime integer. Consider the system of expressions $\sum_{n}^{\infty} a_i p^i$ ($a_i = 0, 1,..., p-1$). Calling the index of the first non-zero coefficient of an expression its "order", then, each class C_i is formed by all the numbers having the same order i. Two numbers belonging to the same class are considered as equivalent, since they behave similarly with respect to division by p. Given two non-equivalent numbers α, β one says that $\alpha < \beta$, whenever the order of the former is lower (in the usual sense) than that of the latter. Clearly the classes C_i are ordered according to their respective indexes. C_∞ contains only the zero element, whose order is ∞. Likewise, C_0 is the class of the units, whose order is zero. Within a given class C_i, two numbers are ordered as follows. Given any two elements,

$$\alpha = \sum_{n}^{\infty} a_i p^i \text{ and } \beta = \sum_{n}^{\infty} b_i p^i$$

then $\alpha \prec \beta$, if $a_n = b_n$; $a_{n+1} = b_{n+1}$; ... $a_{n+m} = b_{n+m}$; but $a_{n+m+1} < b_{n+m+1}$ (in the usual sense).

The set N stipulated by axiom Λ_5 is the infinite set of those p-adic numbers whose expression breaks off after a finite number of factors. This set is clearly an ordered subset of the natural numbers, and it is therefore countable. Moreover, given a number in that set, $a_n p^n + a_{n+1} p^{n+1} +...+ a_r p^r$, then its immediate predecessor (with respect to $\prec$) is

$$a_n p^n + a_{n+1} p^{n+1} + ... + a_{r-1} p^{r-1} + (a_r - 1)p^r + (p-1)p^{r+1} + (p-1)p^{r+2} + ...$$

which is itself not a member of N. It is easy to see that Λ_7 is also satisfied by the system. Finally, the lowest element of the class C_n is p^n, while the highest one is $(p-1)p^n + (p-1)p^{n+1} + ...$ Fraenkel thus concluded that each class of equivalent p-adic numbers constitutes a perfect, nowhere dense set with an initial and final element, which therefore has the power of the continuum.

So much for the order-type of the system S. Fraenkel defined now two operations by postulating a system Π of twelve independent axioms. The first nine involve the standard requirements for fields: Fraenkel took them from Dickson's definition of field, with some slight changes.[32] Given the aims of postulational analysis as practiced by Dickson, this system is neither the earliest, nor the most illuminating, nor the clearest one for fields; it is just the least redundant one from the logical point of view.

The last three axioms of the system Π refer to properties involving both the two operations and the above defined order-properties of S. Thus, axiom Π_{10} may be formulated as follows:

Π_{10} Let ε be a unit with respect to multiplication, and assume ε belongs to the class C_n. If $k\varepsilon$ denotes the sum $\varepsilon + \varepsilon + ... + \varepsilon$ (k times), then, in the sequence 2ε, 3ε, 4ε,..., there exits a first multiple of ε, $p\varepsilon$, such that $p\varepsilon$ belongs to C_{n-1}. p is called the ground number (*Grundzahl*) of the system, and it can be proven to be unique, since in fact ε can be proven to be the only unit of the system S.

Axiom Π_{11} expresses a similar property which may be formulated as follows:

Π_{11} Let α be any non-zero element of S and let γ be another element of S, belonging to the class C_n, and satisfying the property that any number of the series γ, 2γ, 3γ, ...,$(p-1)\gamma$ does not belong to C_{n-1}. Then from $\beta < \gamma$ it follows that $\alpha \cdot \beta < \alpha \cdot \gamma$.

Finally axiom Π_{12} states that if α and β are any two elements of S such that $\beta < \alpha$, then $\alpha + \beta \sim \alpha$.

Fraenkel also derived some immediate consequences from these axioms. Thus, for any non-null element α of S,

$$p\alpha = \alpha + \alpha + ... + \alpha = \alpha \cdot \varepsilon + \alpha \cdot \varepsilon + ... + \alpha \cdot \varepsilon = \alpha \cdot (\varepsilon + \varepsilon + ... + \varepsilon) = \alpha \cdot p\varepsilon$$

but $p\varepsilon < \varepsilon$, hence by Π_{11}: $p\alpha < \alpha$.

Now let m be the smallest integer for which $m\alpha < \alpha$. Suppose p is not a multiple of m, i.e., $p = mr + n$, with $0 < n < m$. In that case $n\alpha \gtrless \alpha > m\alpha$. Hence, by Π_{12}, $(n + m)\alpha = n\alpha + m\alpha \gtrless \alpha$, $(n + 2m)\alpha \gtrless \alpha$, ... $(n + rm)\alpha = p\alpha \gtrless \alpha$, which contradicts the above result. It follows that m is necessarily a divisor of p.

---

32. Dickson 1905.

Notice that in the systems Γ and Λ there is no mention whatsoever of the ground number p. This is also the case for axioms Π_1-Π_9. The ground number appears for the first time in axiom Π_{10}. Thus it is only here that a differentiation may first be introduced between systems with prime ground number p, or p-adic systems, and those with composite ground number g, or g-adic systems. In other words, it is only by means of the last three axioms of system Π that one can establish the difference between systems containing divisors of zero and systems which do not contain such divisors. An important basic result based on this fact differentiation is the following:

> If α is a non-null element of a p-adic system, then $p\alpha <\alpha$. In particular, $p\alpha$ is the first element of the sequence of multiples 2α, 3α, ... which is smaller than α. In a g-adic system, if $m\alpha$ is the first multiple of the sequence which is smaller than α, then m is a divisor of g (probably g itself). (Fraenkel 1912, 66)

According to Fraenkel, this result epitomizes the differences between the p-adic and the g-adic systems and evinces the much simpler structure of the former when compared to that of the latter. At the same time, however, it makes clear that *there is room for meaningful research on g-adic systems.* In his future work, Fraenkel would indeed pursue this line of research, which would eventually lead him to develop his theory of rings.

We thus see how Fraenkel's treatment of the technical details of his system of postulates for g-adic systems allowed this subtle, yet significant shift of interests. It would eventually open a direction of research that in retrospect may seem obvious, but which in actual fact attracted no particular attention before Fraenkel's own contribution. Since Hensel's research had been directly motivated by number-theoretic concerns, and in fact remained strictly within that framework, the mere existence of zero-divisors in g-adic systems was a limitation that discouraged research of such systems. Fraenkel's specific technical analysis suggested to him the convenience of overlooking this limitation.

The remaining sections discussed many additional results—both results which are common to p-adic and g-adic systems, and results which are valid for one kind but not for the other. In particular, Fraenkel introduced three additional axioms that establish the uniqueness of the p-adic and g-adic systems. Thus, Fraenkel introduced for the first time in the German mathematical literature the term "categorical" to denote such systems, a term that had originally been introduced by Veblen in one of his postulational articles.[33]

---

33. See Veblen 1904.

The results attained by Fraenkel himself in his new axiomatic research on p-adic systems, together with the earlier success of Steinitz's work on abstract fields, suggested a new direction of research, that Fraenkel undertook in his following work: the elaboration an abstract theory of domains similar to Steinitz's fields, but possessing zero-divisors.

4.5 Abraham Fraenkel: Abstract Theory of Rings

In his doctoral dissertation, published in 1914, Fraenkel introduced for the first time the axiomatic definition of a ring and discussed systematically the basic properties of this mathematical entity. In his 1912 paper, he had taken a "concrete" mathematical entity—the system of p-adic numbers—and sought to characterize it in minimal axiomatic terms. The system of numbers was here the focus of interest, while the axioms, just the means to improve our understanding of the former. In this sense, Fraenkel had pursued a task very close to Hilbert's own axiomatic concerns, although in a domain originally not envisaged by the latter. The opening pages of Fraenkel's 1914 paper would seem to bring him closer to postulational works of the kind that were published in the USA during the early years of the century. The definition of an "abstract" concept by a system of postulates appears on first sight as the main concern; the study of the postulates themselves, rather than that of the entity which they define, would seem to attract all of the attention. But in fact, after introducing the system of postulates that define abstract rings, and after applying to this system the standard techniques of postulational analysis, Fraenkel immediately proceeded to study the rings themselves, following the model put forward by Steinitz in his study of abstract fields. In Steinitz's own work, it must be added, there was no "postulational analysis" of the axiom system defining fields. Fraenkel's analysis of the system of postulates in 1914, gives a feeling of continuity from his previous work on g-adic systems to this new research. In his later works on rings, nothing of this sort of postulational analysis even appears. Moreover, contrary to what had been the case in his article on g-adic systems, Fraenkel did not provide in 1914 the most logically simple system of postulates to define abstract rings, but rather a clear and useful one. This section discusses in some detail Fraenkel's publications on the theory of abstract rings.

In Hensel's 1913 book, rings were defined as domains satisfying all the axioms of fields, except the last one, namely the "axiom of unrestricted and uniquely determined division" (*Das Gesetz der unbeschränkten und eindeutigen Division*). This axiom accounts simultaneously for the existence of an identity element for multiplication, for the existence of an inverse with respect to multiplication for every element, and for the non-existence of zero-divisors. In the system of axioms for rings that Fraenkel advanced in 1914, these facts were separated into simpler axioms. Nevertheless Fraenkel's system is somewhat more cumbersome and less general than the one used nowadays for rings. The differences between Fraenkel's definitions and present-day ones may in most cases be traced back to their roots in his work on *g*-adic systems.

Fraenkel began with a system of abstract elements endowed with an equivalence relation. This relation is called an "equality" (*Gleichheit*). The identity (*Identität*) is only a particular case of an equality, but there may be other "equalities" defined on a domain. Furthermore, if any operation is defined on the elements of the system, then it is assumed that, regarding this operation, any two "equal" elements are to be considered as one and the same. Now since the elements of the set in question are completely abstract, this assumption is in fact redundant in defining the ring; nevertheless, it is easy to see its origin in Fraenkel's 1912 article. In the framework of *p*-adic numbers the demand was actually meaningful, since *p*-adic numbers may be represented variously, and it was important for Fraenkel to stress the fact that these various representations should not be seen as different elements of the system considered.

On a system R endowed with an equality relation, Fraenkel assumed two abstract operations to be defined: addition and multiplication. Regarding the first operation he demanded that it satisfy the axioms of a group, whereas for the second he demanded that it be associative, that it be distributive with respect to addition, and that R contain at least one unit relative to it. Under these assumptions it is possible that R contains divisors of zero; an element which is not a divisor of zero is called a regular element of the ring.

Fraenkel added two axioms which do not appear in the standard modern definition of rings. These are:

R_8. Every regular element must be invertible with respect to multiplication in the ring.

R_9. For any two elements a,b of the ring there exists a regular element $\alpha_{a,b}$ such that $a.b = \alpha_{a,b} \, b.a$ and a second regular element $\beta_{a,b}$ such that $a.b = b.a.\beta_{a,b}$.

Axiom R_8 implies, that the set of regular elements of the ring is a field, and thus obviously, even the immediate case of the "ring of integers" is not taken into account by Fraenkel's definition. In fact, axiom R_8 describes a situation typical of the system of g-adic numbers, in which the units constitute a group regarding multiplication. Moreover, a unity in a g-adic system is not divisible by g, and therefore it cannot be a zero-divisor.

Fraenkel's 1914 paper deals mainly with the factorization properties of divisors of zero and with additive decompositions of the elements of the abstract rings in terms of some elementary divisors of zero. A ring is called separable (*Zerlegbar*) if whenever three elements a,b,c are given, such that c divides a^2b, then it is always possible to write c as a product $c=c_1{}^2c_2$, where c_1 divides a and c_2 divides b. Fraenkel observed that an exact analog of this condition holds for the rational integers, but it does not always hold for integers in an arbitrary field of algebraic numbers. This divergence, as we have seen, constituted the point of departure for Dedekind theory of ideals. Dedekind addressed the failure of unique factorization in certain domains of algebraic integers by imbedding the theory of integers in a more general one. Fraenkel, on the contrary, confronted with a similar problem in the framework of abstract rings, addressed it with a completely different strategy; instead of providing a broader framework of discussion in which to attain a generalized formulation, he restricted his treatment of factorization problems to separable rings. This fact further sharpens the difference between Dedekind's *Ordnungen* (§ 2.2.4) and Fraenkel's abstract rings. Fraenkel himself did not see any direct connection between his abstract rings and Dedekind's *Ordnungen*. As a matter of fact, in Fraenkel's treatment of factorization in abstract rings there is not even a clue to the connections between rings and concepts such as ideals or modules. From the point of view of "modern algebra", modules and ideals are intimately linked, and in fact, subordinate to abstract rings. This was neither the case in Dedekind's conception in the late nineteenth century, nor in Fraenkel's pioneering work on abstract rings, as late as 1914. In Hilbert's *Zahlbericht* ideals appear in connection to rings of algebraic integers (§ 3.2). It was only after 1920, with the work of Wolfgang Krull and Emmy Noether, that the theory of ideals became organically integrated into the theory of abstract rings.

Rather than providing a framework for studying decomposition of ideals, Fraenkel introduced a special kind of decomposition property, which was not to be extensively developed in later research on rings, but which is a direct extension of Steinitz's line of thought into the domain of rings. Given two elements a,c of R, if a divides c and c divides a, then they are called equivalent (*unwesentlich verschieden oder äquivalent*, p. 147); otherwise they are called essentially different (*wesentlich verschieden*). A zero-divisor is called a prime divisor, whenever it contains no proper divisor, except for regular elements. A ring is called simple (*einfach*) if all of its prime divisors are equivalent to each other. With this terminology, Fraenkel proved a main decomposition theorem for rings which may be formulated as follows (p. 172):

> If a separable ring R contains n essentially different prime divisors $p_1, p_2,..., p_n$, then there exist exactly n univocally-determined, simple rings $R_1, R_2,..., R_n$, satisfying the following conditions:
> I. The simple rings $R_1,..., R_n$ contain only elements of R.
> II. The intersection of any two of the above n simple rings contains only the null element.
> III. The product of two elements of the ring, belonging to two different rings R_i, is always null.
> IV. Every element of R may be written in a unique way as a sum of n elements of R, each belonging to one of the n rings R_i.

This theorem implies, Fraenkel claimed, the possibility of reducing any separable ring into simple rings. Thus, simple rings play in Fraenkel's theory of rings a role similar to that played by prime fields in Steinitz's theory. Of course there are important differences between the two concepts (e.g., fields contain only one prime sub-field), but there is also a basic functional similarity between them, namely, both play the role of building stones of their respective theories. In fact, like in the case of prime fields in Steinitz's theory, a full structural knowledge of separable rings is attained by establishing the properties of simple rings, and by enquiring how these properties are transmitted through the different kinds of extensions. In this article Fraenkel did not pursue this point further, but he did so in his next two works.

Fraenkel developed further his theory of rings in two additional works. In 1916 he published the *Habilitationsschrift* he had submitted to the University of Marburg, and in an article published in 1921 he further developed some of the issues considered in 1916. Both works were prepared during the wartime, while Fraenkel was still mobilized.

In the introduction to his *Habilitationsschrift* Fraenkel claimed that the central achievement of his dissertation had been the proof that the algebraic properties of any separable ring may be reduced to the consideration of a finite or an infinite number of "simpler rings", i.e., rings that in essence contain only one prime divisor of zero. The next task would be to extend, in the framework of abstract rings, the whole range of questions that Steinitz had introduced for fields. In particular, the question of how all the possible algebraic and transcendental extensions of a given ring are characterized, would be elucidated. Although Steinitz had addressed the question for any possible field, Fraenkel—in order to avoid unnecessary complications—considered here only finite and infinite rings of "finite degree" (i.e., separable rings containing only a finite number of essentially different zero-divisors). These are—wrote Fraenkel in the introduction—the rings needed in order to cover all the arithmetical and algebraic applications of the theory. In saying this, Fraenkel had two specific applications in mind: the formulation of a Galois theory for separable rings and the determination of all possible types of finite rings.

In defining rings in 1916 Fraenkel did not bother to produce a minimal system of independent postulates; instead, the reader interested in one such system was referred to Fraenkel's 1914 article. This time it was more important to provide a workable, rather than a logically irredundant system of postulates. As before one starts with an abstract set R endowed with an "equality" (*Gleichheit*) and two operations. The postulates are, in the present version, similar to those accepted nowadays, but a postulate of commutativity for the product is also included. Worthy of special attention is the last axiom of the system.[34] It establishes that R contains at least one regular element, and that if a is a regular element, and b any element of R, then there exists at least one element x in R, such that $ax = b$. In a footnote Fraenkel explained, that postulating the existence of at least one regular element excluded the case of the trivial ring having only zero elements, as well as of other, non-trivial systems, e.g., the system of all classes of congruence modulo g^m, which are divisible by g. As in his earlier version, the ring of integers is also excluded from Fraenkel's 1916 definition. In fact, Fraenkel did not even mention the integers in the framework of his theory. On the other hand, this last postulate implies the existence of a neutral element for the product.

This time, Fraenkel formulated more clearly the relations between the various algebraic concepts involved in his theory: the elements of a ring constitute a group with regard to addition, while the regular elements constitute a group with regard to the product. If a ring, as defined by Fraenkel here, contains no zero-divisors, then it is obviously a field. Thus, all the results valid for fields are also valid for rings, except for those depending on the existence of

---

34. Fraenkel 1916, 4.

division. Also those results on fields are not valid for rings, which depend on the fact that the product of two factors in a field is zero if and only if one of the factors is zero. Loewy had mentioned in his book the other side of the same coin, namely that in the system of integers, though not in itself a field, the product of any two integers is zero, if and only if one of them is zero. Fraenkel was clearly not envisaging the system of integer numbers when he devised his theory of rings. This partially explains why he did not elaborate upon the connection between the theory of rings and the theory of ideals.

The main problem addressed in Fraenkel's publications of 1916 and 1921 is that of the extensions of rings. Fraenkel translated all the concepts that Steinitz had introduced for fields, taking all the necessary precautions needed for rings. He defined prime rings—in a similar way as prime fields had been defined by Steinitz—and classified the possible kinds of prime rings. In order to classify the different kinds of extensions, it is necessary to consider the systems $R(x)$ of rational functions in a single variable, with coefficients in a ring R. Fraenkel studied those systems, stating that they are, in fact, rings. He then deduced several properties of $R(x)$ that depend on those of R. Typical is the theorem stating that (p. 16):

> If the ring R is simple, then $R(x)$ is also a simple ring, and the only prime zero-divisor of $R(x)$ is equivalent in $R(x)$ to p, the only prime zero-divisor of R.

Likewise, Fraenkel defined an Euclidean algorithm for $R(x)$, which was to be the main tool for studying the ring-extension of a given ring. But also in this context there is a noteworthy gap between the problems addressed by Fraenkel and those that were later to become the main problems of the abstract theory of rings. In fact, the main achievement of Emmy Noether's early work on rings was her unified approach to problems of factorization in algebraic number theory and in the theory of polynomials. Her main source in algebraic number theory was the work of Dedekind, whereas in the theory of polynomials it was Hilbert's work, as well as those of Emanuel Lasker and of Francis Sowerby Macaulay. Lasker proved in 1905, that any ideal of polynomials may be decomposed into "primary" ideals and, later on, Macaulay proved that this decomposition is essentially unique. He also provided an algorithm for actually performing the decomposition (see § 4.7 below). The kind of problem they dealt with was not even mentioned by Fraenkel.

To conclude this account of Fraenkel's work, it is worth mentioning that Fraenkel himself, after his article of 1921, totally abandoned the study of rings and moved into the field with which his name has come to be associated ever since, namely, set theory. Among Fraenkel's important achievements in set theory are the independence results for the axiomatic systems defining sets; there is a remarkable proximity of interests between these later concerns and those initially pursued by him, namely, those of postulational analysis.

4.6 Ideals and Abstract Rings after Fraenkel

Fraenkel initiated the study of abstract rings, defining them as domains similar to fields, but containing zero-divisors. Examples of such domains had been known ever since mathematicians studied the congruence classes of the integers, yet within the framework of algebraic number theory they were considered anomalous and devoid of interest precisely because of the existence of zero-divisors. Fraenkel undertook a postulational analysis of the g-adic domains, and this analysis rendered plain and precise the watershed between domains with and without zero-divisors. Fraenkel thus decided to pursue the research of the former, as parallel to the research on fields—i.e., domains without zero-divisors—previously undertaken with great success by Steinitz.

The rather tortuous path that led to this research on abstract rings, the problems chosen by Fraenkel in the framework of that research, and—no less importantly—the problems left untouched by him, have much to tell us about the gradual rise of the structural approach to algebra. In particular, these facts illuminate the gap between the mere technical ability to formulate concepts in abstract terms and the conscious recognition and systematic study of the species of "algebraic structures" as the main task of algebra. The significance of Fraenkel's work lies in its having added a further, non-immediate and non-evident, instance of an entity which should be recognized as belonging to that species.

But at the same time Fraenkel did not see, or at least did not pursue, the full structural picture. Fraenkel did not define the abstract concept of ring in its broader sense. Neither did he choose to address problems that were bound to open a fruitful path of research based on rings as a generalizing framework for several, theretofore separate mathematical issues. The task of realizing the potentialities involved in the idea of ring as a natural framework for dealing with the existing theories of factorization, both in the framework of algebraic number theory and the theory of polynomials, was first undertaken separately by Masazo Sono, Wolfgang Krull and Emmy Noether.

Masazo Sono published several results on rings and ideals of rings, proving that the theorems on unique factorization of ideals—such as had been proved by Dedekind for ideals of algebraic integers—do not always hold in more general rings.[35] Wolfgang Krull (1899-1970) was the first to present the results of the theory of ideals in the framework of the theory of abstract rings.[36] In his articles Krull declared, like Fraenkel before him, that his main task was the translation of Steinitz's achievements of field theory into rings, namely, into the domain of fields where division is not in general defined.[37] Krull explicitly mentioned Fraenkel's earlier work: he stressed the fact that the latter's definition of rings allowed no more than rings that behave essentially like Z_m. According to Krull, the limitations inherent in Fraenkel's work were a consequence of not having included ideals in the treatment. Krull also published a very influential monograph containing a systematic exposition of the abstract theory of ideals.[38] This monograph was published after van der Waerden's *Moderne Algebra*, who also dedicated some chapters to the study of ideals in abstract rings. However, the main inspiration for Krull's work was not provided by van der Waerden, but rather by Steinitz and Emmy Noether. Krull applied the "structural program" of Steinitz to the theory of ideals, which had, on the one hand, a long history going back to Kummer's work and, on the other hand, a recently established re-formulation: the abstract one. This abstract formulation was a contribution of Emmy Noether and it will be analyzed in the next chapter.

Having analyzed the roots of ideal theory in algebraic number theory and the early works on abstract ring theory, the present chapter closes with a brief account of a further direct source of Emmy Noether's theory of ideals—the works of Lasker and Macaulay on the factorization of polynomials forms.

4.7   Polynomials and their Decompositions

Consider a collection M of polynomials in n variables $x_1,..., x_n$ satisfying the two following conditions:

---

35. Sono 1917-8.
36. See Krull 1922; 1923; 1924.
37. Krull 1922, 81.
38. Krull 1935.

1. If F is a member of M and α is any polynomial on n variables, then αF is also in M.

2. If F_1, F_2 are in M also their sum $F_1 + F_2$ is in M.

Denote such a collection M a module of polynomials.[39] A basis of the module M is a finite sub-collection F_1, ...,F_k, such that any member of M may be written as $X_1 F_1 + ... + X_k F_k$, where X_i represent arbitrary polynomials. As was seen above (§ 3.1.), one of Hilbert's important contributions in the framework of his work on invariants during his early career had been the proof of the fact that every collection of polynomials has a *finite* basis. Further, Hilbert proved his *Nullstellensatz*, according to which given a finite set of polynomials F_1, ...,F_h, and another polynomial F, such that F vanishes in all the common roots of F_1, ...,F_h, then there exists a natural number k, and h polynomials A_1, ..., A_h such that

$$F^k = A_1 F_1 + ... + A_h F_h.$$

In the framework of the theory of abstract rings, an ideal A of an arbitrary ring R is called a primary ideal if given two elements a,b of R, such that their product ab belongs to A but a does not belong to A, then there exists an integer k such that b^k belongs to A. This concept arose in the study of ideals of polynomials. Primary ideals of polynomials were first explicitly defined by the German mathematician and World Chess Champion, Emanuel Lasker, in a famous article of 1905. Since the conceptual context of this article is one in which Hilbert's *Nullstellensatz* plays such a central role, it is likely that the latter provided the inspiration for the formulation of primary ideals.

Emanuel Lasker (1868-1941) studied from 1888 to 1891 in Berlin, Heidelberg and Göttingen, and later under Max Noether in Erlangen, where he received his doctorate.[40] His mathematical education was thus within the tradition of the classical theory of algebraic invariants. Lasker's mathematical contribution is mainly associated with his 1905 paper, dealing with systems of polynomial forms. Lasker indicated in the introduction to his paper, that it is based on a very limited number of specific items of mathematical knowledge: the basic concepts of algebra and arithmetic, factorization properties of "forms", determinants, and other standard issues. Interestingly enough, Lasker

---

39. This corresponds to what is called nowadays an ideal of polynomials. At this stage we use the term "Module" like in the texts discussed in the present section.

40. The classical Lasker biography is Hannak 1991 (1953).

pointed out that in spite of the reduced number of research tools needed in his research, he intended to rely on the notation provided by the "new mathematics", meaning by this, for instance, the tools of invariant theory![41]

And indeed, Lasker's link with tradition is more than just an empty statement. The paper opens with a long section where Lasker worked out many results on polynomial forms and invariants. Only in the second section did he introduce more general concepts like "modules" and "ideals." First, Lasker gave a lengthy historical account of the results leading to his work. In this framework he mentioned as his main sources the works of Kummer and Dedekind, Kronecker's contributions to the theory of ideals, Hilbert's work on invariants, and the contributions of Max Noether and other mathematicians to algebraic geometry (which will not be examined here for lack of space).[42]

As a very central principle in his work, he also mentioned what he denoted as the *"Prinzip von Schönemann"*. Theodor Schönemann (1812-1868) was a *Gymnasium* teacher at Brandenburg. He published some works on congruences and was also among the first mathematicians in Germany to work on Galois theory.[43] What the Schönemann's principle meant for Lasker was, simply, that the concept of "congruence" as introduced by Gauss in his *Disquisitiones Arithmeticae* behaves as the usual equality regarding all the relevant properties. The insistence on the centrality of this principle runs parallel to Fraenkel's above mentioned definition of a relation of "equality." It is not exceptional to find in articles of this time (and even later) this recurrent tendency to explain, that—from an abstract point of view—it does not matter whether we use the usual equality or any other equivalence relation when defining abstract objects in algebra. It is useful to keep this in mind when evaluating the degree of penetration of the abstract approach in algebra in the early twentieth century.

Lasker adopted in his text many ideas and notational details from Dedekind. He also developed to some extent, as Dedekind had done before him, an "algebra of modules" by defining for modules the four operations of arithmetic and exponentiation. It should be recalled that Dedekind had advanced a concept similar to primary ideals (*einartige Idealen*) and that he had suggested

41. Lasker 1905, 21: "Obwohl die Mittel der Untersuchung auf diese Weise sehr beschränkte sein sollen, so wird doch die Notation und Symbolik der neueren Mathematik, z.B. der Invarianten-theorie, benutzt werden."

42. See § 3.1., footnote 19.

43. See Schönemann 1846. Cf. Wussing 1984, 119-120.

that in a more general theory of ideals than the one developed by him there may be cases for which these ideals are not powers of prime ideals, as it is the case for algebraic numbers (§ 2.2.3). In spite of the fact that Lasker mentioned so many of Dedekind's ideas, it is remarkable that this particular one, singular ideals, does not appear in his account of Dedekind's work. This provides further evidence for the then still existing gap between the theory of polynomial forms and the algebraic theory of numbers. Were it not for the fact that these two disciplines were considered apart, Lasker could have perhaps seen the direct connection between Dedekind's singular ideals and his own primary ideals, as it is seen today from the point of view of the abstract theory of ideals.

Taken as a whole, even this section—in which Lasker introduced modules and ideals—remains within the traditional late nineteenth-century outlook of the theory of polynomial forms. In particular, the main factorization theorem is proved by means of specific factorization properties of polynomials and not by looking at the properties of their modules. In the first paragraphs of the section, Lasker mentioned the difference between a "module of forms" and an "ideal of forms", according to whether the coefficients of the forms—namely, polynomials in n variables—are taken from an arbitrary field or from the ring of rational integers. A "prime ideal" is defined, as usual, as one to which a product belongs if and only if one of the factors does. The immediate example in the case of polynomials is obtained by considering the set of all polynomials vanishing at a given point. This example has obvious motivations from algebraic geometry. "Primary ideals" are then defined as the central concept of the theory of modules. Before seeing Lasker's definition, some additional concepts must be introduced.

A module is said to "contain" a point $a_1, ..., a_n$ if all the polynomials belonging to the module vanish at that point. The spread[44] (*algebraisches Gebilde*) of a module is the set of all points contained in the module. A spread is called irreducible, when no product of two polynomials of a module vanish in a point of it, unless one of the factors do. Now, it is clear that the spread of a prime module is, by definition, irreducible and that any prime module is determined by its irreducible spread. Moreover, any module containing the irreducible spread defined by a prime module contains the prime module as well. Consider now a module M whose spread is irreducible (i.e. M contains a prime module P). M is called a primary module if whenever a product of forms

---

44. This is the English term used in Macaulay 1913.

belong to M then either one of the factors belongs to P or the other belongs to M. Thus a prime module is the simplest case of a primary module.

The main theorem of Lasker's article states that every module is representable as an intersection of a finite number of primary modules and a module containing no point (p. 51). His proof uses results from invariant theory which are sometimes more succinctly written by means of the "algebra of modules" introduced in the previous sections.

A further theorem (X, p. 56) states that for every form f belonging to the prime ideal P associated with a primary ideal M, there exists an integer h such that f^h is an element of M. The proof of this theorem relies on a lemma which is equivalent to the ascending chain condition. The latter was to be explicitly introduced in its most general formulation some years later by Emmy Noether. The lemma states that given an infinite sequence of ideals M^1, M^2,... there exists a number n so that for $N > n$, the module M^N is contained in $(M^1, M^2,..., M^n)$, the $l.c.m.$ of the n modules $M^1,..., M^n$. This is proven using Hilbert's basis theorem. Lasker did not mention, however, the reciprocal theorem, namely, that the above mentioned condition implies the existence of a finite basis. In the framework of polynomials, this statement would have been, of course, superfluous. For abstract rings, on the contrary, it is a significant result. As will be seen in the next chapter, it would be one of the important discoveries of Emmy Noether.

The decomposition of ideals of polynomials into primary ideals is usually known as the Lasker-Macaulay theorem. In fact, Lasker's results were complemented in a 1913 paper by the English mathematician Francis Sowerby Macaulay (1862-1937). Macaulay graduated from Cambridge and was a high-school teacher who also published original and important work, especially in algebraic geometry around 1900. His results on "modular systems" were published in an article in the *Mathematische Annalen* in 1913 and in a Cambridge Tract in 1916.[45]

There are some minor differences between Lasker and Macaulay as to terminology and presentation of the decomposition theorem, but what actually concerns us here is what Macaulay added to Lasker's work. This addition is twofold: he proved the uniqueness of the decomposition and provided an algorithm to actually perform it. Both these questions are settled through

---

45. Macaulay 1913, 1916. See Baker 1938, and the much more detailed and balance description in Gray 1997, 35-39.

Macaulay's differentiation between *isolated* and *embedded* primary modules in a decomposition.

A primary module appearing in a decomposition is called *isolated* if its spread is not contained in the union of the spreads of the remaining primary components. Otherwise it is called *embedded*. Macaulay showed that the isolated modules in a decomposition are unique but the embedded ones are not. Lasker's decomposition included a further module containing no point. This module is absent in Macaulay's decomposition since he considered only homogeneous systems in which all polynomials contain at least the origin. Macaulay described a transformation by means of which every system of polynomials may be transformed into a system of homogeneous polynomials, thus enabling him to adopt all of Lasker's results.

The preceding account of the process leading to the definition and early research of abstract rings as well as of some developments in the theory of polynomial forms completes the picture of the background to the work of Emmy Noether. Her achievement will be seen to consist, to a large extent, in a synthesis of much of the work on number theory and on the theory of polynomials that was described in this and in the previous chapters. Such a synthesis did not necessitate any substantial enlarging of the existing body of knowledge that went far beyond the results described above. It implied, however, a significant change in the meaning attributed to the concepts and to the results involved, and this implied a significant change in the images of knowledge. The next chapter will describe Emmy Noether's work, while stressing the shift in images of knowledge that characterized her achievements. Thus an examination of her work will allow understanding the last stage in the shaping of the structural trend in algebra.

Chapter 5 **Emmy Noether:
Ideals and Structures**

The preceding chapters described the gradual emergence and adoption of structural elements in algebraic research, and in particular in the research of factorization properties of numbers and polynomials, beginning with the works of Dedekind, through Hilbert and up until the separate works of Steinitz, Fraenkel, Lasker and Macaulay. The culmination of this slow and involved process found its most brilliant expression in the algebraic works of Emmy Noether. The present chapter concludes our account of the consolidation of the structural image of algebra by examining some of her relevant works in the 1920s.

The adoption of the structural image of algebra in its fully articulated conception—as should be clear to the reader at this stage—cannot, by its own nature, be associated to a single idea, nor to a single publication, nor even to the work of an individual mathematician. By the time of publication of Noether's papers, which will be analyzed in the present chapter, several other mathematicians—especially in Germany—were working on related problems, and approaching them from a similar point of view. Van der Waerden's *Moderne Algebra*—the paradigmatic embodiment of the structural image of algebra— was written under Noether's pervasive and decisive influence, but not only under hers. The contemporary works of Emil Artin, Otto Schreier, John von Neumann, and others, provided a direct source of inspiration for some of its chapters.[1]Moreover, a host of articles written still in the twenties by Noether's students and colleagues put forward similar techniques and points of view for a large number of problems in different areas of algebra.[2] Emmy Noether can nevertheless be selected as the most striking representative of this trend, not only because of her systematic application of the methodological principles

---

1. See van der Waerden 1975.
2. See Brewer and Smith (eds.) 1981, 115-163.

basic to the consolidation of the structural image, but also because of the wide variety of algebraic domains she considered in her work. No less important than her own work was her direct influence on the works of many others.

With this picture in mind, the present chapter is intended neither as a comprehensive account of all the works that contributed to the final shaping of the structural image (following the introduction of abstract fields and abstract rings by Steinitz and Fraenkel respectively), nor as a full account of Noether's own mathematical work. Rather, this chapter attempts to indicate—by focusing on technical details of her work—how in Noether's decisive articles on factorization in abstract rings, all the elements that inform the structural image are brought together and combined in an illuminating manner. The innovative insights implied by these articles suggested to her, to her students and colleagues, and to their followers thereafter, the expected gains of addressing the various disciplines of algebra from a unified perspective in which the notion of an algebraic structure lies at the focus of interest.

5.1   Early Works

Emmy Noether (1882-1935) was born in Erlangen to a Jewish family.[3] Her father, Max Noether, was himself a distinguished mathematician. Emmy Noether studied mainly in her native city, except for a semester spent in Göttingen in 1903-04. She began her own research in the domain of invariant theory. Her early works belong to the classical algorithmic trend of the school of Clebsch and Gordan.[4] In 1915 Emmy Noether moved to Göttingen. She had been invited by Klein and Hilbert to join their institute as a specialist in differential invariants, and especially in order to assist Hilbert in his current work on mathematical physics. In 1920, however, Noether's focus of interest had moved to algebra, and most of her work was being done in that discipline. Very soon she became the leading algebraist in Göttingen, and algebra became one of the main specialties of the Göttingen mathematical community. An

---

3. Several biographical notices on Emmy Noether have been published, which provide a detailed picture of her human and mathematical stature: Alexandrov 1935; Brewer & Smith 1981; Dick 1981; Kleiner 1992; Kramer 1974; Tollmien 1990; van der Waerden 1935; Weyl 1935. See also Tollmien 1991 for an account of Noether's difficulties as a woman in the German mathematical establishment. Some of the issues discussed in the present chapter are also considered in articles collected in Brewer & Smith 1981. See especially Gilmer 1981.

4. See for instance, her doctoral dissertation, Noether 1908.

entire generation of young researchers all over the world enjoyed her guidance.[5]

While still in Erlangen, working on the theory of invariants, Noether's
methodology began shifting towards a more abstract approach. Under the
guidance of Ernst Fischer (1875-1959), Noether's work on invariant theory
moved away from Gordan's style and came closer to Hilbert's. In an article
written in 1915 and published one year later, Noether dealt with a generalized
domain of transcendental integers defined by means of abstract properties.
Although Noether explicitly stressed the importance of the "abstract" aspect
of her work,[6] her conception of an "abstract" property was here somewhat different from her own understanding of this in later years. From this work it is
clear that Noether was well acquainted with the works of Steinitz and of
Weber on fields, and thus with the possibility of abstractly formulating concepts. At that time, however, she had not yet adopted such an approach.

From Noether's preserved correspondence with Fischer we know that her
interest in modules and ideals goes back to 1917, when she began lecturing on
that topic in Göttingen.[7] However, her first publication in the field appeared
only in 1920 as a joint paper with Werner Schmeidler (1890-1969).[8] This was
a study on differential operators which received little attention at the time of
its publication. In retrospect one can see that this article contained the germs
of many ideas later developed by Noether herself and by her followers working on algebra.[9]

The theorem of unique decomposition of a polynomial into irreducible
factors is easily generalized from the case of one indeterminate x to the case
of n variables. By contrast, in the theory of differential operators, whereas a

---

5. In the words of Hermann Weyl (1935, 60): "In my Göttingen years, 1930-1933, she was
without doubt the strongest center of mathematical activity there, considering both the fertility of
her scientific research program and her influence upon a large circle of pupils." Among Emmy Noether's direct students we can mention leading algebraists such as Max Deuring (1907-1984), Hans
Fitting (1906-1938), Friedrich Grell (1903-1974), Jakob Levitzki (1904-1956), Kenjuri Shoda
(1902-1977), Otto Shilling (1911-1973) and Ernst Witt (1911-1991). On Noether's relations with
her students see Kimberling 1981, 39-44. Several other leading mathematicians co-worked with her
and her influence was decisive on their own works, among them: B.L. van der Waerden, Helmuth
Hasse, Wolfgang Krull, Paul Alexandrov, Oscar Zariski. See Parikh 1991, 75-76.

6. Noether 1916, 195.

7. Cf. Kimberling 1981, 12; Tollmien 1990, 161-162.

8. Noether & Schmeidler 1920.

9. More details on this issue appear in Jacobson 1983, 12.

corresponding formulation of the unique decomposition theorem is valid, its generalization to partial differential operators is not. In their joint paper, Noether and Schmeidler attempted to show that if the decomposition of the differential operator is formulated in terms of *l.c.m.* of ideals, then, the desired generalization will indeed be possible. This reformulation is not straightforward at all, however, since it requires, in the first place, the possibility of defining "ideals" in a domain in which multiplication is not commutative. This is seen as follows. A differential operator is an expression of the form.

$$a_0\frac{d^n y}{dx^n} + a_1\frac{d^{n-1}y}{dx^{n-1}} + \ldots + a_n x = A(y)$$

The formal analogy of this expression with a polynomial on a variable x is manifest. Now given a second operator

$$b_0\frac{d^n y}{dx^n} + b_1\frac{d^{n-1}y}{dx^{n-1}} + \ldots + b_n x = B(y) \ ,$$

then, the addition of the two operators is defined in the obvious way, whereas their product $AB(y)$ is defined as follows

$$a_0\frac{d^n}{dx^n}B(y) + b_1\frac{d^{n-1}}{dx^{n-1}}B(y) + \ldots + a_n x = AB(y) \ .$$

Write now $(d/dx)^n$ instead of d^n/dx^n, and further μ instead of d/dx. The analogy with polynomials—as well as the limitations of this analogy—become even more perspicuous: we obtain polynomials on μ, whose multiplication is albeit non-commutative. This is more clearly seen by taking the derivative of a linear factor $a=a(x)\cdot y$,

$$\frac{d}{dx}(a(x) \cdot y) = a(x)\frac{d}{dx}(y) + a'(x) \cdot y$$

or, expressed in terms of μ,

$$\mu \cdot a = a \cdot \mu + a' \tag{*}$$

The general setting in which the paper is worked out, then, is that of the domain of polynomials in which multiplication, though in general non-commutative, satisfies at least condition (*). In order to treat factorization in such domains Noether and Schmeidler introduced "one-sided ideals" (*einseitige Moduln*). A set A of polynomials is called a one-sided ideal if the following two conditions hold:

1. If M and N are polynomials in A, then their sum also belongs to A.

2. If M belongs to A and F is any polynomial, then FM belongs to the ideal for any F. (This is the case of a right-ideal, whereas in the case of a left-ideal it is required that MF belongs to the ideal).

A remarkable feature of this paper, then, is that right- and left-ideals are defined here for the first time, but not in what would be considered a classical algebraic context.

Residue classes of one-sided ideals are defined in the standard way. The set of these classes constitute a group with respect to addition but not so with respect to multiplication. The study of this additive group lies at the heart of the paper.

The study of factorization properties undertaken by Noether and Schmeidler in their joint paper focuses on the relationship between the multiplicative decomposition of certain one-sided ideals and the corresponding additive decompositions of the quotient groups determined by them. Thus, it is proven that a module may be decomposed as the *l.c.m.* of a system of mutually exclusive prime ideals[10] if and only if the respective quotient groups may be decomposed into sums of certain "irreducible" subgroups.[11] This decomposition of groups corresponds to the decomposition of a group into irreducible direct factors. The paper thus established the conditions for existence and uniqueness of finite decompositions in given domains of non-commutative polynomials in terms of properties of the corresponding quotient groups. This recognition of the tight connection between decomposition properties of polynomials and groups, stressed the conceptual proximity of these two mathematical entities such as few algebraic results had done before. In her future work Noether would further elaborate the approach implied by this recognition.

Dedekind's concepts and methods clearly dominate the background of this joint paper. This is manifest in the notation as well as in the formulation of the

---

10. Noether & Schmeidler 1920, 13. A system of ideals is said to be a system of mutually exclusive modules (*total teilerfremdes System*) if, whenever the system is divided in two classes, and one separately takes the *l.c.m.* of the ideals belonging to each class, then the two resulting *l.c.m.*'s generate the whole system of ideals.

11. Noether & Schmeidler 1920, 20.

concepts and in the general methodological approach. The paper pays much attention to the parallel analysis of three different domains of mathematical discourse: the domain of operators, the ideals and the group of rest classes. The paper discusses in detail their interrelations and identifies the concept of isomorphism as the essential tool for working out these interrelations.[12] Yet, much like in Dedekind's own work, the main concepts used here are not abstract axiomatic constructs, but rather concrete mathematical entities. There are repeated references to the "usual rules of algebra" (meaning by that, the algebra of real numbers) as the basis of the discussion A domain is characterized according to whether or not it satisfies all those "usual rules of algebra" or, as in the case of the operators, all those rules except one (e.g., commutativity for the product).[13]

The content and significance of Noether and Schmeidler's 1920 joint article can be summarized as follows: it approached the study of factorization properties of a given domain (differential operators), while relying on a limited analogy with a better-known, second domain (polynomials), and while stressing the close connection of these properties with those of a specific collection of groups. Thus the article stressed the conceptual parallel between these three domains. At the same time, however, the conceptual hierarchy that dominated the classical image of algebra remained here unaltered: properties of polynomials are derived from the basic, known properties of the systems of real and complex numbers. The crucial step implicit in reversing this hierarchy was only taken by Noether in 1921. In that year she published the paper which marked the beginning of her research on the abstract structure of rings and of the widespread acknowledgment of her genius.

5.2   Idealtheorie in Ringbereichen

Factorization properties of algebraic numbers had been studied throughout the nineteenth century using several newly introduced concepts such as modules, ideals, and fields. From the works of Gauss onward, through Kummer, Kronecker, Dedekind and, finally, Hilbert, an impressive numbers of results

---

12. Noether & Schmeidler 1920, 2.
13. Thus, for instance on p. 6, Noether and Schmeidler speak in the same sentence of "einem beliebigen abstrakt definierten Körper P" satisfying "alle Regeln der Algebra, insbesondere also auch das kommutative Gesetz der Multiplikation mit eindeutiger Umkehrung," and an integral domain J that satisfy all rules of algebra, except commutativity of the product.

had been proved, and new techniques had been thoroughly elaborated; the theory of algebraic fields was by the turn of the century a well-established mathematical discipline. This theory, as conceived before Noether's work, derived its results from the known properties of the real and complex numbers. Steinitz's 1910 article had initiated the study of abstract fields as a subject of independent interest, but it did not change the basic assumptions of the theory.

In the theory of polynomials, on the other hand, specific factorization theorems had also been proven and special techniques had been developed in the works of Hilbert, Lasker and Macaulay. These results were also derived from the peculiar properties of polynomials, which themselves derived from those of the real and complex numbers. Ideals were found to play an important role in elucidating phenomena of factorization both for algebraic numbers and for polynomials, yet these ideals were conceived as specific collections either of numbers or of polynomials.

When Emmy Noether undertook in 1921 the unification and generalization of this entire body of mathematical knowledge, she was almost forty. She had already made significant original contributions, but her most important works were still ahead. The conceptual framework used by Noether in putting forward her unification was based on the abstract concept of ring. This concept was first defined by Fraenkel, but in his own work no connection at all had been made between it and the problem of factorization in terms of ideals. Noether was the first to identify the relevance of this connection. In fact, the very concept of ring was still so alien to contemporary mathematicians in 1921, that in her article Noether felt it necessary to prove its most elementary properties—as the need arose—throughout the paper. Thus, for instance, she proved in a footnote such elementary results as the uniqueness of the identity element in a ring. One can find this result, for instance, in the opening chapter of Hensel's book of 1913. But the need to prove it again in detail in this paper seems to indicate that the concept of ring had theretofore passed essentially unnoticed as a useful one for algebra.

Noether's 1921 paper opens with a clear statement of purpose:

> The aim of the present work is to *translate the factorization theorems of the rational integer numbers and of the ideals in algebraic number fields into ideals of arbitrary integral domains and domains of general rings.* (Noether 1921, 25. Italics in the original.)

In order to show how this translation is achieved, Noether presented the factorization of integers into primes and their uniqueness properties in a formulation that differed slightly from the usual one. Thus, Noether considered four different aspects of the unique factorization of an integer. Given an integer a, one can write it as a product of primes:

$$a = p^1 p^2 \dots p^n = q_1 q_2 \dots q_n$$

Ideals of polynomials may be decomposed into primary components in such a way, that a prime ideal is associated to each primary component (§ 4.7 above). A similar result may be obtained for prime numbers, if one focuses on the powers of the primes q_i as the factors of decomposition, rather than on the primes themselves. Accordingly, the properties of the decomposition of integers may be written as follows:

(1) Two different factors q_i, q_j are relatively prime and they cannot be further decomposed into factors having this property. This is the first sense in which the factors are irreducible. Moreover, since the factors are relatively prime, the product of the q_i's represents also their *l.c.m.*

(2) Given q_i and q_k, and an integer b such that q_k divides $b^2 q_i$ then it follows that q_k necessarily divides b.

(3) Given that q divides $b^2 c$ but q does not divide b, then q necessarily divides a power of c. However, the product of two factors q_i and q_k does not satisfy this property anymore. In this sense, the above mentioned decomposition into "primary" components is said to be an irreducible factorization.

(4) Each factor q_i in the above decomposition cannot be represented as the *l.c.m.* of two proper divisors of q_i.[14]

Now the theorem of unique factorization of integer numbers may be formulated in terms of the primary components q_i. To each q_i there is a uniquely associated prime number p_i and a natural number σ, such that $p_i^\sigma = q_i$. The theorem is stated as follows:

In any two different factorizations of a rational integer numbers into irreducible, greatest primary factors q, the number of factors, the corresponding prime numbers (up to their sign), and the exponents coincide. (Noether 1921, 25)

Substituting the word "number" by "ideal generated by a number", yields the theorem for ideals in algebraic number fields, as Dedekind had formulated it. The translation of the theorem from prime to primary components may seem trivial when applied to integers, and as giving less precise information

---

14. Noether 1921, 25: In (1) Noether uses the term *paarweise teilerfremd*, while in (2), any two numbers satisfying that property are called *relativprim*. Since these terms are used here only for the sake of illustration, I use here the current terminology.

about the decomposition than the usual formulation of the theorem. But it was precisely this formulation and the concomitant recognition of these simple basic properties as the key to understanding factorization phenomena that opened the way to Noether's unification.

In defining abstract rings, Noether mentioned the earlier work of Fraenkel. Fraenkel's definition, she claimed, imposed unnecessary restrictions on the ring.[15] The four above mentioned decompositions appear as one and the same in the more restricted class of rings allowed by Fraenkel's theory. Noether's own definition of a ring uses general, abstract axioms, similar in essence to those used until today. The 1921 article, however, focused on commutative rings Σ (in general, not a ring with unity) satisfying the "finiteness" condition (*Endlichkeitbedingung*), i.e., rings in which every ideal has a finite basis. Right from the beginning Noether proved the equivalence of the finiteness condition and the ascending chain condition (*a.c.c.*).[16] She conceded that the *a.c.c.* had been identified and used (albeit in particular cases) prior to her work by both Dedekind and Lasker, but these authors had used it in a rather limited way.

In the few next pages of the present section we describe in some detail the technical aspects of Noether's decompositions theorems. This is necessary in order to understand the peculiar way in which the new, structural considerations, and in particular, the chain conditions, replace the original, number-theoretical or polynomial arguments in her work. This change in the conceptual priorities of the mathematical entities involved, as has insistently been stressed in the foregoing chapters, represents a crucial turning point necessary for the definite consolidation of the structural image of algebra.

Noether's first decomposition theorem considers "irreducible ideals", i.e., ideals that cannot be written as the intersection of two different ideals properly containing it. If a given ideal M of Σ may be represented as the intersection of a finite number of ideals of Σ, $M = [B_1, B_2, ..., B_n]$, we say that the representation is "shortest" (*kürzeste*) if none of the B_i's contains the intersection of other ideals appearing in the representation. If, in addition, none of the ideals may be

---

15. These restrictions were mentioned above in § 4.6.

16. There are alternative formulations of the finiteness condition which lead to proofs shorter than those proposed by Noether's. For instance, in van der Waerden (1930 Vol. 2, p. 22), the "maximality condition" states that if a ring satisfies the *a.c.c.*, then every set of ideals has a maximal ideal. The "principle of divisor induction" states that in a ring with the *a.c.c.*, every property is satisfied by any ideal A, whenever it is satisfied by all ideals properly containing it.

replaced by another ideal properly containing it, then the decomposition is said to be "reduced." Noether did not actually speak of intersection but of the *l.c.m.* of the ideals, and, as a matter of fact, she did not in general use in this context the standard set-theoretical notation accepted nowadays. Thus, if an element f belongs to an ideal M, Noether writes $f \equiv 0(M)$, which reads "f is divisible by M" (*f ist durch M teilbar*). Similarly if all the elements of an ideal N are also contained in M, Noether writes $N \equiv 0$ (M), which reads "N is divisible by M." Moreover, M is called a "proper divisor of N" (*echter Theiler von N*) when M contains all the elements of N and there is at least one element in M which is not in N. Clearly, all this stems directly from Dedekind's work (§ 2.2.2 above). It was seen that Dedekind was aware of the tension between this notation and the inclusion relations of the ideals taken as sets. Even in Krull's 1935 important monograph on ideals one still find traces of this contradictory use of notation. Krull proposed to abandon the usage introduced by Dedekind since he considered it "clear from the historic point of view, but untenable from the point of view of set-theory."[17] Following the publication of van der Waerden's textbook and of Krull's monograph the unified notation became universally accepted, so that an ideal N is a "partial set" of the ideal M only when N is a "partial ideal" of M.

The first lemma proved by Noether (*Hilfsatz I*) states that if an ideal may be written as the *l.c.m.* of a finite number of ideals containing it, then there exists at least one reduced decomposition of that ideal. The lemma is proved using the *a.c.c.* Thus, if $M = [B_1, B_2, ..., B_n]$ is the given decomposition, and U_i is the intersection of all the ideals appearing in the decomposition, which are different from the B_i (for fixed i), then clearly $M = [B_i, U_i]$. If there exists no ideal containing B_i which can be written in the last decomposition instead of B_i, then B_i may be taken to stand in a reduced decomposition of M. If that is not the case, there exists a second ideal $B_{i/1}$ containing B_i, which can be placed in the reduced decomposition of M. Applying the same argument successively one obtains either a reduced decomposition of M or an infinite sequence of ideals B_i, $B_{i/1}$, $B_{i/2}$, ... each properly containing the former. This last possibility, however, clearly contradicts the *a.c.c.* Thus, the theorem can be proved without any recourse to properties derived from the nature of the elements, and only based on an inclusion property of their collections.

---

17. Krull 1935, 149: "... historisch klar, aber mengentheoretisch unmöglich."

In an example dealing with polynomials, Noether showed that two reduced decompositions built according to the above procedure may be different. However, it may be also shown that the number of ideals appearing in each possible reduced decomposition is invariant.

Noether then proved a second decomposition theorem:

II. Every ideal is decomposable as the *l.c.m.* of a finite number of irreducible ideals. (Noether 1921, 33)

This is proven using an argument similar to the one above, and relying again on the *a.c.c.* Theorems I,II are then combined into the following:

IV. Every ideal is representable as a reduced *l.c.m.* of a finite, invariant number of irreducible ideals. (Noether 1921, 36)

In the next section Noether introduced primary ideals. D is a primary ideal of Σ if from $AB \equiv 0\ (D)$ and $A \not\equiv 0\ (D)$, it follows that there exists an integer n such that $B^n \equiv 0\ (D)$. Naturally, if $n=1$ for every ideal B, then D is a prime ideal. For any primary ideal D, there exists a unique prime ideal P containing D, such that a power of P is contained in D. The lowest such power is called the exponent of D. Using again the *a.c.c.*, and by an argument similar to the one used above, Noether proved the following theorem:

VI. Every irreducible ideal is primary (but there are primary ideals which are not irreducible). (Noether 1921, 39)

Next, from theorems IV and VI she obtained the following:

* Every ideal is factorizable as a reduced intersection of primary ideals.

As for uniqueness of this last decomposition, she proved not only that the number of such primary ideals is an invariant of the ideal, but also that:

VII. In a reduced factorization of a given ideal as intersection of primary ideals, the collection of the associated prime ideals is an invariant property of the given ideal.

The above results can be summarized as follows:

In a ring with *a.c.c.* every ideal is representable *as the reduced intersection of a finite number of indecomposable ideals (which are also primary); the number of such ideals and the collection of associated prime ideals is invariant for every given ideal, though probably the specific primary ideals used in the factorization are not.*

One can in this way associate an invariant finite collection of prime ideals to any ideal in a ring with *a.c.c.*, and the invariance of this collection is proved relying precisely by using the latter condition.

One can go further and construct the decomposition as a reduced intersection of greatest primary ideals (*kleinstes gemeinsames Vielfaches von größten primären Idealen*), that is, the primary factors can be chosen in such a way that the intersection of any two of them is not a primary ideal anymore. In that case the decomposition will remain invariant in the sense described above, namely, the number of components and the associated prime ideals will be invariant (Theorem IX). Moreover, Noether proved as a corollary, that every prime ideal is irreducible. We saw when examining Kummer's work, that in certain domains there are irreducible numbers which are not prime (§ 2.2.1). This was a main motivation for Dedekind's work in the theory of algebraic numbers, and his theory of ideals succeeded in providing the adequate generalizations required to overcome the limitations of Kummer's theory. In this regard, Noether's study attained an even higher degree of generality concerning the concepts of irreducible, prime, and primary ideals.

The third type of decomposition considered by Noether involves relatively prime, irreducible ideals. An ideal A is relatively prime with respect to an ideal B, when for any ideal T such that TA is contained in B, it follows that T is contained in B. This is not a symmetric property; if A is relatively prime to B and B relatively prime to A, then A and B are called mutually prime (*gegenseitig relativprim*). If an ideal cannot be written as the intersection of two mutually prime ideals it is called *RI* (*relativprim-irreduzibel*). Noether proved a theorem of decomposition into *RI* factors. Although this proof was not as influential on future research as the other three, describing its details is useful for understanding later developments. As in the former case, one must prove the existence of an essentially unique decomposition of any ideal as an irreducible intersection of mutually prime ideals (theorem XII).

Noether started by proving some auxiliary results which are also necessary for the proof of the invariance of the so-called "isolated primary ideals", which will be described below. If an ideal R and each of the ideals of a collection $S_1, S_2, ...,S_l$ are mutually prime, then both R and the intersection of the ideals in the collection are also mutually prime. An important auxiliary theorem states that:

XI. Given two ideals A and B contained in a ring R, the ideal A is relatively prime to B, and B is different from R, if and only if none of the prime ideals associated with A is contained in a prime ideal associated with B. (Noether 1921, 46)

With the help of this result, Noether proved the existence and invariance of the decomposition of an ideal as the intersection of irreducible relative-prime ideals. This is done as follows:

Let $M = [D_1,...,D_n]$ be an irreducible decomposition of M into greatest primary ideals as obtained in the former section. Let $\{B_i\}$ be the set of prime ideals associated with the primary ideals $\{D_i\}$. We can separate $\{B_i\}$ into disjoint equivalence classes $\{G_1,...,G_m\}$ with the property that no ideal in a class contains another ideal from a different class, and no class can be further divided into sub- classes satisfying this property. Noether called these classes *Gruppen*. Given any ideal, the class to which it belongs includes as members all the other ideals of the ring that are either contained in that ideal or contain it. This division into classes is uniquely determined for every ideal in a ring.

Let now R_i be the intersection of all those ideals belonging to a given class G_i. Since $M = [D_1,...,D_n]$ it follows that also $M = [R_1,...,R_m]$. On the other hand, for any two different ideals R_k and R_j it is clear, by definition, that their respective associated prime ideals are not contained in each other. Hence, by the auxiliary theorem, the R_i's are mutually prime. The number of ideals R_i is equal to the number of classes G_i, and therefore, it is invariant for a given M. To see that the decomposition is unique, suppose that two different decompositions of M into greater primary ideals are given: $M = [D_1,...,D_n]$ and $M = [D'_1,...,D'_n]$. By theorem IX the collection of prime ideal associated with both sets of primary ideals is identical; therefore, the classes G_i will also be identical. Noether's third decomposition theorem appears next as:

XII. Every ideal is uniquely decomposable as an irreducible intersection of mutually prime ideals.

As Noether indicated in the footnotes, the existence of such a decomposition could have been proved directly, in a similar way as theorem II was proved. The above described proof, however, yields directly the invariance. And, what is more important for the present discussion, this proof reveals more than any other the close connection between the uniqueness properties of the decomposition and the relative inclusion properties of the prime ideals. Oystein Ore would later exploit in detail this connection in his work on the foundations of abstract algebra (§ 6.3 below).

Now, as was said above, while the number of irreducible ideals and the particular prime ideals are invariantly determined by decomposition, this is in general not the case concerning the specific irreducible (respectively: the primary) ideals, whose intersection represents the given ideal. In some particular cases, however, even these latter ideals are also invariantly determined; this is considered by Noether in section 7, on "uniqueness of isolated ideals." Let $M = [R,L]$ be a reduced decomposition of M. R is called an isolated factor of the reduced decomposition whenever no prime ideal associated with R is contained in a prime ideal associated with L, in other words, when R is relatively prime to L. Using the auxiliary theorem, Noether proves a further theorem:

XIII. In every reduced decomposition of an ideal as intersection of irreducible (resp.: greatest primary) ideals, the irreducible (resp. primary) components associated with isolated prime ideals are uniquely determined. (Noether 1921, 50)

The details of the proof are similar to the preceding ones.[18]

This result is one of the important innovations of Noether's article. In fact, neither the invariance of the associated prime ideals nor that of the isolated factors had appeared in the works of Lasker. Macaulay, on the other hand, had introduced the isolated ideals but as a concept built on properties of the spread of a variety. Noether's result is therefore not only a generalized formulation of the known theorem on polynomials, but in fact a new result which could not have been attained in the particular case of polynomial rings.[19]

The fourth and last decomposition theorem of Noether holds for rings with identity and *a.c.c.* Two ideals A and B are called comaximal (*teilerfremd*)[20]

---

18. It should be noted, however, that although the theorem of decomposition into relative-prime factors did not appear in most algebraic texts published after Noether's article, the invariance of isolated factors is proved in many of those texts, and the proof given there relies on a theorem equivalent to Theorem XI. This is the case. e.g. in van der Waerden's *Moderne Algebra*, § 88.

19. Noether added a footnote in this section on invariance of isolated primary factors, explaining that Macaulay had already formulated the uniqueness theorem for ideals of polynomials, although without giving the proof. See Noether 1921, 50: "Dieser Satz ist für Ideale aus Polynomen im Fall der Zerlegung in größte primäre schon ohne Beweis von Macaulay mitgeteilt; seine Definition der isolierten und nicht-isolierten primären Ideale kann als irrationale Fassung der unten mitgeteilten angesehen werden."

20. The English translation of van der Waerden's book is somewhat confusing regarding this term. What Noether calls *relativprim* is translated there as "relatively prime to" (Vol. 2, p.25), while *teilerfremd* is translated as "relative prime." The latter is the common usage for integer numbers, and it may be natural to translate it also to this domain. However, we shall abide here by the translation of Gilmer 1981 which seems clearer for this specific context.

when their g.c.d. equals the ideal generated by the unit-element of the ring, or in other words, when every element of the ring may be written as a sum of an element of A and an element of B. When an ideal cannot be written as an intersection of two comaximal ideals, it is called comaximal-irreducible (*teilerfremd-irreduzibel*). It is easy to prove that two comaximal ideals are mutually prime.[21] The theorem deals with comaximal-irreducible ideals and the proof is similar to that of the third decomposition theorem. First, construct the invariant relative prime irreducible ideals R_i as in the previous proof. Group them into disjoint classes such that two ideals belonging to different classes are comaximal and that no class may be further divided in this way.

After having separated the ideals into t different classes $G_1,...,G_t$ let T_i be the *l.c.m.* of all the ideals included in the class G_i. The set of ideals $T_1,..., T_t$ is used to represent M as desired: $M = [T_1,...,T_t]$ with the T_i's pairwise comaximal- irreducible. Once more, it is straightforward to prove that the grouping into classes G_i is uniquely determined. It follows that the representation is uniquely determined as well. Finally, one can show that given two comaximal ideals, their *l.c.m.* is equivalent to their product.[22] The fourth decomposition theorem is finally formulated as follows:

> XV. Every ideal is uniquely expressible as a product of a finite number of pairwise comaximal irreducible ideals. (Noether 1921, 53)

As in the case of the third decomposition, Noether remarked in a footnote (# 34) that the existence of the decomposition could have been proven directly. The proof presented in the article has the advantage of allowing a closer inspection of "the structure of the comaximal-irreducible ideals."[23] Of course, this sentence may be understood in a vague, informal sense, but it is interesting

---

21. Cf. Noether 1921, 51: If A and B are comaximal, then there exist by definition, an element a in A and an element b in B, such that their product equals e, the identity element of the ring. Take now an ideal T, satisfying $TA \equiv 0$ (B). Obviously also $Ta \equiv 0$ (B), and therefore $Te = T \equiv 0$ (B). In other words, if TA is included in B it follows that T is also included in B, and therefore, by definition, A is relatively prime to B. By symmetry considerations, it is clear that also B is relatively prime to A.

22. Cf. Noether 1921, 53: If $M = [T_1,...,T_t]$ and L_i denotes the intersection of all ideals T_j, different from T_i itself, then L_i and T_i are comaximal. Therefore, there exist elements l_i in L_i and t_i in T_i, such that $e = l_i + t_i$.

Take now any element f in the intersection $[L_i,T_i]$

$$f = f.e = f.t_i + f.l_i$$

But f belongs to T_i, so that $f.l_i$ belongs to $T_i.L_i$. Likewise $f.t_i$ belongs to $f.t_i$ belongs to $T_i.L_i$ and, finally, f itself belongs to $T_i.L_i$. On the other hand, the product $T_i.L_i$ is included, by definition, in the intersection $[T_i,L_i]$.

that this is the only place where the word structure appears, and it denotes an aspect of the inner arrangement and, in particular, the inclusion properties of the ideals in a given ring. Moreover, the classes G_i used in the proofs of the third and fourth decompositions are termed by Noether "groups" (*Gruppen*). By this, she obviously did not mean groups in the usual, strict sense of the term, but a rather general collection of certain classes of ideals.

Noether next considered the validity of the decomposition theorems for non-commutative rings. In her joint paper with Schmeidler, Noether had considered ideals in non-commutative rings of polynomials. In "*Idealtheorie*" she reformulated in abstract terms some of the results that had been obtained there. In order to do this, she first defined a module over a (not necessarily commutative) ring and showed that an ideal in a ring is a particular case of a module. She then reformulated in terms of modules all the concepts introduced in the opening sections for ideals, and also the first decomposition theorem: Any module with *a.c.c.* may be decomposed as a reduced intersection of a finite number of irreducible modules, and the number of components is invariantly determined. The proof does not use the commutativity of multiplication in the ring and is, therefore, true for the non-commutative case. One cannot, however, translate to the case of modules concepts such as prime or primary ideal, since their definition presupposes commutativity. Noether's fourth decomposition does not involve the product of the ring and therefore it holds independently of commutativity. Its proof, however, relies on the third decomposition, and therefore a completely different proof must be given if one is interested in extending its validity to the non-commutative case. But as she already mentioned in a footnote, the decomposition itself (not the uniqueness) can be directly proved by an argument similar to that of the proof of the first decomposition. Noether explained the peculiarity of these two decompositions:

> Whereas all these theorems are grounded only on the concepts of *divisibility* and of *l.c.m.*, the following theorems of uniqueness are based essentially on the concept of *product* and therefore admit no direct translation. (Noether 1921, 56)

Thus, Noether's increasing interest in the non-commutative goes hand in hand with her increasing understanding of the internal logic of factorization properties, and in particular, with the recognition of those properties that depend uniquely on the inclusion properties of the sub-domains of the rings

---

23. Noether 1921, 52: "Der hier gegebene Beweis gibt zugleich Einblick in die Struktur der teilerfremd-irreduzibelen Ideale."

investigated rather than on their two operations. This interest will translate into important results in her future works.

These are the four factorization theorems that Noether proved in her 1921 paper. All four were already known for the domains of polynomials, but the previously existing proofs relied, as Noether herself remarked, on the fact that every polynomial is uniquely representable as a product of irreducible polynomials. The latter result depended itself on properties of polynomials that are directly derived from those of the systems of real and complex numbers. Noether's main insight was thus to understand that the factorization of ideals may indeed be formulated *independently* of this property of polynomials, and that what determines it is, in fact, what may be properly called a "structural property", namely, the *a.c.c.* Moreover, Noether clearly stated the uniqueness conditions of the different factorizations. In the introduction she mentioned a series of earlier articles in which uniqueness had been discussed, but always in a partial, non-systematic fashion.

The decomposition theorems of this paper are all "multiplicative" ones: namely, they present the ideals as products or *l.c.m.* of other ideals. However, Noether mentioned the existence of other, "additive" decomposition theorems—such as those proved in her joint paper with Schmeidler—and pointed out that a translation of the latter kind of decompositions into the former would be possible using the concepts introduced in this article, and some additional ones.[24] These additional concepts were meant to define properties of the systems of residue classes defined by the ideals in the ring. The latter happen to constitute themselves a ring satisfying the same general properties as the original one. In fact, the original ring might itself have been considered as a special case of a residue system.[25] Noether's structural concerns appear clearly conveyed by these kinds of remarks. Moreover, they condition the direction of her research, since they suggest both the kinds of questions that should be addressed and the preferred way to answer them.

Noether chose here decomposition questions as her main focus of interest, but these are addressed from a peculiar perspective. Not only are the subjects of her research abstractly defined by means of axioms, but, moreover, contrary to the classical approach in which the systems of numbers have a privileged

---

24. Noether 1921, 27-28.
25. Noether 1921, 28.

status and the properties of other algebraic constructs may be derived from those of the number-systems, in the approach advanced here by Noether there is no essential difference between a given algebraic system and the building blocks that constitute it. Faced with these systems the question arises, how their properties (in this case, decomposition properties) are passed over back and forth, from the original structure into its building blocks (the quotient systems). These kinds of questions had already been worked out by Steinitz in the case of fields, but in extending Steinitz's point of view to rings, Noether was in fact establishing a guideline for research with much more general implications, namely, that these questions are relevant for a more general class of interesting, abstractly defined algebraic systems.

According to van der Waerden, Noether's 1921 article signified the great success of Hilbert's approach to the theory of polynomial forms, since it contained a proof of the Lasker theorem based only on the finiteness condition. Noether's work implied that the theorem is valid for all rings in which every ideal has a finite basis, thus reinforcing the point of view put forward by Hilbert.[26] While this evaluation is clearly correct, it seems to overlook a more general change in perspective that Noether's own work implied: whereas in Hilbert's work the finiteness condition was derived from the properties of the concrete mathematical entities, Noether's use of the condition was meant as an abstract, implicit definition valid for all those entities for which the factorization theorems hold.

And yet the point of view put forward by Emmy Noether in her 1921 article could be further elaborated beyond the results appearing here, and a more thoroughly axiomatic treatment of the problem of factorization in abstract rings could still be advanced. This was done by Noether herself in 1926.

5.3   Abstrakter Aufbau der Idealtheorie

The next major paper by Emmy Noether on abstract algebra appeared in 1926. Noether reconsidered in it the problem of factorization, but this time within a much more maturely conceived axiomatic framework. In "*Idealtheorie*", the *a.c.c.* was assumed at the outset and no clear connection was established between individual axioms and theorems. Noether now started from a commutative ring R and proved a series of decomposition theorems by succes-

---

26. Van der Waerden 1933, 402.

sively introducing additional axioms, as the need arose for particular arguments.

Taken together, the five axioms introduced by Noether in her 1926 article add up to define what is known nowadays as a "Dedekind ring", namely a ring in which every primary ideal is a power of a prime ideal. The five axioms are formulated at the beginning of the paper as follows:

I. R satisfies the *a.c.c.*

II. Every proper descending chain of ideals in R, each of which contains a given non-zero ideal, is a finite chain.

III. There exists a unit for the multiplication in R.

IV. There are no divisors of zero in R.

V. The field of fractions of the ring R is integrally closed (i.e., each element of the field of fractions, which is an integer with respect to R, belongs in fact to R).

The first two axioms are chain conditions, while the last three deal with multiplicative properties. Since the very definitions of prime and primary ideals are based on the commutativity of the multiplication in the ring, Krull classified this paper, in his 1935 monograph on ideal theory, as belonging to what he called the "multiplicative theory of ideals", as opposed to the "additive" one. The additive theory of ideals approaches the ring R as an Abelian group together with a domain of operators R. This approach was directly inspired by the works of Kronecker, Lasker and Macaulay. The multiplicative approach, on the other hand, was inspired by Dedekind's ideas; it examines the decomposition properties of the ideals from the perspective of the "algebra of ideals" defined by Dedekind in his second version of the theory of ideals. In particular, it relies on the multiplicative properties of the "invertible ideals" (*umkehrbare Idealen*) defined therein (§ 2.2.3 above). Thus, this work of Noether was a crucial link in the passage from the operational conception of algebra to the structural one. On the one hand, it is still strongly connected to Dedekind's original conception, but on the other, it clearly bears many of the central features of the non-formal idea of "mathematical structure." It also develops many of the central ideas upon which Ore would later attempt to base his formal concept of **structure**.

Before proving any factorization theorems, Noether opened the article with some general considerations about rings. Noether recalled in a footnote (#6, p. 29) the abstract definition of a ring as a set endowed with two operations satisfying certain properties. Once again one finds here the somewhat surprising and redundant requirement that a "equality relation" (*Gleichheitsre-*

lation) be defined on the set. Further explanations added by Noether are strongly reminiscent of those invoked by Fraenkel in his papers on rings. This "equality relation" is tightly connected with the passage from the original ring to its sub-domains or to its extensions, and therefore it seems that, in Noether's view, it played an important role in elucidating the structure of the ring actually investigated. Thus, Noether wrote in the same footnote:

> If, however, the underlying definition of "equality" is not the set-theoretical identity, then, in order that the same definition of "equality" might be applied to each sub-system, every such sub-system (sub-ring, module, ideal) must contain, together with a given element, all the elements "equal" to it. (Noether 1926, 29)

Any mention of "element" in a concept or a statement should be understood as denoting a class of equivalent elements under the given "equality relation." Sometimes this is the usual identity relation, but in some other instances it is not. Therefore—Noether added—in the final analysis any such equality relation can be grasped as the usual identity of set-theory by considering the classes as the elements themselves. Noether mentioned in this footnote one of her former students, Robert Hölzer, who died shortly after finishing his dissertation. She credited him with the ideas expressed in the above remarks, but we have already seen them appear earlier in various contexts.

This concept of "equality relation" is mentioned again only in section 4 which constitutes a brief digression from the main issue of the paper: it contains an abstract treatment of the isomorphism theorems for modules over an arbitrary ring. Thus, the passage to the quotient module (*Restklassenmodul*) is realized in this section by a change in the *Gleichheitsrelation*: given a module *M*, and a sub-module *U*, the quotient module is defined as a new module *M'* identical to *M*, except that "congruence modulo *U*" is taken now as the "equality relation" (p. 40). Noether also formulated the theorems themselves in terms of this concept. It should be noticed that the isomorphism theorems had been were already widely known in separate contexts for group theory and for ideals. In particular cases they had also been established by Dedekind for certain kinds of modules. Noether gave here a general formulation of the theorem and several simple proofs valid for all algebraic domains. She sketched the proofs of the first and second theorems of isomorphism and suggested how these could be formulated exclusively in terms of inclusions and intersections of submodules while explicitly overlooking the operations defined among the elements of the module. This avoidance is not casual; it corresponds to Noether's desire to attain as great a generality as possible in her proofs and formulations.

In her 1926 article Noether considered abstract rings, not only as an appropriate framework for formulating general theorems of factorization, but increasingly as an object of intrinsic interest that raised structural questions worthy of detailed study. As with her 1921 paper, it is convenient to see in some detail in the next few pages, the technical details of this study, which show how these general structural concerns are translated into meaningful mathematical results.

In the first sections of the paper Noether defines again the basic concepts of the theory. Take a ring T with identity and having no divisors of zero, and consider a fixed sub-ring R of T containing the identity. An R-Module is defined as a sub-set of T satisfying the usual conditions for the addition and for the product by an element of R. An ideal is a module all of whose elements belong to R. Now, consider an extension ring S of R, fully contained in T. Noether calls such a ring an *Ordnung*. It may appear strange at first sight that Noether introduces such a definition, since an *Ordnung* is nothing but a ring. But recall that this concept had been originally defined in Dedekind's theory of ideals to denote those systems of numbers for which the factorization theorem of ideals hold; the system of algebraic integers is one such *Ordnung*, but not the only one (§ 2.2.3). Dedekind had suggested that there is a general theory of *Ordnungen* of which the theory of algebraic integers is just a particular case. Noether undertook to develop here that suggestion of Dedekind, within the framework of abstract rings. By sticking to Dedekind's original train of thought, Noether stressed the original meaning attached to the concept of *Ordnung* by Dedekind himself; she directly translated his idea into the setting of abstract rings and their extensions. Then she added a concept that can be seen in retrospect as redundant to her conceptual system, although it was not to Dedekind's.

Now, given an element α of T, the *R-Ordnung* generated by α is, as usual, the intersection of all *R-Ordnungen* containing α. Denote by U_n the *R-Ordnung* generated by $\alpha^0, \alpha, ..., \alpha^{n-1}$. Clearly U_n is contained in U_{n+1}. Now, an element α is called an integer with respect to R, if the sequence of ideals $U_1, U_2, ..., U_n, ...$ is finite. Noether proved that this definition is equivalent to Dedekind's general definition of integer in a domain T of algebraic numbers: α is an integer in T, if there exist coefficients $r_i \in R$, such that $\alpha^n + r_1 \alpha^{n-1} + ... + r_n = 0$. This is a remarkable step in Noether's reformulation of Dedekind's theory of ideals in the context of a structural theory of rings, not so much for the direct results to which it leads, but especially for the conception of algebra that it reflects. In fact, here Noether went one step further in fulfilling Dedekind's own methodological guidelines. She presented the tools of the theory—such as ideals and modules— as well as the concept of integer itself, in terms of sets (which are now abstract sets, rather than collections of rational numbers) and their inclusion properties.

Noether also showed that the set S of all integers of T is an *Ordnung*, and that, moreover, if α is an integer with respect to an *Ordnung* S then it is integer with respect to R. If all the integers of T relative to R belong to R, then R is said to be integrally closed. It is easy to see how Noether simply translated, into the new abstract framework, all the concepts with which Dedekind solved his original problem for the ring of algebraic integers.

This is also exemplified in the following section of the "*Aufbau*" where Noether dealt with finite modules, namely, modules for which a finite basis exists. Noether proved the following theorem:

> If R is a commutative ring with unit satisfying the *a.c.c.*, and M is a finite R- module, then also M satisfies the *a.c.c.* (Noether 1926, 34)

Notice that the theorem is conceived, formulated and proved "structurally": it considers an abstractly defined domain and establishes the conditions under which certain properties of the basic domain are passed over to a new domain built on it. Of special significance is the fact that the proof relies on a certain correspondence between the lattice of sub-modules of M and that of the ideals of R. This lattice expresses in a succinct fashion the inclusion relations of the sub-domains of the two domains in question.

This "structural" approach is explicitly worked out when Noether explains how the properties expressed as axioms I to V carry over from certain basic rings to their finite extensions. In particular, if axioms I-V hold for a given ring R, they hold for the domain S of all integers in T relative to R. At this stage it is clear how to proceed:

> It is thus enough to prove for subordinate fields of numbers and of functions that the basic domains satisfy axioms I-V: integer numbers, single-variable polynomials, functional domains of many-variables polynomials. (Noether 1926, 37)

In the case of fields, Steinitz had established that the most basic fields are the prime fields associated with any field. These are the building blocks of field theory. The factorization theorems constitute a natural step in the process of determining the building blocks of abstract ring theory and of realizing the structural program in it.

Noether next considered the theorems that follow from Axiom I (the *a.c.c.*) alone. Although it was not specified in the opening section together with the other assumptions, Noether assumed here that a well-ordering can be defined on the elements of R. She also assumed that a well-ordering on the system of ideals of R may be deduced from the well-ordering of the elements. Recall that Steinitz had separately considered those theorems depending upon the assumption of the axiom of choice (§ 4.2 above). Naturally, no truly abstract treatment of an algebraic domain can be undertaken without this kind of consideration, and here Noether for the first time explicitly refers to it.

The first decomposition proven under those assumptions in 1926 is the factorization into indecomposable ideals:

> I. Any ring satisfying the *a.c.c.* is decomposable as a finite intersection of irreducible ideals.

Recall that the proof given in "*Idealtheorie*" relied on the *a.c.c.* in a preliminary theorem in order to show that any representation of an ideal as intersection of a finite number of ideals containing it can be substituted by an equivalent, *reduced* one. This preliminary step is skipped here using the well-ordering.

If M itself is not irreducible, let $M = [A,B]$ be a representation of M as intersection of two ideals containing it. A contradiction to the *a.c.c.* is obtained directly by assuming that A and B are not irreducible. In fact, if that is the case, then M is properly contained in an ideal for which theorem I is not valid. Choose A_1 (via the well-ordering) as the smallest among such ideals and proceed to form an infinite chain. Especially relevant for the present discussion is the footnote added here by Noether. She draws the reader's attention once again to the system of sub-sets as a valuable source of information. Theorem I, she claimed, has a "purely set-theoretical character." By that she meant that the theorem can be formulated exclusively in terms of properties of inclusions of sets and "independently of any operation" (*unabhängig von allen Verknüpfungen*) defined among the elements of the system. In fact, given a set M, and a system Σ of sub-sets of M satisfying a certain chain condition (indeed, similar to the *a.c.c.*) one can formulate the following representation theorem

> * Every set belonging to Σ is representable as the intersection of a finite number of sets belonging to Σ. (Noether 1926, 46, note 27)

The expression "purely set-theoretical considerations", in Noether's usage, does not refer to concepts nowadays related to the theory of sets (membership, power, etc.). It denotes arguments for proof in algebra, which do not rely on the properties of the operation defining the domain under inspection, but rather properties of the inclusions and intersections of sub-domains of it. In her works on algebra, such arguments only appear implicitly in certain proofs, or in footnotes like the one quoted above. However, the expression "purely set-theoretical considerations" will appear again in her later works. We shall return to it in § 5.4 below.

As in her 1921 article on ideals, Noether proved here that irreducible ideals are primary. Also this fact is a direct consequence of the *a.c.c.* It follows that any ideal admits a "shortest representation" as an intersection of a finite number of primary ideals associated with different prime ideals (Theorem III, p. 47).

Noether next considered the consequences of adding axioms II-V to the *a.c.c.* Assuming all the five axioms allows proving that primary ideals are powers of their associated prime ideals. As a consequence, the main decomposition theorem yields:

> VI. If a ring R satisfies axioms I-V, then every ideal in R is uniquely representable as intersection of a finite number of powers prime ideals. (Noether 1926, 53)

In the closing section of the article Noether proved the equivalence of the double-chain condition (i.e. the simultaneous occurrence of the ascending and descending chain conditions) with the existence of a composition series. All her arguments are formulated here in terms of modules in general, and the "pure set-theoretical character" (in the sense stated above) of the proofs is preserved throughout. There is no mention of the operation among elements of the module, and only arguments based on the interrelation of the ideals and sub-modules involved. Moreover, as Noether explained in a footnote, since the modules are in fact Abelian groups under addition, the composition series are indeed principal series.[27] All the theorems in this section are formulated in

---

27. A sequence $E \leq H_n \leq ...H_1 \leq G$ of sub-groups of a given group constitute a composition series when each subgroups is normal relative to its immediate predecessor and when the factor groups H_i/H_{i+1} are all simple groups. Clearly, it might be the case that H_{i+1} is normal in H_i, but not in G. If, however, all the H_i's are normal in G, then the series is called a principal series. For Abelian groups, every composition series is also a principal series.

terms of composition series, and thus, remain valid for the case of non-Abelian groups.

The converse is also true: the existence of a composition series implies the double chain condition. As a by-product of this result, Noether also proved a general version of the Jordan-Hölder theorem using induction on the length of the series in the standard way. In a footnote (# 34, p. 58), she mentioned Dedekind's 1900 article on lattices and the proof of this theorem that appears there (§ 2.3 above). She also added some remarks concerning the validity of the theorem for more general cases. The attempt to find the most general formulation and proof of this theorem also turned out to be a central motivation for Ore's work on the foundations of abstract algebra (§ 6.5 below).

So much for Noether's article of 1926 and the structural treatment of concepts that characterize it. We proceed to discuss now other ideas from her subsequent work.

5.4   Later Works

During the years 1920-1926 Noether concentrated her efforts on developing the abstract theory of ideals. The next stage of her research centered on the study of non-commutative algebra and representation theory. As was seen in the preceding sections, Noether's innovations stemmed, partially a least, from a desire to provide proofs of factorization theorems, whose validity is independent of the commutativity of the product in the domain considered. In fact Noether believed that a thorough study of the non-commutative cases would be of great significance for better understanding the commutative one.[28] This trend was initiated in her joint paper with Schmeidler, and also appeared in some of the cases of factorization that she developed in her articles. When she returned to non-commutative algebra in 1927, she could count on an arsenal of new methods and ideas that she had developed by herself, and that were applicable to this kind of study. Using the concepts of modules, sub-modules, direct sums, homomorphisms, etc. Noether was able to unify many results that had accumulated over the past years. In particular, she reformulated Wedderburn's theory of decomposition for a general ring in terms of ascending and descending chain conditions, thereby expanding the validity of results that had

---

28. See for instance the openings remarks to her invited address to the 1932 International Congress of Mathematicians (Zürich), in English translation, in Brewer & Smith (eds.) 1981, 167.

formerly been proven only for algebraically closed fields of characteristic zero to fields of an arbitrary characteristic.[29]

Up to what degree did Noether intend to generalize her results? What was the most general level that she thought convenient for developing relevant algebraic results? Beyond her published work, some unpublished ideas she seems to have been working on in her later years may indicate that Noether was aiming at a rather more general abstract framework than she actually succeeded in bringing into publication. Shortly after her death, Pavel Alexandrov (1896-1985), by then president of the Moscow Mathematical Society, delivered a memorial talk in honor of Emmy Noether. She had been in close contact with many Russian mathematicians, some of whom were deeply influenced by her.[30]Alexandrov himself had made important contributions, in collaboration with Heinrich Hopf (1894-1971), to the foundations of algebraic topology. In his obituary talk on Noether, as well as in various other opportunities, Alexandrov stated very clearly that many central ideas of his joint work with Hopf had been suggested to them by Noether. In particular, Alexandrov referred to the idea of describing the invariants of a topological space in group-theoretical terms, rather than in the by then customary Betti numbers.[31] A detailed account of the gradual adoption of the structural approach in topology would be well beyond the scope of the present book. Nevertheless, Alexandrov's remark would suggest that any such account should indeed make reference to Noether's central contribution to that discipline.

But, more relevant to our account, Alexandrov's obituary also suggests that Noether might have been working on an early generalization of the con-

---

29. Her most important results were published in Noether 1929; 1933. See Curtis 1999, 214-223; Lam 1981; Scharlau 1999. On Wedderburn, see Parshall 1985.

30. Cf. Kimberling 1981, 24.

31. See also Hopf 1966; Hirzebruch 1999. Dieudonné 1984 describes this episode and also mentions the early use of homology groups in topology by W. Mayer (in Mayer 1929). In his article, Mayer—who worked at the time in Vienna—did not credit Noether with the original conception of this idea (Hopf did so in several places: e.g., Hopf 1928). However, Dieudonné suggested, "by that time the spirit of 'modern algebra' had spread to many German universities" and therefore "it is not unlikely that it could also have reached Vienna" (Dieudonné 1984, 6). It is worth remarking that in a sequel to this article of Dieudonné, Mac Lane (1986a, 306) wrote that Dieudonné's claim "accords with a favorite view that Mathematical ideas originated in Göttingen and then spread to lesser places. It was not always so simple." Thus Mac Lane described the intensive activity in combinatorial topology that existed in Vienna since 1907 and concluded: "Probably ideas passed back and forth in both direction between Vienna and Göttingen (and Berlin, Moscow and Paris)."

cept of algebraic structures. Alexandrov and Noether met several times during their careers, and both considered their interchanges of ideas to have been fruitful for their respective works. Concerning one such meetings, Alexandrov recalled:

> My theory of continuous partitions of topological spaces arose to a large extent under the influence of conversations with her in December-January of 1925-1926, when we were both in Holland. On the other hand, this was also the time when Emmy Noether's first ideas on the set-theoretic foundations of group theory arose, serving as the subject of her course of lectures of 1926. In their original form these ideas were not developed further, but later she returned to them several times. The reason for this delay is probably the difficulty involved in axiomatizing the notion of a group starting from its partition into cosets as the fundamental concept. But the *idea* of set-theoretic analysis of the concept of group itself turned out to be fruitful, as shown by recent works of Ore, Kurosh and others. (Alexandrov 1935, 9. Italics in the original).

What were these ideas of Noether concerning the "set-theoretic foundations of group-theory"? We have no direct evidence to decide what Alexandrov meant by that. However, as was seen above in § 5.2, in her 1921 paper on ideals Noether used the adjective "set-theoretical" to refer to theorems and proofs that are expressed "independently of any operation" (*unabhängig von allen Verknüpfungen*) defined among the elements of the system considered, and which are formulated exclusively in terms of properties of inclusions of its sub-systems. Such an approach would certainly correspond to the problem, mentioned by Alexandrov, of "axiomatizing the notion of a group starting from its partition into cosets as the fundamental concept." One might thus tentatively conjecture that these are indeed the ideas that Alexandrov was referring to on that occasion.

Even if these ideas of Noether were not published, the fact that her lectures of 1926 (according to Alexandrov's testimony) dealt with them indicates that Noether saw in their elaboration a meaningful project. In fact, many direct accounts of her students describe how she used to lecture on those issues on which she was currently working. In those lectures many ideas were exposed in unfinished form, and the interchange with her students was often crucial in producing the final shape of her research. This is perhaps the reason why her lectures are often described as rather unclear but highly thought-provoking.[32] Be that as it may, even if Noether herself did not elaborate these ideas into a putative full-blown theory, they did provide a basis for later works of Ore and of the Russian group-theorist Alexander Kurosh (1902-1971). This latter fact would also be in agreement with Alexandrov's account. These ideas will be considered again when discussing Ore's program in § 6.6 below.

Noether's conception of an algebraic structure is known to us only implicitly, through the images of knowledge manifest in her published works. An attempt to formulate an axiomatically based, general theory of algebraic structures does not appear among them, but one wonders what would have been her attitude towards such are reflexive theory. The first such theories, as considered in the second part of this book, appeared very close to the time of her death. Unless direct evidence will eventually be found, the above discussion may be taken as conjectural evidence that she might have been interested in them.

5.5 Emmy Noether and the Structural Image of Algebra

Emmy Noether's work on abstract ideal theory constitutes a main turning point leading to the new, structural image of algebra. On the one hand, her ideas are still close enough to those elaborated by Dedekind within the framework of algebraic number theory, to Hilbert's research on invariants and algebraic number theory, and to the works of Lasker and Macaulay on polynomials. Therefore, they enable a direct identification of her immediate roots and motivations. On the other, her works display all those features that characterize modern abstract algebra as a discipline of structures. Noether fully adopted the methodological guidelines established by Dedekind together with his specific achievements—especially in algebraic number theory— and sought to combine the latter with similar results that had been attained in the theory of polynomials. Like Dedekind before her, she strove to formulate precise concepts as the legitimate basis for explaining the similarity and the proximity of apparently distant theories.

In the images of knowledge of both masters every theorem of mathematics has an essential reason, whose clear elucidation is a primary task of research. It is not enough to justify a result by a series of diverse *ad-hoc* arguments or by rude calculation. Rather, it is necessary to isolate and present the actual essence underlying that result. This can only be done after having put forward

---

32. Thus, we have Saunders Mac Lane's description: "I attended one course of Emmy Noether's on representation theory and found her lectures enthusiastic but obscure." (Quoted from Mehrtens 1979, 157.) Cf. also Lam 1981, 146.

the proper concepts which help telling the essential from the mere auxiliary. This view, which we saw thoroughly implemented in Dedekind's works, comes to the fore also in Noether's works analyzed here. It is encapsulated in the following formulation attributed to her:

> If one proves the equality of two numbers a and b showing first that $a \leq b$ and then $a \geq b$ it is unfair; one should instead show that they are really equal by disclosing the inner ground of their equality.[33]

This was a leading idea of Dedekind which Noether wholeheartedly adopted and developed further on.

Nevertheless, her work differs from Dedekind's in the greater generality of her results and in the much clearer axiomatic presentation of the ideas. Dedekind had opened the way to generalized factorization theorems by introducing new concepts, built by focusing on certain characteristic collections of numbers (and, in his joint work with Weber, on collections of functions). Noether advanced one step forward. She abandoned the restrictive framework of systems of numbers and reformulated Dedekind's concepts in terms of collections of *abstract elements* of an abstract ring. Dedekind's recurrent use of sets of numbers and their inclusion properties as the conceptual ground of many of his proofs was fully absorbed by Noether and re-elaborated into the systematic use of the technical device of the *a.c.c.* The centrality of the latter had been formerly identified by Hilbert, but—at variance with him—Noether now conceived and applied it in its abstract formulation. The increasing interest of her later years of research in the non-commutative case gave this tendency its most extreme expression. In fact, in the non-commutative case it is somewhat limitative to rely on the properties of the operations defined on the individual elements of the abstract ring. Decomposition theorems in this case are thus best proved purely in terms of inclusion properties of sub-domains. Proofs of this kind should reveal, in Noether's view, the real "structure of the ideals of the ring."

Noether's abstractly conceived concepts provide a natural framework in which conceptual priority may be given to the axiomatic definitions over the numerical systems considered as concrete mathematical entities. With Noether, then, the balance between the genetic and the axiomatic point of view begins to shift more consciously in favor of the latter. This new balance was a necessary condition for the redefinition of the conceptual hierarchies, and for

---

33. Quoted from Weyl 1935, 148.

the establishment of a new image of knowledge. In the latter, the idea of structure dominates algebraic research and the various number systems are particular instances of it. Nevertheless, Noether's axiomatic conception, perhaps because of her own deep acquaintance with the classical aims of concrete algebraic research, remained very close to Hilbert's own. For Noether, the axiomatic analysis of concepts is only one of two complementary aspects, rather than the exclusive essence of mathematical research. Thus she was quoted as saying:

> In mathematics, as in knowledge of the world, both aspects are equally valuable: the accumulation of facts and concrete constructions and the establishment of general principles which overcome the isolation of each fact and bring the factual knowledge to a new stage of axiomatic understanding.[34]

Steinitz's work had highlighted the potential insights that abstractly defined algebraic theories can offer. It stressed their meaning not only concerning relatively subordinate mathematical ideas (such as permutations and continuous transformations of the plane), but also the numerical systems which constitute the very heart of higher mathematics, and which hitherto provided a conceptual basis for the whole of algebra. Noether elaborated further this trend of ideas. She pursued the study of abstract rings as an object of interest in itself and used it as the main conceptual framework of algebra. Her work had a greater overall impact on algebra than Steinitz's, if only because, having appeared about ten years later, it showed that Steinitz's program applied not only for the particular case worked out by him, but for many other significant cases as well. Group theory was thus the first algebraic discipline to be abstractly investigated, and field theory the first discipline that arose from the research of numerical domains into an abstract, structural subject. The study of ideal theory in an abstract ring consolidated the idea that a more general conception lay behind all this: the conception that algebra should be concerned, as a discipline, with the study of algebraic structures.

Beyond the intrinsically mathematical virtues of Noether's work, it also seems clear that the great influence she was able to exert can be explained by the quantity and the quality of her Göttingen students. It seems unlikely that such a circumstance could have come about in a different institutional environment. Obviously, neither Dedekind, nor Steinitz, Fraenkel, Lasker or Macaulay—regardless of their personal abilities to create a stable group of stu-

---

34. Quoted in Alexandrov 1935, 4.

dents around them and to communicate to them their own ideas—had an opportunity to do so in conditions similar to those enjoyed by Noether in Göttingen.[35]

But at the purely mathematical level, Noether did not simply combine the achievements of her predecessors; she came forward with a mathematical work which was essentially different from theirs. Of course she proved many new theorems, and developed new theories. But, in addition, she also helped establish new images of algebra that were to dominate the stage for many decades. The idea of an algebraic structure tacitly dominates Noether's work more deeply than that of any of her predecessors, from Dedekind on. At least concerning this basic difference in the images of algebra of Dedekind and Noether, one cannot but disagree with Noether's famous assessment: "*Es steht alles schon bei Dedekind.*"[36]

To conclude, Noether's influence on van der Waerden's *Moderne Algebra* and—through the impact of the latter—on modern algebraic research at large can be now more clearly characterized. Van der Waerden adopted many results of Noether and presented them in a systematic way. Of course, Noether—original as her thought was—was not the only important algebraist from whom van der Waerden took his ideas. He also included methods and results of Artin, Krull and other important mathematicians who were working on the same issues and were influenced by (and probably also influenced) Emmy Noether. In van der Waerden's presentation, different mathematical domains are considered as individual instances of algebraic structures, and therefore undergo similar treatments: they are abstractly defined, they are investigated by recurrently using a well-defined collection of key concepts, and a series of questions and standard techniques is applied to all of them. Some of these features had already appeared in Steinitz's and in Fraenkel's work, but Noether's research on ideals definitely contributed to give legitimation and interest to the possibility of applying them systematically. Van der Waerden also included issues that had not been developed by Noether but which found a natural place among the other algebraic domains. The most outstanding example of this was

---

35. In fact, it must be pointed out, that Fraenkel, working in his new domain of set-theory was indeed able to create a highly successful and influential school at the Hebrew University in Jerusalem after his arrival there in the mid-thirties.

36. Noether's assessment has been repeatedly echoed. Interestingly for the present discussion, van der Waerden quotes it as the opening motto of his 1975 article "On the sources of my book *Moderne Algebra.*"

group theory, which appeared in van der Waerden's book for the first time as an algebraic theory of parallel status to field theory, ring theory, etc.

Noether's work implied significant innovations in both the body and images of algebra, and van der Waerden's textbook helped disseminate those innovations worldwide. Moreover, the impact of van der Waerden's book triggered a process that eventually led to relocating the whole discipline, in its new garb, as an intermediate graduate student's level subject (and later on—mainly through the influence of Bourbaki's treatise—perhaps even undergraduate level). In Noether's own lectures and in those of her immediate followers, the issues considered in van der Waerden's book had been seen as a rather advanced and specialized domain. However—it should be remarked once again—this transformation was not accompanied by an explicit explanation of what it entailed at the level of the images of knowledge. The rise of the structural approach in algebra was neither preceded nor dictated by a formal, reflexive mathematical elucidation of the concept of algebraic structure or, more generally, of mathematical structure. It was not long, however, before such reflexive attempts were underway. The origins and the initial steps of some of those attempts constitute the main subject of the second part of this book.

Part Two: Structures in the Body of Mathematics

The preceding chapters described the gradual transformation of algebra into the discipline dealing with algebraic structures, with special attention to the gradual penetration of structural concerns into research on the theory of ideals, beginning with Richard Dedekind and up until the work of Emmy Noether. Our account of the rise of the structural approach to algebra stressed the watershed marked by the publication in 1930 of van der Waerden's textbook, and by the innovative image of algebra it put forward. Nevertheless, as was also stressed, at no place in his book did van der Waerden include an articulated explanation of the essence of this new conception or of the expected advantages of its adoption. The idea of structure appeared implicitly in the text, exhibiting through its actual applications both its nature and its advantages.

Later, a-posteriori explanations of the meaning and the applications of this new image of algebraic knowledge appeared in various contexts. They appeared, in the first place, in the introductions to textbooks written after *Moderne Algebra*, that adopted a similar approach.[1] Likewise, several expository articles were written over the years explaining the main traits of the new spirit of algebra. But besides this kind of explanation, several reflexive theories were formulated that attempted to elucidate, in strict mathematical terms, the idea of a mathematical structure and its significance within the whole edifice of mathematics. Among the most elaborate such theories were Oystein Ore's theory of **structures**, Bourbaki's theory of *structures*, and the theory of categories and functors, first developed in the USA by Samuel Eilenberg and Saunders Mac Lane (1909-).[2] A detailed analysis of these three theories, their

---

1. Such as for instance: Birkhoff & Mac Lane 1941; Chevalley 1956; Dubreil & Dubreil Jacotin 1964; Hall 1966; Kurosh 1963.

2. To avoid confusion, in the second part of the book we adopt the convention of referring to Ore's concept as **structure** (boldface), and to Bourbaki's technical term as *structure* (italics). The general, non-formal notion, intended with its usual meaning, is simply referred as structure.

motivations, early evolution and further elaboration or lack of it constitute the main subject of this second part of this book.[3]

The increased adoption of the structural approach in algebra, and later in other mathematical domains, raised new questions concerning the nature of mathematical knowledge and the role of the idea of structures in it. The creation and elaboration of reflexive theories of structures were motivated to a large extent by the desire to address those questions. But at the same time, these theories may also be seen in retrospect as part of a more general trend characteristic of the first decades of the present century, during which reflexive theories enjoyed impressive success in elucidating central questions regarding the nature of mathematical entities and mathematical practice. Metamathematical theories, such as proof theory and mathematical logic, had been steadily providing significant insights—supported by mathematical proofs—concerning the nature and scope of mathematical knowledge. This success could but encourage the formulation of similar, reflexive theories, aimed at elucidating related meta-questions about mathematics, such as those raised by the recent rise of the structural approach.

Before proceeding to consider the reflexive theories that constitute the main subject of the second part of this study, we first discuss some views set forward in 1939 by Mac Lane, with regard to the nature of the new, structural image of algebra. Mac Lane, one of the leading figures in algebraic research in the USA, was not only one of the creators of the theory of categories, but also played a major role in its further development and widespread adoption.[4]

Mac Lane first became acquainted with modern algebra, and particularly with group theory and Galois theory, through Ore's lectures at Yale back in 1929-30. Under the influence of these lectures, of his reading of Otto Haupt's 1929 textbook on algebra, and of Ernst Steinitz's theory of fields, Mac Lane decided to write a master's thesis that was a rudimentary attempt to generalize

---

3. It should be stressed, however, that there were at least two further, minor attempts to develop mathematical theories, which can be seen as elaborate axiomatic treatments of a generalized concept of mathematical structure. These attempt were advanced, separately, by the Odessa-born, French mathematician Mark Krasner (1912-1985) in an article of 1938 (see Krasner 1938), and by the Portuguese José Sebastião e Silva (1914-1972) in an article originally published in Italian in 1945 (See the English translation in Sebastião e Silva 1985). These two works had no tangible influence, nor were they further pursued by the authors themselves (except for two short notes in the case of Krasner: Krasner 1939, 1950). These two attempts are therefore not discussed here. However, on a possible renewed interest in Sebastião e Silva's ideas, see Da Costa 1986.

4. For a biographical note on Mac Lane, see Putnam 1979.

the theory of fields into a kind of universal algebra.[5] His actual research career, however, began as a Ph.D. student in Göttingen in 1930-31, at a time when the structural trend was finally being consolidated. Mac Lane was then a post-doctoral fellow at Yale, precisely when Ore was working on his **structural** program. Since the 1940s Mac Lane worked for many years in collaboration with Samuel Eilenberg (who was also a Bourbaki member) contributing with many important innovations to algebra; the creation of the theory of categories was only one among their many joint undertakings.[6] Thus, from 1930 onwards, Mac Lane had been an active witness to the development of abstract algebra in all its branches, in particular homological algebra and category theory. Over all those years Mac Lane published important research works in several branches of mathematics and he also published expository articles about mathematics. Moreover throughout his career he expressed a definite interest in basic questions about the essence of algebra and of mathematics in general. Although foundational questions were certainly not Mac Lane's main field of research throughout his long career, he did address such questions several times from various perspectives and he also wrote abundantly on the issue.

In an article written for a conference held in 1939 at the University of Chicago, Mac Lane surveyed "Some recent advances in algebra", while attempting to give a general picture of the "central problems considered and the types of answers obtained" over the past years of algebraic research. His survey amounted to an overview of the changes in the body of algebraic knowledge. Yet, Mac Lane also stressed what in his view were the dominant images of algebra at the time. Accounts such as this were of vital importance to the mathematical community, given the deep changes that algebraic research had undergone in the two previous decades. Beyond the individual description of the main current problems of algebra, Mac Lane also attempted to answer the then pressing question: "What is algebra?". His answer stressed the "structural character" of current research, while explaining what he meant by this. He thus wrote:

> Algebra concerns itself with the postulational description of certain systems of elements in which some or all of the four rational operations are possible: fields, linear algebras, Lie algebras, groups. the abstract or postulational development

---

5. Cf. Mac Lane 1988, 329.

6. Their joint work has been collected in Eilenberg & Mac Lane 1986. See also Mac Lane 1976a.

of these systems must then be supplemented by an investigation of their "structure." (Mac Lane 1939, 17)

This account, however, necessitates explaining what is meant by investigating the "structure" of an algebraic system. Mac Lane did so by inventorying
the kind of possible answers and the expected, legitimate questions of current
research, in a manner similar to our analysis of the structural character of van
der Waerden's algebra in § 1.3 above. Thus, under "structural" questions Mac
Lane included the following:

(a) the number and interrelations of the subsystems of a given system, either subsystems just like the whole system (lattice of subgroups), or of subsystems with
especially characteristic properties (sets of integers, maximal orders, ideals, subfields of an algebra, etc.);
(b) the group of automorphisms of a system, and connections between the subgroups of this group and the subsystems of the given system (Galois theory, class
field theory);
(c) the construction of all systems of specific types out of simpler systems of the
same or other types (the construction of cyclic algebras and matrix algebras, the
reduction of a given surface to a irrationally equivalent surface without singularities, construction of Lie algebras);
(d) alternatively, the description of given systems as subsystems of larger systems (complete fields, power series fields);
(e) criteria of invariants to determine when two explicitly but differently constructed systems are abstractly the same or *isomorphic* (the canonical generation
of a cyclical algebra; the genus as an invariant defined by the differential of a
function field). (Mac Lane 1939, 17-18. Italics in the original)

Most of the features mentioned here by Mac Lane as characterizing the
structural character of the new algebra were already mentioned above in connection with the evolution of algebra from Dedekind to Noether. Thus, regarding (a) it was seen that Dedekind's definition both of the system of integers
within an arbitrary field and, of course, of the ideals themselves were key steps
in his treatment of factorization. So was Noether's treatment of the ideals as a
distinguished type of subrings. Regarding (b), the role of Galois theory in the
rise of the structural approach has been evident throughout. An important
example of (c) (and hence of its reciprocal (d)) is of course the relation
between the properties of a field or a ring and those of their respective ring of
polynomials. It was seen how this question was investigated in increasingly
structural terms in the works of Hilbert, Lasker, Macaulay and Noether. Features (c) and (d) constitute the leading idea in Dedekind's work, and more so

in Noether's theory of factorization of ideals. They were also mentioned (§ 1.3 above) as among the typical questions pursued in van der Waerden's structural analysis of the various algebraic systems.

These features also served as starting points for the reflexive, structural theories that will be described in the forthcoming chapters. Thus Ore's program, as will be seen below, is nothing but the full elaboration of the view that the whole edifice of algebra can be reconstructed starting with feature (a) alone. Ore's program also comprises among its concerns, questions of the kind mentioned in (c), (d) and, naturally, also in (e). In fact, Mac Lane mentioned Ore's early works on the foundation of abstract algebra, which began to appear in 1935, and among the central concerns of current research in algebra he explicitly included the question of whether or not the nature of a group is completely determined by the lattice of its subgroups (pp. 6-7)—a particular example of (a). Bourbaki's theory of *structures*, on the other hand, purports to deal with questions such as described in (c) and (d).[7]

Mac Lane did not mention among the central concerns of current algebraic research the task of elaborating a reflexive mathematical theory that will deal with a general concept of "algebraic structure." However, he himself had announced back in 1934 a forthcoming research paper in which he intended to develop an algebraic formal theory focused on the study of the properties of "structures." In this announcement Mac Lane did not explicitly state how his research would proceed, but it is quite clear that his line of thought was closely connected with that of Ore, with whom he had been working at Yale. Thus Mac Lane explained that "algebraic varieties" [sic] (groups, fields, etc.) "concern essentially a system composed of a number of functions" and that "an abstract mathematical theory consists of some system and a number of axioms and theorems about this system." Mac Lane left the terms "systems" and "structures" vague, but he added:

> From this standpoint, we can prove a number of interesting theorems concerning the interconnections of these relations and including as special cases many well known theorems of algebra. One such theorem is the generalization to systems of the "second isomorphism theorem" for groups. Many algebraic theorems may be viewed as special cases of the general theorem that two isomorphic systems have the same structure. (Mac Lane 1934, 53)

---

7. The centrality of Galois theory and the kind of pursuit described in (b) generated further attempts to generalize the idea of structure, such as those of Krasner and Sebastião e Silva mentioned in footnote 3 above.

Clearly the last sentence in this quotation becomes meaningful only if "structure" is given a specific, formally defined meaning such as Ore did in his own work. This is in all likelihood what Mac Lane seems to have meant here. This kind of issue is not mentioned separately in Mac Lane's 1939 review, but as will be seen in chapters 8 and 9 below, generalizing questions about the nature of mathematics were always among his concerns. Typical of his attitude, however, is a tendency to exercise caution when generalizing, as attested by the concluding remarks to his definition of algebra appearing in the review of 1939. There he remarked:

> *Algebra tends to the study of the explicit structure of postulationally defined systems closed with respect to one or more rational operations.* This summary does not account well for the use of topological operations in algebra ... [and for] the reduction of matrices to canonical forms ... As with many hyper-generalizations our statements fit the facts only when the facts are slightly distorted. (Mac Lane 1934, 53. Italics in the original)

Summarizing the above paragraphs we can say that in 1939 Mac Lane described algebra (with some reservations) as the "structural investigation of algebraic systems", and what he meant by this was explained by inventorying the main problems, techniques and expected answers of current algebraic research. In§ 9.2 the evolution of Mac Lane's images of mathematics after 1940 will be re-examined.

Mac Lane's views exemplify how algebraists were explaining to themselves and to other mathematicians the essence of the new image of algebra.[8] The reflexive attempts to elucidate the idea of an algebraic structure had to account at least for some of the features mentioned by these algebraists. Having seen in detail in Part One how the structural approach gradually entered algebra, and how algebraists conceived it in the 1930s, we can now turn to a more detailed discussion of the main reflexive attempts to elucidate the mathematical meaning of the idea of a mathematical structure.

---

8. Additional expository articles in the same trend include Hasse 1930; Macaulay 1933; Richardson 1940. Hasse's article was specifically addressed to analysts and it explicitly meant to "pave the way to a more sympathetic understanding of modern algebra." Richardson's article is an early English review of the new perspective offered by van der Waerden's book and by recent research on ideal theory. It is interesting because of its many oddities regarding nomenclature and certain basic concepts.

Chapter 6   Oystein Ore: Algebraic Structures

The first reflexive attempt to develop a formal theory of structures that we consider here is the research program initiated by the Norwegian mathematician Oystein Ore, beginning in 1935 at Yale. Ore's program was an attempt to develop a general foundation for all of abstract algebra based on the concept of lattice, which he denoted with the term **structure**. The leading idea behind this attempt was that the key for understanding the essential properties of any given algebraic system lay in overlooking not only the specific nature, but even the very existence of any elements in it, and in focusing on the properties of the lattice of certain of its distinguished subsystems. Ore intended to derive, in this way, all the main, general theorems valid for all algebraic domains.

The basic notions and the images of mathematics underlying Ore's attempt were deeply rooted in the development of ideal theory from Dedekind to Noether, discussed in detail in Part One. The very concept of lattice was first studied at the turn of the century independently by Dedekind and Ernst Schröder (§ 2.3 above). Dedekind's articles on *Dualgruppen* were a direct by-product of his own work on ideals. Beginning in 1930 Ore edited and published the complete works of Dedekind, together with Emmy Noether and Robert Fricke. Ore's own 1924 dissertation had dealt with the theory of fields.[1] He was thus well acquainted with the latest developments in algebra, and in particular with Dedekind's theory of ideals as well as with Noether's ideas on the latter.

The early works on lattices by both Dedekind and Schröder aroused only feeble interest among contemporary mathematicians. In the 1930s the concept reemerged in the independent works of various mathematicians working in fields as diverse as projective geometry and abstract algebra. Garrett Birkhoff (1911-1996) and Oystein Ore (1899-1968)—in separate works and, very

---

1. For further biographical details on Ore see Anon. 1970. See also Birkhoff 1977, 70-71.

likely, unaware of each other's research—were the main revivers of the theory in its abstract version of the 1930s. In the initial stages of their respective works, both researchers considered the theory as a plausible unifying conceptual framework within which many results that had recently been obtained simultaneously in the diverse branches of algebra could be derived in a generalized formulation. Although Birkhoff's is the name usually associated with the new beginnings of lattice theory, Ore's influence was no less important in this regard. Ore's interest focused—more than Birkhoff's—on the possibility of relying on the theory as an abstract foundation for all of algebra. Although "lattices" became the term generally accepted, "**structure**" was used in many instances with the meaning intended by Ore.

The first mathematical symposium dedicated exclusively to the theory of lattices took place in 1938 in Charlottesville, Virginia.[2] This symposium signified that lattice theory had come of age as an autonomous mathematical research discipline. The year 1938 also saw the semicentennial anniversary of the American Mathematical Society. E.T. Bell was invited to address the society with an historical account on the first fifty years of algebra in America. In Bell's view, the most significant trend of algebraic research in the USA since the activities of the society began was the search for ever more general concepts, enabling to prove theorems of an ever increasing generality. Thus Bell wrote:

> If there is any clue through the tangled jungle of elaborated theories and special theorems for the past fifty years, it is perhaps only this steady progression from the particular to the less particular ... many aspects of the algebra of groups, rings and fields which were hitherto obscured by a multitude of special details, are now seen to be simple consequences of underlying general concepts. (Bell 1938, 6)

Among the "underlying general concepts" referred to by Bell, Ore's concept of **structure** provided the most outstanding example.[3] Whether this

---

2. See the proceedings in the *Bulletin of the AMS*, Vol. 44 (1938), pp. 793 ff.

3. Bell dedicated a whole chapter in his popular *Development of Mathematics* to describe the rise of the concept of **structure**. In a chapter entitled "Towards Mathematical Structure: 1801-1910", Bell described the process leading to the consolidation of this concept as he could have done for any other mathematical concept. He wrote (Bell 1945, 245-246): "Structure ... was the final outcome of this accelerated progress from the particular to the general ... The entire development required about a century. Its progress [which] is typical of any major mathematical discipline of the recent periods ... [culminated] with the formulation of the postulates crystallizing in abstract form the structure of the system investigated." Bell used the term explicitly to mean Ore's concept, and he made no distinction whatsoever between a formal and a non-formal meaning of the term.

name, or "lattice" as proposed by Birkhoff, would be the one that would finally survive was still to be settled. But Bell was certain of the central role that this notion was bound to play as a unifying notion in the coming years of mathematical activity.[4] This expectation turned out to be overconfident, however, and in fact, by 1940—the time of publication of Bell's popular book on the history of mathematics, in which he reiterated his confidence in the unifying powers of **structures**[5]—Ore's own enthusiasm for his program had already waned considerably. Lattice theory had become a mature research discipline, but **structures** could no longer strive to fulfill the far-reaching, unifying expectations attached to them in the beginning.

Ore's program developed as part of the re-birth of lattice theory in the 1930s, and therefore it was also influenced by the many sources from which lattice theory in general emerged. Thus, a more faithful description of the rise of Ore's program should take into account the broader framework of the development of lattice theory. Here, only those aspects of Ore's program will be considered which touch directly upon his attempt to formalize the general idea of an algebraic structure. The present chapter thus examines Ore's research on **structures** and in particular those aspects in which the unifying intentions of his work are manifest.[6] Before discussing Ore's work itself, however, we must first consider one aspect of its immediate background which was not mentioned so far, namely, the interest in direct decomposition theorems in abstract algebra during the first third of the present century.

6.1 Decomposition Theorems and Algebraic Structures

Factorization theorems and their relationship to chain conditions—such as described in Part One for the case of ideal theory—embodied a main research concern for all branches of algebra during the first two decades of the twentieth century. Parallel to the decomposition of algebraic systems as intersections of more basic systems of a similar kind, a second type of decomposition theorems developed in algebra at about the same time: the so-called "direct

---

4. Cf. Bell 1938, 7.

5. Bell 1945, 245-246.

6. As already said in § 2.3, a comprehensive account of the rise of lattice theory appears in Mehrtens 1979. A shorter account, which however discusses some applications of lattices not covered by Mehrtens, appears in Rota 1997. Mehrtens's account includes, of course, a description of Ore's work, although the emphasis is somewhat different to the one intended here.

decompositions." Direct decompositions were studied especially within group theory, and the mathematician who took the lead in this research was Robert Remak (1888-1942). Remak wrote his doctoral dissertation under Frobenius and published it in 1911.[7] He was also known for his contributions to the geometry of numbers, universal algebra and mathematical economics.[8] Some of his group-theoretical ideas are important for understanding Ore's work and they deserve a brief discussion here.

Given a group G and two of its subgroups U,B, the group G is called the direct product $G = U \otimes B$ if and only if G is generated by products of elements of U and B and, in addition, the two following properties hold: (1) for any pair of elements u,b, with $u \in U$ and $b \in B$, the identity $ub = bu$ holds, and (2) the intersection (U,B) contains only the identity element of G. G is called "directly indecomposable" (*direkt unzerlegbar*) if it cannot be written as a non-trivial direct product of two of its subgroups. In his dissertation Remak proved that any two direct decompositions of a given group are essentially identical, i.e., that they have the same number of pairwise isomorphic direct factors. This theorem had formerly been proven by Frobenius and Stickelberger[9] for the commutative case. Remak's proof generalized their result.

Very much like had been the case with the Jordan-Hölder theorem (see above § 1.1), Remak's theorem was soon reformulated in more general contexts. These reformulations stressed the connection of the theorem with chain conditions. Two years before Remak published his proof, Henry MacLagan Wedderburn had published a similar one.[10] Remak claimed that he was not aware of Wedderburn's work and that, moreover, the latter's proof contained a gap which was in need of improvement. In 1912 the Russian mathematician Otto Schmidt (1891-1956) published a simplified proof of Remak's result.[11] Later, in 1928, Schmidt widened the scope of Remak's theorem by proving that several decomposition theorems of finite groups (Remak's theorem amongst them) are valid for the infinite case as well, whenever the group in

---

7. An improved version of his dissertation was published in Remak 1924.

8. On Remak, see Biermann 1988, 209-212; Merzbach 1992, Siegmund-Schultze 1998, esp. 121-123. In May 1940 Remak was deported to Auschwitz from Holland. In 1942 his wife heard of him for the last time.

9. Frobenius & Stickelberger 1879.

10. See Wedderburn 1909.

11. Schmidt 1912; 1913.

question satisfies certain chain conditions.[12] Schmidt's proof, in turn, strongly relied on techniques introduced earlier by Krull.

In 1925, Krull had defined generalized Abelian groups—a concept equivalent to what is called nowadays an Abelian group with operators. He intended to generalize the concept of Abelian group, so that it might be used in the study of differential operators and groups of matrices. Krull proved that many known theorems of group theory are also valid for generalized Abelian groups. He analyzed the role played by the integers (seen as operators) in the decomposition theorems of Abelian groups, and extended the validity of those theorems by translating that role to the systems of external, generalized operators of the generalized Abelian group. In particular, one of Krull's main theorem was a generalized version of Remak's decomposition theorem.[13] Since then the theorem has been associated with the names of Krull-Remak-Schmidt, or alternatively with Krull-Schmidt. At any rate, what concerns us here is the fact that, parallel to the many theorems dealing with representations of algebraic systems in terms of intersections, like in the case of Noether's results, algebraists were also dealing with decomposition theorems in terms of direct products of more basic subsystems, and in particular, they were trying to find their most general formulations. This trend was manifest in the structural image of algebra embodied in *Moderne Algebra*. For Ore, it became one of the central traits of his foundational program for abstract algebra.

6.2 Non-Commutative Polynomials and Algebraic Structure

Oystein Ore was born in Christiania (now Oslo) in 1899. He received a Ph.D. in 1924 with a dissertation on field theory. Like many other young European mathematicians at that time, Ore was brought to the USA by James Pierpont, and from 1927 on, he taught at Yale.[14] Ore's main field of research was algebra, but he also published several works on graph theory and topology, as well as a book on the history of number theory and two well-known mathematical biographies: of Niels Henrik Abel and of Girolamo Cardano.[15]

In the years 1930-34, Ore's research efforts concentrated on non-commutative fields and non-commutative polynomials. Ore studied the possibility of

---

12. Schmidt 1928.
13. Krull 1925, 186.
14. On Pierpont's role in building Yale's department of mathematics see Dorwart 1989.
15. Ore 1948, 1957, 1958, respectively.

extending the validity of many unique factorization theorems of polynomials from the commutative to the non-commutative cases. Surprisingly enough, in spite of being acquainted with Noether's work on ideals, Ore dismissed the possibility of conducting his own research along the lines advanced by her. Ore did not treat the theory of polynomials as a particular case of Noether's theory of abstract ideals, but rather within their specific, more reduced context. In his proofs he relied on particular properties of polynomials, as Lasker and Macaulay had done for the commutative case at the beginning of the century. This early position of Ore makes his later generalizing efforts especially appealing for historical inquiry.

In 1931 Ore published an article on the resolution of a system of linear equations, with coefficients taken from a non-commutative field.[16] Among earlier works on the same question, which Ore mentioned in the introductory section of his article, was an article by Noether.[17] Among others, Ore gave a full characterization of all the integral domains of which a field of fractions can be constructed, a task that Van der Waerden had mentioned in his book (Vol. 1, § 12) as a key open question. Thus, Ore was clearly well-aware of the scope and the projections of the new approach to algebra and of the important issues at stake in current algebraic research. His choice of a more 'classical' approach to deal with the problem considered in this article of 1931 and in his articles on non-commutative polynomials, was not a product of ignorance, but a conscious decision. On the other hand, it is worth pointing out that in his definition of ring, Ore also demanded that an "equality relation" be satisfied. This same relation had already appeared in Fraenkel's definition (§ 4.5 above). From a logical point of view, this demand is redundant, as we saw, and thus Noether and van der Waerden did not include it in their respective definitions of rings. Apparently, then, Ore may have been taking his ideas on rings directly from Fraenkel.

Ore published his research on non-commutative polynomials in two German papers,[18] which he later summarized in an English version.[19] In the introductory sections to the English summary, Ore listed several works that had previously addressed the issue of non-commutative polynomials. Noether and Schmeidler's article of 1920 appears among them. As noted earlier, several

---

16. Ore 1931a.
17. Noether 1929.
18. Ore 1932; 1932a.
19. Ore 1933a.

important ideas that Noether developed later in her abstract treatment of the theory of ideals had already been foreshadowed in this earlier article (see § 5.1 above), and Ore referred here to this fact. He thus wrote:

> One could have deduced this theory [i.e., of non-commutative polynomials] using the modules studied by Noether and Schmeidler. One would then have to study the residue classes of non-commutative polynomials, these form a generalized Abelian group according to the terminology introduced by Krull, and there exists a correspondence between the structure of this generalized Abelian group and the corresponding polynomial such that the properties of a polynomial can be deduced from the general theorems on generalized Abelian groups. I have preferred, however, to build up the theory directly, that is to use only the properties of polynomials themselves as, for instance, in the ordinary polynomial theory. This seems preferable for various reasons. It makes the theory independent of the more general theory and it brings out more clearly some of the specific properties of polynomials. (Ore 1933a, 480)

In order to better understand the factorization laws of the specific domain at issue here—Ore claimed—Noether's line of research should be abandoned. The theorems of factorization for polynomials should be derived directly from their particular properties, and not from the properties of Krull's generalized Abelian groups, which *conceal*—in Ore's view—the essence of the proofs.

Like Lasker and Macaulay in their works before Noether's unification, the central tool of Ore's research on polynomials was the Euclidean algorithm of division. However, since he was working here in a non-commutative domain, Ore had to redefine the algorithm and to adapt it to this particular case. The aim of the new algorithm should be similar to that of the old one, namely, to provide a standard device for calculating the *g.c.d.* of two arbitrary polynomials. Given two polynomials $A(y),B(y)$ Ore denoted their *g.c.d.* (*Durschnitt*) as usual by $D(y) = (A(y),B(y))$. In case $D(y) = y$ then the polynomials are called relatively prime. The *l.c.m.* (*Hülle*) is denoted by $M = [A(y),B(y)]$ and is defined as the smallest monic polynomial which is divisible by both A and B.

Ore proved several theorems about the relationship between the above concepts and the generalized Euclidean algorithm. In particular, he introduced the concept of "transformation" of polynomials. Given a polynomial $A(y)$, the transformation of $A(y)$ by a second polynomial $B(y)$ is defined as a third polynomial $A_1(y)$, as follows:

$$A_1(y) = a_0 b_0 [A(y),B(y)] \times B(y)^{-1}$$

where a_0, b_0 are the leading coefficients of $A(y)$ and $B(y)$ respectively. This transformation is denoted by $A_1(y) = BA(y)B^{-1}$. Among the main theorems proven algorithmically by Ore are the following two:[20]

Th. 9: $C[A(y), B(y)]C^{-1} = [CA(y)C^{-1}, CB(y)C^{-1}]$

Th. 10: $C(A(y) \times B(y))C^{-1} = C_1 A(y) C_1^{-1} \times CB(y)C^{-1}$,

where $C_1(y) = AC(y)A^{-1}$.

After the one-sided operations on polynomial were defined, as well as the *g.c.d.*, the *l.c.m.* and the transformation, Ore proceeded to elucidate the factorization properties of the non-commutative polynomials. However, whereas in the first part of the paper Ore proved the theorems in terms of the operations directly defined on the polynomials as was his declared intention, the theorems of the second part were all proven using the *g.c.d*, the *l.c.m.*, and the transformation. Relying on them, Ore proved several factorization theorems, quite similar to the four decomposition theorems of Noether's *"Idealtheorie"* (§ 5.2 above). Nevertheless, in his English article of 1933, Ore declared once again that he intended to exploit directly the particular properties of polynomials as such, instead of deriving the theorems from more general principles. In fact, in the English version Ore relied even more strongly on the properties of the *g.c.d.*, the *l.c.m.* and the transformation than he had done in the German version of his article, but he consciously refrained from using concepts or terms taken from Noether's abstract theory of ideals. In fact, the very concept of ideal was not mentioned at all in any of Ore's early articles.

Ore proved four different theorems of unique decomposition of polynomials into more basic ones: prime, maximal completely reducible, etc. In the English version, between the third and fourth decomposition, Ore remarked:

> In the preceding discussion we have deduced some of the more important structure theorems for non-commutative polynomials. They depend on various notions of indecomposability which lead to a number of decomposition theorems. (Ore 1933a, 505)

Thus, as late as 1933, after Noether's work on ideals and the publication of van der Waerden's book, in spite of his own knowledge of the current trends in algebra, and in spite of the limited role than the specific properties of polynomials actually played in the proofs formulated in his own results, Ore

---

20. Ore 1932, 230-231.

ascribed to the particular theorems of the theory of polynomials a higher mathematical value than he did to generalizing theories such as Noether's. Ore saw in the former the legitimate way to unveiling the true essence of the "structure of the polynomials." At this stage Ore used the expression "structure of polynomials" as a non-formal one, denoting those decomposition properties of polynomials which derive from the generalized Euclidean algorithm defined on them. The expression denoted here neither a clear and defined notion which can be used for all domains of algebra nor a formal concept which appears as part of a formal, deductive theory. It is therefore rather surprising to see the sharp turn that affected Ore's conception after 1935, leading him to adopt a totally opposed view. He would now attempt to produce a single formal concept from which the theorems valid in all domains of algebra could be deduced in a unified and generalized formulation. Using his concept of **structure**, Ore tried to prove the main decomposition theorems of algebra, including those which he previously insisted on proving separately for the sake of their better understanding.

6.3 **Structures and Lattices**

Birkhoff and Ore were the leading figures behind the re-elaboration of lattice theory in the 1930s.[21] For Birkhoff, the concept of lattice itself was the main focus of interest. Ore's interest in lattices was directly motivated by his desire to a general, formal foundation for abstract algebra. This difference in emphasis is the main reason why the present chapter focuses on Ore's work as opposed to that of Birkhoff. Nevertheless, it is relevant to start here with a brief account of Birkhoff's works.

Birkhoff's early articles on lattices attracted very little attention. Among his main sources of ideas he mentioned van der Waerden's book, Remak's work on decompositions of groups, and Dedekind's works on *Dualgruppen*.[22] Birkhoff defined the role he envisaged for lattice theory within the whole picture of mathematics by comparison with group theory: whereas the latter ana-

---

21. Other mathematicians that contributed to the early stages of research on lattice theory, besides Ore and Birkhoff, were Karl Menger (see Menger 1928) and Fritz Klein-Barmen. The works of the latter concentrated on the axiomatic formulation of the theory (Klein-Barmen 1929, 1931). Klein-Barmen introduced the German term *Verband* for lattices. As far as we know, all those mathematicians worked independently and unaware of one another's work. Cf. Mehrtens 1979, Ch. 3 and Birkhoff 1977, 28 & 70.

22. Birkhoff 1933.

lyzes the abstract aspects of the phenomena of symmetry—he wrote—the former will enable a similar treatment of a phenomenon of no less importance in mathematics, namely the phenomenon of order in its broadest sense.[23] Birkhoff discussed the connections between lattice theory and ideal theory,[24] as well as the possibility of providing a lattice-theoretical foundation of universal algebra.[25] His principal result was a theorem connecting the lattice of all subsystems of any given algebraic system with the lattice of the equivalence relations that can be defined on it. Birkhoff proved that a one-to-one mapping can be defined between them. We will return to this issue below.

Ore started publishing his theory of **structures** immediately after Birkhoff, and the first full presentation appeared in 1935 in the *Annals of Mathematics*. Here we find a detailed description of his current conception of algebra, a conception which was diametrically opposed to that of his earlier works. Ore believed that in light of the many theorems that had recurrently appeared over the last decades in the different algebraic domains, it was natural to expect the existence of a single, general concept from which one could derive equivalent theorems simultaneously valid in all those domains. Ore mentioned the work of Krull on generalized groups as an example of an earlier attempt in the same direction. His own current proposal implied a further generalization of the algebraic domains discussed: while the abstract approach to algebra had come about when ignoring the *nature* of the elements involved in the operations and when concentrating on the properties of the operations themselves, the second-order generalization, according to Ore, should proceed by overlooking the very *existence* of elements and of operations among them and by focusing on a different level of reference. This level is what Ore called the **structural** aspect of the algebraic domain, which he described as follows:

> In the discussion of the structure of algebraic domains, one is not primarily interested in the *elements* of these domains but in the relations of certain *distinguished sub-domains* ... For all these systems there are defined the *two operations* of union *and* cross-cut satisfying the ordinary axioms. This leads naturally to the introduction of new systems, which we shall call *structures*, having these two operations. The elements of the structure correspond isomorphically with respect to union and cross-cut to the distinguished sub-domains of the original sub-

---

23. See Birkhoff 1938.
24. Birkhoff 1934.
25. Birkhoff 1935.

domain while the elements of the original domain are completely eliminated in the structure. (Ore 1935, 406)

He also remarked that it is only in integrally closed rings that the theorems of algebra are naturally written in a multiplicative formulation, but that these theorems might actually be written without reference to the operation that defines the domain.[26]

Ore presented two alternative definitions of a **structure**. The first one is based on the definition of an abstract order relation "<" among members of a given system Σ. This relation is required to satisfy the properties of the usual set-theoretical inclusion. Union and cross-cut are defined as an abstract operation, in terms of inclusion, as follows:

To every couple of elements A,B there exists an element $D = (A,B)$, the cross-cut of A and B, satisfying the property that
$$D \leq A, \quad D \leq B$$
and $D_1 \leq D$ for every other element D_1 having the same property...
There exists an element $M = [A,B]$, the union of A and B such that
$$M \geq A, \quad M \geq B$$
and $M_1 \geq M$ for every other M_1 with the same property.[27]

Now, every system endowed with two operations satisfying these properties is called a **structure**. Ore showed that these abstract operations satisfy the usual properties of union and cross-cut:

$$(A,B) = (B,A) \qquad [A,B] = [B,A]$$
$$(A,A) = A \qquad [A,A] = A$$

(α)

$$(A,(B,C)) = ((A,B),C) \qquad [A,[B,C]] = [[A,B],C]$$
$$[A,(A,B)] = A \qquad (A,[B,C]) = A$$

The properties are written in two parallel columns in order to emphasize the "duality principle" of **structures**, namely, that every true statement of the theory remains true whenever one substitutes simultaneously every appearance of cross-cut by union (and vice-versa) and < by > (and vice-versa).

An alternative definition of **structures** is obtained by taking properties (α) as starting point. The two operations of cross-cut and union may be defined on

---

26. Recall that a ring R is integrally closed, if every element in the field of fractions of R, which is integer respect to R, is in fact an element of R. In § 5.3 above, it was seen that among the five axioms defining a "Dedekind ring", the demand appears that the ring R be integrally closed. In Dedekind rings, every primary ideal is the power of a prime ideal.

27. Ore 1935, 408.

any system Σ, and they will constitute a **structure** whenever the defined operations satisfy axioms (α). Accordingly, given two elements of the system A and B, one can say that $A > B$ whenever the following two equivalent relations hold:

$$[A,B] = A \qquad (A,B) = B$$

Starting from these two alternative definitions, Ore proposed to derive all the general theorems for which special versions existed in each of the branches of algebra, and especially for decomposition theorems. He thereby sought to lay down the conceptual foundations of a new approach to abstract algebra. These foundations—Ore thought—should not be taken as a mere sterile, formal exercise on the possibility of presenting a unified outlook of already known results. Thus he wrote:

> It is of course quite interesting to examine to what extent this is possible, but the real usefulness of the idea appears through the various new results to which it leads. (Ore 1938a, 801)

The "distinguished" subdomains on which Ore's attention focused are, for instance, normal subgroups for group theory or ideals for ring theory. Ore's first important insight was to notice that the peculiar algebraic properties of these distinguished subdomains may indeed be expressed in strictly **structural** terms, since, as it happens, their intersections and unions obey an additional axiom which Dedekind had called the "modular axiom" (§ 2.2.4 above), and which Ore would call the "Dedekind axiom", namely that if $C > A$, then

$$(C,[A,B]) = [A,(B,C)] \tag{1}$$

Thus Ore's research was to concentrate on "Dedekind **structures**", namely **structures** satisfying the "Dedekind axiom."

The first stage in realizing Ore's program was to redefine some central, recurrent concepts of algebra (e.g., cosets and isomorphism) in **structural** terms. The ascending, as well as the descending chain conditions were of central importance for this purpose. A **structure** satisfying both the $a.c.c.$ and the $d.c.c.$ is called an Archimedean **structure**. If $A > B$ and there are no further elements between A and B, then A is called "prime over B." A chain of elements in the **structure** $A = A_0 > A_1 > ... > A_n = B$ is called a "principal chain of length n" if each of its components is prime over the following one. A **structure** Σ is called a "Dedekind **structure**" if it satisfies the Dedekind axiom: if A,B,C belong to Σ and the inequality $A < C < [A,B]$ holds, then $C = [A,(B,C)]$.

Ore proved that this axiom is equivalent to condition (1) above which corresponds to Dedekind's original formulation (see above § 2.2.4).[28]

A further, somewhat stronger axiom, satisfied by the so-called "arithmetical **structures**", asserts the distributivity property:

$$(A,[B,C]) = [(A,B),(A,C)].$$

When $A > B$, Ore defined the quotient A/B as the collection of elements of the **structure** which are smaller than A and greater than B. Obviously, A/B constitutes in itself a **structure**. Given any two structures, a mapping from the first to the second is called a homomorphism if it preserves unions and crosscuts. Ore, like van der Waerden in the early editions of his book, implicitly considered only morphisms which are also surjective, and thus, an isomorphism was defined as an homomorphism which is one-to-one. Ore proved a central theorem that had already appeared in Dedekind's work on *Dualgruppen*: If A and B are elements in a Dedekind **structure**, then the **structures** $[A,B]/A$ and $A/(A,B)$ are isomorphic.[29] More specifically, the theorem states that given the principal chain $(A,B) < A_1 < ... < A_n = A$, between (A,B) and A, then the chain

$$B < [B,A_1] < [B,A_2] < ... < [B,A_n] = [B,A]$$

is a principal chain between B and $[B,A]$. The most elementary principal chains are of the kind

$$[A,B] > A > (A,B) \quad \text{and} \quad [A,B] > B > (A,B).$$

These two chains are obtained from one another by means of a "prime transformation." Ore formulated and proved here an abstract version of the Jordan-Hölder theorem for "Dedekind **structures**":

> When there exists a finite principal chain between A and B in the Dedekind structure Σ, then all principal chains between A and B have the same length and one can be obtained from the other by successive prime transposition. (Ore 1935, 419)

---

28. The proof is obtained by substituting $A = A_1$, $B = B_1$ and $C = (A_1,[C_1,B_1])$.

29. See Ore 1935, 418. M. Ward (1939) proved the reciprocal of that theorem: "Let Σ be a **structure** in which for every pair of elements A and B, the quotient **structures** $[A,B]/A$ and $B/(A,B)$ are isomorphic. Then if either the ascending or descending chain condition holds in Σ, the **structure** is Dedekindian."

A further concept, to which Ore attached central importance in his work, arose directly from Noether's work on ideals. Let A, A' be two elements in a Dedekind **structure**. Denote by $P \not\approx Q$ the fact that either $P > Q$ or $P < Q$ and suppose there exists a chain $A' \not\approx A_1 \not\approx ... \not\approx A_n \not\approx A$. In this case one says that A and A' are connected. Obviously this is an equivalence relation and one can therefore define equivalence classes or, more precisely, connectivity classes in any **structure**. Such classes, albeit unnamed, played a central role in Noether's proofs of her third and fourth decompositions theorems of 1921 (§ 5.2 above). As was said above, in spite of the significant influence of Noether's work in general, these particular proofs, and the partition into connectivity classes used therein, were soon forgotten. But Ore saw in them a useful concept which he put to use in his own work. Thus, for instance, he used them to prove the validity of the Jordan-Hölder theorem in non-Archimedean **structures**, i.e., **structures** with infinite principal chains (Th. 9, p. 424).

As he advanced into the paper, Ore abandoned the original **structures** and focused the discussion on the quotients A/B, whose collections constitute themselves **structures**. In fact—as Ore showed—all the chain conditions satisfied by the original **structure** are also satisfied by its quotient **structures**.

Given two quotients A/B and B/C, define their product by:

$$A/C = A/B \times B/C$$

Take now two quotients with a common denominator $U = A/B$, $T = C/B$ and define the quotient U' by:

$$U' = TUT^{-1} = [U,T] \times T^{-1} = [A,C]/C \qquad (*)$$

U' is called the "transformation of A by C." Clearly, this concept connects us directly to Ore's earlier work on non-commutative polynomials. Thus, the remaining sections of the article were devoted to translating into the realm of **structures** those decomposition theorems that Ore had obtained earlier for non-commutative polynomials. Ore also formulated some additional decomposition theorems, but this time in **structural** terms, refraining from using elements of the original domains and relying only on the properties of union, cross-cut, inclusion and transformations.

If in the transformation (*), one demands that $(A,C) = B$, then one says that U' is obtained from U by a "similarity transformation." In that case, one can prove that U and U' are isomorphic **structures**. Ore reformulated the Jordan-

Hölder theorem in terms of transformations. If A and B are Dedekind **structures**, the theorem states that:

> When there exists a finite principal chain between A and B, then the quotient
> $U = A/B$ is representable as the product of prime quotients
> $$U = T_1 \times \ldots \times T_r$$
> Any other such representation will have the same number of prime factors similar to the T_i in some order. (Ore 1935, 436)

Likewise, the **structural** framework allows Ore a very general formulation of the Schreier theorem, which in itself is a generalization of the Jordan-Hölder theorem for the case in which no finite chains between A and B necessarily exist.[30] The **structural** version of the Schreier theorem reads as follows:

> When
> $$U = T_1 \times \ldots \times T_r = R_1 \times \ldots \times R_s$$
> are any two finite product representations of a quotient in a Dedekind structure,
> then it is possible to decompose these factors further in such a way that in the
> resulting finite products both sides have the same number of factors and the factors are similar in pairs. (Ore 1935, 436)

In 1936 Ore published the second part of his foundations of abstract algebra. He went on to prove additional decomposition theorems, while stressing the advantages of his approach: not only does it enable general proofs for the theorems of algebra, but it also leads to the solution of an additional important problem, namely, the exact demarcation of the validity domain of each theorem. Thus, for instance, Ore proved that the very *existence* of a decomposition of a Dedekind **structure** into indecomposable Dedekind sub-**structures** implies the uniqueness of that decomposition.[31] This result differed from former formulations and proofs (including Noether's) which assumed the *a.c.c.* as a necessary condition for uniqueness.

So much for the two articles that opened the series to which we refer here as Ore's program. Before going on to consider the next articles published by Ore and by other mathematicians in the same direction, it seems necessary to discuss a further point concerning the background to Ore's work. Although many theorems proven before 1935 reinforced the views put forward by Ore

---

30. See Schreier 1928 for the original proof. The standard accepted proof of this theorem includes a small correction first suggested in Zassenhaus 1934. For a more recent formulation and proof of the theorem see Jacobson 1980, 104-108.

31. Ore 1936, § 1.3.

when developing his program, at least one result can be seen as completely refuting it. As early as 1928 Ada Rottländer, a doctoral student of Ernst Fischer (who had also been Emmy Noether's teacher, and the one who introduced her to the abstract trend in algebra), had dealt with the question of the group and the lattice of its subgroups. Given two isomorphic groups G, G', then their respective lattices of subgroups are obviously isomorphic. The isomorphism $f: G \rightarrow G'$ preserves inclusions and conjugation, i.e., if K, H are two subgroups of G and $K' = f(K), H' = f(H)$, then $f(K^{-1}HK) = K'^{-1}H'K'$. Rottländer addressed the reciprocal question: let G, G' be two groups such that there exist an inclusion- and conjugation-preserving isomorphism of lattices between the respective lattices of subgroups of G and G', are the two groups themselves isomorphic? Rottländer proved that if one of the groups is Abelian then the answer is positive. She also gave an example of two non-Abelian groups for which the answer to the above question is negative.

Rottländer's article was published in the *Mathematische Zeitschrift*. It is therefore unlikely that Ore was unaware of it, although he did not explicitly mention the article in any of his publications. On the face of it, Rottländer's article represents a glaring contradiction to one of the central tenets of Ore's program. In fact, like Noether before him, one of Ore's central drives in pursuing generalizing frameworks in algebra was the desire to deal with commutative and non-commutative cases simultaneously. If the theorems of algebra can be proven without recourse to the operation defining the domain at issue, then it is not even necessary to mention the two different cases. Rottländer's result brings to the fore a peculiar difficulty involved in the non-commutative cases. Were it not for the non-commutative case, then Rottländer could have formulated a correspondence theorem between the structure of the group and that of its lattice of subgroups in all the generality envisaged by Ore's program several years later. This, however, was not the possible.

How did Ore see the prospects of his program in view of Rottländer's negative result? The best one can do, for lack of documentary evidence, is to advance a conjecture. It seems that for Ore the decomposition theorems represented the main (and probably the only) target of algebraic research. Rottländer's result does not show that it is impossible to study all the decomposition properties of algebraic domains from a **structuralist** perspective and, therefore, it does not refute Ore's program a-priori. In this case Ore's program should be formulated not as an attempt to provide abstract foundations for all of algebra, but only for the decomposition theorems within it.

6.4   Structures in Action

Several mathematicians, including Ore and some of his students, published many research works between 1936 and 1943, in which they tried to work out the **structural** program as a meaningful mathematical project. We describe now cursorily the contents of those works.

In a series of articles following his first two programmatic ones, Ore attempted to make his ideas known to a wider audience of algebraists and mathematicians in general. In the first place, he presented the main lines of his program in an address to the International Congress of Mathematicians held at Oslo in 1936, where he was invited to lecture in one of the plenary sessions.[32] This invitation makes clear that, at least at that time, the theory aroused some interest. At that opportunity Ore claimed that the guidelines of his program, although originating within algebra, should not be limited to that domain alone, and he envisaged that they would be applied in additional fields of mathematics as well. It is worth stressing that at this special event, Ore chose to open with an informal account of what he saw as the essence of the new approach to algebra. In his description, the axiomatic formulation of theories is only one side of this new approach. As one of the central problems of modern algebra Ore discussed the above mentioned problem of determining the exact domain of validity of each theorem in the discipline. The fact that Ore used the privileged stage offered to him to mention these two aspects of algebra indicates how far they were from being taken for granted by the mathematical community at large still in 1936 and, on the other hand, the high degree of importance that he himself conceded to them.

Ore returned to the same issues in 1936 in a French booklet in which he presented his results. In this booklet Ore added a general review of the present state of algebra as he saw it, with special emphasis on decomposition theorems. As an example of an important theorem of algebra in **structural** version Ore formulated the main correspondence theorem of Galois theory as follows:

> There exists an inverse **structure**-isomorphism between the collection of subfields of a normal extension K of a field k, and the collection of the sub-groups of the group of automorphisms of K which leave k invariant.[33]

---

32. See Ore 1936b.

33. Ore 1936c, 50: "Il existe une isomorphie de structure (inverse) entre les sous- corps d'une extension normale K d'un corps k, et les sous-groupes du groupe d'automorphismes de K par rapport à k."

After having proven many general versions of the decomposition theorems in his first articles, Ore went on to attempt a purely **structural** characterization of the concept of group. This he did in two articles, whose explicit aim was to "provide a foundation of the theory of groups, as far as it is possible, directly upon their subgroups, and overlooking their elements."[34]

Ore published two further articles in which he gave an exhaustive account of the abstract aspects of the Jordan-Hölder theorem. In the first of them, written in 1937, he focused on the properties of composition series, as opposed to principal series that he had considered in earlier works.[35] Later, in 1943, Ore proved a very general version of the theorem valid for partially ordered sets.[36] The theorem of Jordan-Hölder very much attracted Ore's attention in the framework of the elaboration of his program, but it is interesting to notice that even after his exhaustive treatment one can still find articles containing further refinements of it.[37]

In implementing his program, Ore envisaged the analysis of algebraic domains defined by one or more internal operations, as well as of any kind of abstract relations. Thus, in a further paper he examined the **structural** properties of abstractly defined equivalence relations.[38] All equivalence relations defined on a given set constitute a **structure**, although not a Dedekind **structure**. Precisely for that reason Ore found a special interest in their study, since they constitute a concrete example of a **structure** with non-standard behavior. The same problem was also addressed in 1939 by the French mathematicians Paul Dubreil and Marie L. Dubreil Jacotin.[39] In their article they pointed out that Birkhoff was the first to show that the collection of equivalence relations defined on a given set is in fact a lattice. Birkhoff even proved that this lattice is isomorphic to the lattice of the subgroups (both normal and non-normal) of a group of permutations.[40] Ore also analyzed the **structural** properties of the set of order relations defined on a given set, while stressing the relevance of this analysis for an eventual **structural** approach to topological spaces. In his

---

34. Ore 1937a; 1938.

35. Ore 1937.

36. Ore 1943b.

37. Cf., e.g., George 1939. An even more general formulation, in categorical terms appears in Pareigis 1970, 174-177.

38. Ore 1942.

39. See Dubreil & Dubreil Jacotin 1939. Dubreil attended the first Bourbaki meetings and was in close relations with the group members. See § 7.1 below.

40. In Birkhoff 1935, 446-452.

view, this analysis would help understanding the discipline in its most general setting.[41] Based on the latter analysis, Ore also presented a generalized version of Galois theory, aimed at discussing "a general type of correspondence between **structures**" which "occur in a great variety of mathematical theories."[42]

A further direction in which Ore's program developed was the search for generalized definitions of group. This direction was explored in two works conducted by Ore in cooperation with his graduate students. In 1937 Ore published together with Bernard Hausmann a research paper on the "Theory of Quasi-Groups." Any algebraic system with a single internal operation was called by Ore and Hausmann "grupoid." They claimed that several isolated results on grupoids had already been obtained, which could be found scattered around the algebraic literature. In their article they proposed to follow the guidelines of Ore's program in order to present those results systematically, and to obtain meaningful new ones. Moreover, they saw in their work an axiomatic analysis of the property of associativity in general:

> Instead of starting from a chosen set of axioms we analyze the conditions necessary in order to obtain specific theories and theorems. In the end this turns out to be mainly an analysis of the associative law and the various possible formulations of the associative law represent the condition for the existence of various analogues to the theorems in ordinary group theory. (Hausmann and Ore 1937, 983)

A quasi-group differs from a group in that the former does not necessarily satisfy the associative law, nor does it necessarily contain an identity and inverse elements. The only axiom that necessarily holds in a quasi-group is the following:

> *Quotient Axiom*: To each pair a and b there exists a unique x and y such that $ax = b$, $ya = b$ (p. 986)

In other words, if one writes down the multiplication table of a quasi-group, then each element appears in a row once and only once, exactly as it is the case for groups. Ore and Hausmann defined the parallels of cosets and normal subgroups for quasi-groups and established several connections between different axioms of associativity and the properties of the **structure** of "normal sub-quasi-groups" of any given quasi-group. In particular the authors

---

41. Ore 1943, 1943a.
42. Ore 1944.

found out the minimal distributivity conditions needed to ensure that the Jordan-Hölder and Schmidt-Remak theorems hold in a quasi-group. The authors also proposed several open problems for further research: to develop a theory of Abelian quasi- groups, to formulate Sylow theorems for quasi-groups, etc.

In the long run, the study of quasi-groups did not turn into a lively domain of research in algebra; nevertheless it should be remarked that several research papers, which followed the lines opened by Ore and Hausmann, were published in leading American mathematical journals during the forties: papers elaborating ideas such as Abelian quasi-groups,[43] simple quasi-groups,[44] and "loops", namely, quasi-groups with identity.[45] More interestingly, several generalizations of the Jordan-Hölder theorem appear in this context. Thus Duffin and Pate proposed an axiomatic analysis of the Jordan-Hölder theorem based on a generalized concept of "normal subgroup", which was a direct elaboration of the ideas of Ore and Hausmann.[46] Likewise Richardson proposed a study of the theorem for all binary operations which are not associative.[47] A very systematic and detailed study of quasi-groups was published by the American algebraist Adrian A. Albert. He not only proved generalized versions of the Schreier refinement theorem and of the Jordan-Hölder theorem, but he also proposed a generalized version of the extension problem for groups, namely, to construct all loops G with given "normal divisor" H and given "quotient loop" G/H.[48] Albert mentioned the article of Ore and Hausmann as a seminal one for this field, although his own proof and concepts do not build on **structural** formulations. Albert's papers were soon followed by a paper in which many of his theorems, in particular the Jordan-Hölder theorem for quasi-groups, were given new **structural** proofs based on a work of Ore.[49]

Similar to Ore's cooperation with Hausmann was the research conducted by Ore together with Melvin Dresher: a **structural** "theory of multigroups."[50] The concept of multigroup is a further direct extension of an early

43. Griffin 1940.
44. Bruck 1944.
45. Bruck 1949.
46. In Duffin 1938 and Duffin & Pate 1943.
47. Richardson 1945.
48. See Albert 1943; 1944.
49. See Smiley 1944. Smiley referred in his article to Ore 1937.
50. Dresher & Ore 1938.

idea of Ore. Any two elements A,B of a **structure** define a quotient A/B whenever $A > B$. On the contrary, in the theory of groups, for instance, B defines a quotient in A if and only if B is a *normal* subgroup of A. The concept of multigroup enables a generalization of the concept of quotient so that any subgroup, not only the normal ones, define a quotient on the group. A multigroup is an algebraic system that satisfies all the axioms of a group, except that the product of two elements does not necessarily yield a single element as a result, but rather a set of elements. In their paper on multigroups, Ore and Dresher discussed the normality conditions of submultigroups and conducted a **structural** research of the decomposition properties of these domains.

6.5   Universal Algebra, Model Theory, Boolean Algebras

Beyond Ore's own work and that of his contemporaries on the abstract foundation of algebra based on the concept of lattice, early work on lattices gave rise to other developments which are more or less directly connected with the reflexive attempt to elucidate the idea of a mathematical structure, and in particular of an algebraic one. They cover a period of time relatively later than Ore's program itself, yet they share a common background and many of the same leading principles. Some of these developments deserve being discussed here. They concern disciplines such as universal algebra, model theory and Boolean algebras.

The term "Universal Algebra" was used for the first time by Alfred North Whitehead (1861-1947) as the title of his 1898 book. Whitehead had proposed to develop a comparative study of various algebraic disciplines. In his study he covered Hamilton's quaternions, Hermann Grassman's *Ausdehnungslehre* and Boole's algebra of logic. However it was not until the consolidation of the structural approach to algebra that a broader interest in such an undertaking became manifest. In fact, the works of Ore, and especially those of Birkhoff on lattice theory are usually considered to contain the first significant contributions to universal algebra.[51]

A universal algebra is a straightforward formalization of the non-formal idea of an algebraic structure. It comprises a non-empty set A and a list of finitary relations abstractly defined on the elements of the set. The immediate interest of theorems on universal algebra concern results on homomorphisms

---

51. Cf. Grätzer 1968, *v*; Jacobson 1980, 52; Mehrtens 1979, 206.

and isomorphisms, quotient and subalgebras, free algebras, as well as decomposition and chain theorems. All these kinds of questions were addressed by Birkhoff, and by Ore and his collaborators during the decade 1935-1945.[52]

A most significant turning-point in the research on universal algebra took place during the 1950s when this research was combined with tools provided by mathematical logic, thus incorporating it into the gradual rise of model theory. Model theory proposed to analyze the relations between systems of abstract postulates and mathematical systems. Thus the roots of this theory lie in the early work on axiomatics of Hilbert (§ 3.3), Peano, Frege and others. The early works on postulational analysis in the USA (§ 3.5) provided a further significant thrust to this development. Later, Birkhoff's lattice-theoretical work on universal algebra has usually been seen as the main contribution to the early stages of model theory.[53] Although it is difficult to separate the early developments of universal algebra from those of model theory,[54] the latter began to be recognized as an independent domain of research since the early 1950s.[55]

The subsequent development of model theory provided a further contribution to the understanding of the nature and classification of mathematical structures. It would be quite beyond the scope of the present book to discuss this contribution in detail.[56] In fact, it would seem more natural to consider the development of model theory in the framework of a history of logic than in one of algebra. In particular, model theory is essentially different from the three reflexive theories considered in this second part of the present book. The latter are characterized by the definition of a specific mathematical entity that is similar in nature to the mathematical structures it attempts to account for. Thus *structures*, **structures**, and categories are each an instance of itself. They attempt to elucidate the generalized, abstract concepts which constitute their subject-matter by formulating a second-order generalization modelled on the same kind of mathematical perspective on which the first generalization was

---

52. The early stages of universal algebra and its close relation to the lattice- theoretical research of Birkhoff and Ore are discussed in Mehrtens 1979, 206-210.

53. Accounts of the early stages in the development of model theory appear in Vaught 1974; Weaver 1993, 671-674.

54. In fact, as Chang and Keisler (1990, 1) point out "the line between universal algebra and model theory is sometimes fuzzy; our own usage is explained by the equation

universal algebra + logic = model theory."

55. Cf. Chang and Keisler 1990, 3; Grätzer 1968, *vi*.

56. For such an account see, e.g., Chang 1974.

performed. Thus a category and a **structure** generalize the idea of a ring in a similar fashion that a ring generalizes the idea of the integers equipped with two operations. In other words these three concepts are both an attempt to understand the structural character of mathematics and a result of it. This is not the case with model theory, which provides a side-view, as it were, of the phenomena of mathematical structure, as seen from the vantage point of view afforded by mathematical logic.[57]

It should also be pointed out here that the subsequent developments of universal algebra increasingly concentrated, since the late 1960s, on phenomena closely connected with category theory and to its main concepts, and one thus finds during this period several research-works addressing issues common to both theories.[58]

A further direction of research connecting lattice-theoretical research with the reflexive attempt to elucidate the idea of mathematical structures is associated with the work of Marshall Stone (1903-1989) on Boolean algebras, and to its subsequent influence on the rise of category theory. In the late 1930s Stone published a series of articles that changed the whole conception of this particular mathematical domain of research. By the time of publication of those articles Stone was already a leading mathematician whose main fields of research included neither of the two main fields directly associated with Boolean algebra, algebra and logic. Rather Stone was especially known for his work on functional analysis. His book on linear operators in Hilbert spaces was the classical textbook of the discipline for many years.[59] Stone's interest

---

57. Without specifically referring to the three main reflexive attempts discussed here, Hourya Sinaceur has described this difference as follows (Sinaceur 1991, 13): "Cherchant les raisons logiques de certains faits mathématiques, [le théoricien des modèles] s'intéresse *aussi* au *langage* utile pour une description formelle de ces faits. Quel que soit leur degré d'abstraction, les structures considérées ne sont pas étudiées pour elles-mêmes, mais en tant que modèles d'ensembles d'énoncés d'un langage logique du premier ordre." (Italics in the original.)

58. Thus in the classical textbook on Universal algebra, Grätzer 1968, *vii*, the readers are told that "category theory is excluded from this book because a superficial treatment seems to present pedagogical difficulties and it would be mathematically not too effective; moreover, those topics that can be treated in depth in a categorical framework ... are to be discussed by Eilenberg in a book (universal algebra and automata theory) and by F.W. Lawvere in a book (on elementary theories, in lecture notes (on algebraic theories), and in an expository article (on the category of sets)." A second standard book on universal algebra (Cohn 1965) included in its second edition (1981) a new supplement on category theory and universal algebra. Jacobson 1980 has a chapter on universal algebra, whose exposition strongly relies on categorical formulations.

in Boolean algebra thus arose from an unusual source, namely, from his research on commutative operators in Hilbert spaces.[60] Stone published his results on Boolean algebras in two articles,[61] which were preceded by two short preliminary reports.[62]

It was seen above in § 3.5, that the algebra of logic was the single issue to which a majority of research papers were devoted during the first two decades of the century in works on postulational analysis in he USA. However, contrary to what may appear on first sight, it is a fact that Boolean algebra (or the algebra of logic) was *not* considered an organic part of modern, structural algebra even in the early 1930s! The absence of a chapter on Boolean algebras in van der Waerden's book is symptomatic of the status of the discipline. Boolean algebra was part of logic, and logic was not seen as an integral part of mathematics in most universities until the early thirties, at least in the USA. It is true that a central importance was accorded to mathematical logic in Göttingen since the time of Hilbert, but in this particular respect Göttingen was more the exception than the rule.[63] In the USA, Alonzo Church (1903-1995) was the first logician to be appointed in a mathematics department. It happened at Princeton in the 1930s.[64] Saunders Mac Lane has reported that, having returned from Göttingen after finishing his Ph.D., he faced a strong opposition to his interest in logic (especially from Ore, then Sterling Professor of Mathematics at Yale).[65] The fact that so many articles on postulational analysis dealt with Boolean algebra corresponds to the fact that the real interest of those articles lay in the systems of postulates as such, rather than in the domain of research that those systems define. Boolean algebras were especially interesting from the perspective of postulational analysis because they could be defined by two different, but equivalent, kinds of systems of postulates: by focusing either on the binary operation or on the order properties.[66] Moreover,

---

59. Stone 1932.

60. A brief historical account on this issue appears in the introduction to Johnstone 1982.

61. Stone 1936; 1937.

62. Stone 1934; 1935.

63. The developments leading from 1907 on to the creation of the first chair of mathematical logic in Germany (for Ernst Zermelo in Göttingen) are described in Peckhaus 1990, 106-110 & 116-122.

64. Nevertheless, logical research had had an earlier chapter in American universities, especially with the works of Charles Sanders Peirce (1839-1940) and of his student Christine Ladd. See Parshall and Rowe 1994, 130-134.

65. Mac Lane 1988, 323-324.

Huntington, who was among the most active authors of that trend, was not a full professor of mathematics, but rather a part-time professor of mechanics at Harvard, and this fact may explain his choice of a field of research which was rather unusual among mainstream working mathematicians.

Stone was among the first to understand and explicitly declare the gains to be expected for the discipline of Boolean algebras if it be incorporated into the picture of modern, structural algebra. When in 1938 the first symposium on lattice theory was organized, Stone was among the invited lecturers. In his lecture, he described the status of Boolean algebra within mathematics in the following terms:

> In the mathematical literature one frequently comes upon evidences of an opinion that Boolean algebras are in some sense fundamentally different from the algebras commonly met in dealing with families of real or complex numbers and their generalizations. Thus Whitehead [in his *Universal Algebra* (1898)] sets Boolean algebras apart as instances of "algebras of nonnumerical genus"; and quite recently Bell (1927) follows him, writing in the introductory remarks of a paper on Boolean algebras that "this is probably the first attempt to construct an arithmetic for an algebra of non-numerical genus." (Stone 1938, 809)

In our attempt to understand the rise, the meaning and the real import of the structural approach to algebra, one can hardly overlook the fact that the study of Boolean algebras and of the algebra of logic—a discipline that had long since been presented in a fully axiomatized formulation, and whose subject-matter is a domain on which two operations are abstractly defined on arbitrary elements—was not part and parcel of structural algebra by 1930. This situation gives further weight to the claim that the abstract axiomatic formulation of theories was neither the only nor even the single most important feature of the structural image of algebra. Structural inquiries pursued in non-abstractly defined domains were already mentioned in earlier sections of this book; here one finds a further example of a domain that—although *defined* very early in thoroughly abstract axiomatic terms—was not immediately absorbed into the structural picture of algebra, because it had not been previously *investigated* in structural terms. Algebra was not the main field of research of Stone. As a first-rate mathematician who was a relative newcomer to this particular field, he was perhaps relieved from thinking in terms of

---

66. Huntington 1904 contains a detailed postulational analysis of both approaches to the definition of the algebra of logic, as well as historical references to earlier works on both directions.

accepted schemes. Consequently, he was able to understand clearly for the first time what in retrospect seems almost obvious to mathematicians: that Boolean algebras are "algebraic structures", just as fields or groups.

In his research, Stone was led by the idea that the discipline of Boolean algebras would considerably benefit from being approached from within the perspective offered by the structural image of algebra. At the same time, he thought that algebra at large would also benefit from the incorporation of Boolean algebras into its whole picture. Stone thus presented a program for Boolean algebra in terms strikingly reminiscent of those used by Ore in the presentation of his own foundational program for abstract algebra. In his article of 1936 (which had been actually written in 1935) Stone wrote:

> If one reflects upon various algebraic phenomena occurring in group theory, in ideal theory, and even in analysis, one is easily convinced that a systematic investigation of Boolean algebras, together with still more general systems, is probably essential to further progress in these theories. (Stone 1936, 37)

Stone left no room for doubt concerning the nature of those "more general systems", so important for the future development of the theory. In a footnote, he cited the works of Birkhoff, Ore and Klein-Barmen, and even those of Dedekind before them, on lattices.

The basis for Stone's intended incorporation of Boolean algebras into modern algebra was provided by his famous insight that a Boolean algebra is nothing but a particular kind of ring: a ring in which every element is idempotent (i.e., that $a^2 = a$ for every element of the ring). The central theorem proven in his 1936 article is the dual formulation of that insight, namely, that given any Boolean ring, it is possible to build an algebra of classes which is isomorphic to the given ring. In a more general formulation, Stone showed how to build all the algebras of classes which are isomorphic to a given Boolean ring. This theorem is parallel to the representation theorem for groups which states that for any abstractly defined group it is possible to find a certain group of permutations isomorphic to the given group. Beyond the intrinsic interest of his results, Stone's work implied the introduction of a topological space of non-trivial interest, defined on a purely algebraic basis. Until this work of Stone, most important spaces considered in topology—even within the most abstract formulation of the discipline—had arisen from purely geometric or analytic contexts.[67]

Stone defined a topology on the collection of prime ideals of a given Boolean algebra A as follows: for any ideal I, define the open set $U(I)$ of all prime ideals $P \subseteq A$, such that $P \not\subset I$. A correspondence may be established between the principal ideals (and therefore the elements of the ring) and the clopen sets of the topology (i.e., those sets which are both closed and open). Therefore, given the topology one can obtain an isomorphic copy of the original algebra. Stone also proved that every topological space obtained in this way is compact, Hausdorff, and totally disconnected. Every topological space satisfying these three properties was called by Stone a "Boolean space." Later, that name would change to "Stone spaces", a concept which developed into an autonomous field of research.[68]

The background of Stone's ideas on Boolean algebras was totally different from that of Ore, but these two mathematicians stressed a similar kind of leading principles in their approach to algebra. This similarity should be born in mind when evaluating their historical influence; while Stone's contribution has come to be considered a central milestone for many branches of twenty-century mathematics, Ore's work has been forgotten. As a matter of fact, however, many of the influential features of the former's work appear also in the latter's. This issue will be retaken below in § 8.3, when analyzing the initial stages of category theory.

In order to complete the picture of the relationship between lattice-theory and topology at the time of publication of Stone's work, it should be mentioned that besides Stone, Henry Wallman exploited results of lattice theory for his research on topology.[69] Likewise, the work of Karl Menger (1902-1983) should also be mentioned in this context. Menger summarized previous works in topology, which attempted to develop this discipline from a perspective similar to that of Ore's program for algebra, i.e., avoiding any mention of the elements of the spaces considered.[70] These works, however, did not have a meaningful impact on later research in topology.

---

67. An interesting example of a topology defined earlier following a purely algebraic motivation appears in a generalized formulation of Galois theory for infinite algebraic extensions, published in Krull 1928. Similar examples appear in the works of Krasner and Sebastiao e Silva mentioned in the introduction to Part Two (footnote 3).

68. For an account of Stone's work on Boolean spaces and its relation to Tarski's unification of logic, Boolean algebras and topological spaces, see Grosholz 1985, 148-151.

69. See Wallman 1938. Other influences of Stone's work are described in Johnstone 1983 and in the introduction to Johnstone 1982.

70. See Menger 1940.

6.6   Ore's Structures and the Structural Image of Algebra

Ore published his program for providing a general foundation of abstract algebra in 1935. In his early articles of 1930-1932, Ore had explicitly expressed the opinion that the most appropriate way to conduct research in any domain of algebra is to exploit the specific defining features of the domain in question. This view was in clear opposition to that advanced by Noether, which favored the introduction of generalizing concepts as the best way to bring out the essence of the theorems proven. Ore was well aware of the latest advances of algebra and, in particular, of Noether's works on ideals and her general views regarding algebra. Ore's early choice of a more classical, concrete approach to algebra derived from reasons other than sheer ignorance of the possibilities opened up by the modern abstract approach. Moreover, it is hard to specify any dramatic developments in the body of algebra over the critical years during which Ore's views changed, and that might account for this change. All the relevant knowledge on which Ore based his program was already available back in 1930. Thus, it is not clear at all what caused the change in Ore's view and why this change happened when it happened. From the available evidence regarding Ore's work there is no direct answer to these questions.

Be that as it may, Ore did publish his generalizing program in 1935, and the issues discussed in Part One of the present book clearly indicate that ideas contained in the works of Dedekind and Noether decisively influenced Ore's views. Ore took from Dedekind the concept of lattice itself, but also the idea that this concept should have meaningful applications in generalizing results of algebra. From Noether, Ore learned the value of the abstract approach as a source for important results, as well as many technical details which became central to his proofs.

But besides particular manifestations of Dedekind's and Noether's influence on Ore, one can state, in a more general formulation, that Ore's program embodies a reflexive attempt to transform into a substantial, formally developed body of knowledge, the images of knowledge shared by these two mathematicians. Ore took very seriously several of the leading non-formal images of algebra tacitly underlying the works of Dedekind and Noether; he formulated them explicitly, in strict mathematical terms within an axiomatic system, explored their deductive consequences, and expected that such an inquiry would suffice to produce what he took to be the central theorems of algebra.

Dedekind and Noether had laid special stress upon the inclusion properties of distinguished sets of elements. This was the case in Dedekind's foundational studies on the natural and irrational numbers, and especially so in his theory of ideals. This was also the case with many of the details of Noether's proofs and, above all, with her insistence on the centrality of chain conditions.

There are also some grounds to conjecture that the specific direction and the very leading notion of Ore's program had formerly been envisaged by Noether herself. We find clues to this in Alexandrov's obituary on Noether quoted above in § 5.4. Alexandrov mentioned some ideas that Noether was working on since 1926, but that she did not get to publish. These ideas concern a "purely set- theoretical" definition of group. Now, in the footnotes to the "*Aufbau*" Noether had explained what she meant by "purely set-theoretical considerations", namely, that the proofs be given "independently of any operation" (§ 5.3). Alexandrov, in turn, mentioned the "difficulty involved in axiomatizing the notion of a group starting from its partition into cosets as the fundamental concept", and the later success of Ore and Kurosh in carrying out some of the ideas suggested by Noether in this context. Until new evidence is gathered on this issue, the conjecture may be thus left open that Noether was working in her late years on ideas such as developed by Ore in his program.

It should also be pointed out that, even after Ore's program as such was abandoned, the question of the relationship between the structure of the group and the lattice of its subgroup continued to attract interest as an open research question in the theory of groups. An exhaustive and systematic exposition of the main results obtained concerning this issue was published in 1956 by the Japanese mathematician Michio Suzuki in his book *Structure of Groups and the Structure of the Lattice of Subgroups*. Suzuki summarized the relation between these two algebraic constructs by claiming that "the structure of the subgroup lattice reflects very strongly the structure of the group."[71] This statement is very close to the basic presupposition behind Ore's program, but certainly not identical. At the time when he was working out his program, Ore would not have been satisfied merely to know that *much* information about a particular algebraic domain is carried by its lattice of subdomains. Rather, he

---

71. Suzuki 1956, 2. On Suzuki's work on group theory, see Aschbacher et al 1999. For a more recent, comprehensive exposition of the current state of research on the relationship between a group and its lattice of sub-groups, see Schmidt 1994.

had thought that *all the important theorems of algebra* could have been derived from the information contained in those lattices.

The latest papers that may be considered as developing ideas connected with Ore's program were published around 1945. It turned out than the program was not fruitful enough and that it did not pose enough interesting problems to capture the attention of many researchers. Moreover, since the early 1940s many American mathematicians turned their attention towards the war effort and, as a result, applied mathematics received much more emphasis than the pure, foundational studies embodied in Ore's program.[72] The program was not only abandoned; it fell into oblivion and today it is not even mentioned as an object of historical interest. The above account of Ore's program shows, however, that it was a significant stage in the rise and development of the structural approach to algebra and a chapter in the history of twentieth century mathematics worthy of attention.

---

72. Birkhoff, in particular, became fully involved in applied research since 1940. See Young 1997.

Chapter 7   Nicolas Bourbaki: Theory of *Structures*

The widespread identification of contemporary mathematics with the idea of structure has often been associated with the identification of the structural trend in mathematics with the name of Nicolas Bourbaki. Fields medalist René Thom, in a famous polemical article concerning modern trends in mathematical education, asserted that Bourbaki "undertook the monumental task of reorganizing mathematics in terms of basic structural components."[1] Thom further claimed that:

> Contemporary mathematicians, steeped in the ideas of Bourbaki, have had the natural tendency to introduce into secondary and university courses the algebraic theories and structures that have been so useful in their own work and that are uppermost in the mathematical thought of today. (Thom 1971, 695)

The identification of Bourbaki with modern, structural mathematics is not always as explicitly formulated as in Thom's quotation, but it has interestingly been manifest in many other ways. Thus for instance, during the years 1956 and 1957 in Paris the "Association des professeurs de mathématiques de l'enseignement public" organized in Paris two cycles of lectures for its members. The lectures, meant to present high-school teachers with an up-to-date picture of the discipline, were delivered by such leading French mathematicians as Henri Cartan, Jacques Dixmier, Roger Godement, Jean-Pierre Serre, and several others. They were later collected into a volume entitled *Algebraic Structures and Topological Structures*.[2] Thus we have the first component of the two-fold identification: "up-to-date mathematics = structural mathematics." The second component—"Bourbaki's structures = mathematical structures"—was not explicitly articulated therein, yet the editors made this second identification plain and clear by choosing to conclude the book with a motto

---

1. Thom 1971, 699.
2. Cartan et al., 1958.

quoted from a famous article in which Bourbaki described "The Architecture of Mathematics" in terms of mathematical structures, as follows:

> From the axiomatic point of view, mathematics appears thus as a storehouse of abstract forms—the mathematical structures; and it so happens—without our knowing how—that certain aspects of empirical reality fit themselves into these forms, as if through a kind of preadaptation.[3]

Historians of mathematics have also accepted very often this identification of "mathematical structures" with the name of Bourbaki. An explicit instance of this appears in Hans Wussing's well-known account of the rise of the concept of abstract groups. In the introduction to his book Wussing wrote:

> The conscious tendency to think in terms of structures has even produced its own characterization of mathematics. An extreme characterization of this kind, advanced by ... Nicolas Bourbaki ... sees mathematics as a hierarchy of structures. (Wussing 1984, 15)

And of course, the identification of mathematical structures with Bourbaki has had a marked influence *outside* mathematics. The best-known example of this is found in the work of Jean Piaget. In his widely-read general exposition of the central ideas of structuralism, one of the chapters discusses the "new structuralist view of mathematics." In this context Piaget mentions Klein's Erlangen Program, because of its successful use of the concept of group structures, as the first victory of the new approach. "However", he adds:

> ... in the eyes of contemporary structuralist mathematicians, like the Bourbaki, the Erlangen Program amounts to only a partial victory for structuralism, since they want to subordinate all mathematics, not just geometry, to the idea of structure. (Piaget 1971, 28)

In the same place Piaget also pointed out the close correspondence between Bourbaki's so called "mother structures" (i.e. algebraic structures, order structures and topological structures) and the first operations through which the child interacts with the world.[4]

However, as was already seen, the term "mathematical structure" has been used and understood in sharply divergent ways by different authors. The question therefore arises, what is the meaning attributed to the term by those authors identifying it with the work of Bourbaki. Moreover, what did Bour-

---

3. We quote here from the English version, Bourbaki 1950, 231. This article of Bourbaki is discussed in greater detail below.

baki actually mean by "mathematical structure", and how do mathematical structures appear in Bourbaki's actual mathematical work?

The present chapter discusses the place of the idea of mathematical structure in Bourbaki's work. As already mentioned, Bourbaki's concept of *structure* was meant to provide a unifying framework for all the domains covered by Bourbaki's work. Ore's **structures**—discussed in the previous chapter—had arisen within, and remained focused on, the relatively more limited framework of abstract algebra. Thus Bourbaki's works, and in particular the concept of *structure*, relate to a much larger territory of pure mathematics concerning both motivations and intended scope of application. Bourbaki's concept of *structure* has rarely been explicitly considered as what it really is, namely, *one among several formal attempts* to elucidate the non-formal idea of mathematical structure, and in fact, a rather unsuccessful one at that. It will be seen that there is a wide gap separating the significance of Bourbaki's overall contribution to contemporary mathematics from the significance of this one particular component of Bourbaki's work, namely, the theory of *structures*. Thus, the present chapter examines Bourbaki's concept of *structure*, its relation to Bourbaki's work at large, and the degree of its success in formally elucidating the idea of "mathematical structure."

7.1 The Myth

Nicolas Bourbaki was the pseudonym adopted during the 1930s by a group of young French mathematicians who undertook the collective writing of an up-to-date treatise of mathematical analysis, suitable both as a textbook for students and as reference for researchers, and adapted to the latest advances and the current needs of the discipline. These mathematicians were initially motivated by an increasing dissatisfaction with the texts then traditionally used in their country for courses in analysis,[5] which were based on the university lectures of the older French masters: Jacques Hadamard (1865-1963), Emile Picard (1856-1941), Edouard Goursat (1858-1936), and others.[6]

---

4. See also Piaget 1973, 84: "From the level of concrete operations—at about 7/8 years—another interesting convergence may be found, that is the elementary equivalence of the three "mother structures" discovered by Bourbaki, and this in itself shows the "natural" character of these structures." See also Gauthier 1969; 1976. The issue of the "mother structures" is further elaborated below. See Aubin 1997, for an illuminating analysis of the close conceptual relationship between Bourbaki, Levi-Strauss and French "potential literature".

5. See Dieudonné 1970, 136; Weil 1992, 99-100.

They also felt that French mathematical research was lagging far behind that of other countries,[7] especially Germany,[8] and they sought to provide a fresh perspective from which to reinvigorate local mathematical activity.

The would-be members of Bourbaki met for the first time to discuss the project in the end of 1934. They stated as the goal of their joint undertaking "to define for 25 years the syllabus for the certificate in differential and integral calculus by writing, collectively, a treatise on analysis. Of course, this treatise will be as modern as possible."[9] The names involved in the project, as well as the details regarding the scope and contents of the treatise were to fluctuate many times in the following decades, but some of the essentials of Bourbaki's self-identity are already condensed in this last quotation: collective work, wide-ranging coverage of the hard-core of mathematics, modern approach.

The first actual Bourbaki congress took place in 1935.[10] The founding members of the group included Henri Cartan (1904-), Claude Chevalley (1909-1984), Jean Coulomb, Jean Delsarte (1903-1968), Jean Dieudonné (1906-1994), Charles Ehresmann (1905-1979), Szolem Mandelbrojt (1899-1983), René de Possel and André Weil (1906-1998).[11] Over the years, many younger, prominent mathematicians joined the group, while the elder members were supposed to quit at the age of fifty. From the second-generation Bourbaki members the following are among the most prominent: Samuel

---

6. On the French tradition of *Cours d'analyse*, based on lectures delivered by leading mathematicians see Beaulieu 1993, 29-30.

7. Thus, Weil 1992, 120, described the situation, as he saw it, in the following words: "At that time [1937], scientific life in France was dominated by two or three coteries of academicians, some of whom were visibly driven more by their appetite for power than by a devotion to science. This situation, along with the hecatomb of 1914-1918 which had slaughtered virtually an entire generation, had had a disastrous effect on the level of research in France. During my visits abroad, and particularly in the United States, my contact with many truly distinguished scholars had opened my eyes to the discouraging state of scientific scholarship in France." Weil 1938 contains a general assessment of the state of scientific activity in France during the 1930s. See also Dieudonné 1970, 136.

8. Israel 1977, 42-43 analyses the differences between the German and the French mathematical schools at the time, based on sociological considerations.

9. Quoted from Beaulieu 1993, 28.

10. Cf. Weil 1978, 537-538.

11. Paul Dubreil (1904-1994) and Jean Leray (1906-1998) also attended the early meetings. See Beaulieu 1994, 243 Guedj 1985, 8; Weil 1992, 100 ff. For an account of the participants in the meetings preceding the actual work of the group see Beaulieu 1993, 28-31.

Eilenberg (1913-1998), Alexander Grothendieck (1928-), Pierre Samuel (1921-), Jean Pierre Serre (1926-). All of these mathematicians were pursuing separately their own individual work (usually being among the leading researchers of their respective disciplines), while the activities of Bourbaki absorbed part of their time and effort.

What was initially projected as a modern textbook for a course of analysis eventually evolved into a multi-volume treatise entitled *Eléments de Mathématique*, each volume of which was meant to contain a comprehensive exposition of a different mathematical discipline. Each chapter and each volume of Bourbaki's treatise was the outcome of arduous collective work. Members of the group used to meet from time to time in different places around France. At each meeting, individual members were commissioned to produce drafts of the different chapters. The drafts were then subjected to harsh criticism by the other members, and then reassigned for revision. Only after several drafts had been written and criticized was the final document ready for publication.[12] Minutes of meetings were taken and circulated among members of the group in the form of an internal bulletin called "La Tribu." Although the contents of the issues of "La Tribu" abound with personal jokes, obscure references and slangy expressions which sometimes hinder their understanding, they provide a very useful source for the historian researching the development of Bourbaki's ideas.[13]

---

12. Bourbaki's mechanism of collective writing has been documented in several places. See, e.g., Boas 1970, 351; Cartan 1980, 179; Dieudonné 1970, 141; Weil 1992, 105. See also the vivid description of Mac Lane (1988, 337): "Debate at Bourbaki could be vigorous. For example, in one such meeting (about 1952) a text on homological algebra was under consideration. Cartan observed that it repeated three times the phrase 'kernel equal image' and proposed the use there of the exact sequence terminology. A. Weil objected violently, apparently on the grounds that just saying 'exact sequence' did not convey an understanding as to why that kernel was exactly this image." On the introduction of this term, see below the opening passages of § 8.2.

13. According to Weil (1992, 100) since the early meetings of Bourbaki an archive was established, of which Delsarte was first in charge. Later it was kept in Nancy and later on in Paris. Some years ago, the "Association des Collaborateurs de Nicolas Bourbaki" was established at the Ecole Normale Supérieure, in Paris. An archive containing relevant documents, probably including many copies of "La Tribu" was created. Unfortunately, it has yet to be opened to the public. Those issues of "La Tribu" quoted in the present article belong to personal collections. Professor Andrée Ch. Ehresmann kindly allowed me to read and quote from documents belonging to her late husband, Professor Charles Ehresmann. This includes volumes of "La Tribu" from 1948 to 1952. Other quotations here are taken from the personal collections of Chevalley and Szolem Mandelbrojt, as they appear in an appendix to Friedmann 1975.

In the decades following the founding of the group, Bourbaki's books became classic in many areas of pure mathematics in which the concepts and main problems, the nomenclature and the peculiar style introduced by Bourbaki were adopted as standard. Bourbaki's actual influence on the last fifty years of mathematical activity (research, teaching, publishing, resources distribution) has been enormously significant.[14] However, even now that the Bourbaki phenomenon is receding into the past, a fair historical evaluation of Bourbaki's influence on contemporary mathematics remains an arduous task.[15] Such an assessment should take into account, in the first place, the diverse degrees of influence which Bourbaki exerted on mathematical research and on mathematical education during different periods of time and in different countries.[16] Second, it should take into account Bourbaki's varying influence on different branches of mathematics. There are certain branches upon which Bourbaki exerted the deepest influence, like algebra and topology; assessing Bourbaki's influence on them would be tantamount to analyzing the development of considerable portions of these disciplines since the 1940s. Here we can only briefly overview the scope of this influence.

The first chapters of Bourbaki's book on topology were published in 1940,[17] following almost four years of the usual procedure of drafting and criticism. This treatise on topology was meant to provide the conceptual basis needed for discussing convergence and continuity in real and complex analysis. Bourbaki's early debates on topology were gradually dominated by a tendency to define this conceptual basis in the most general framework possible, avoiding whenever possible the need to rely on the traditional, most immediately intuitive concepts such as sequences and their limits. This effort helped understanding, among others, the centrality of compactness in general topol-

---

14. As Mac Lane 1988, 338, wrote: "A whole generation of graduate students were trained to think like Bourbaki."

15. Beaulieu 1989 contains the most detailed and perhaps only comprehensive historical study of Bourbaki's work written to the present. It concentrates on the first ten years of activity. For more recent works on Bourbaki see: Borel 1998, Cartier 1998, Chouchan 1995, Mashaal 2000.

16. To the best of my knowledge, beyond scattered remarks, there are no detailed studies of Bourbaki's influence on research and teaching of mathematics in individual countries or regions. For Bourbaki's influence on shaping mathematical tastes in American universities, see Lax 1989, 455-456. On the influence of Bourbaki on mathematical education in the USSR see Sobolev 1973. Cf. also Israel 1977, 68.

17. The English version appeared as Bourbaki 1966. For a detailed discussion of Bourbaki's book on topology see § 7.3.3 below.

ogy.[18] It also yielded a thorough analysis of the various alternative ways to define general topological spaces and their central characteristic concepts: open and closed sets, neighborhoods, uniform spaces.[19] Moreover, an important by-product of Bourbaki's discussions was the introduction of filters and ultrafilters as a basis for defining convergence while avoiding reliance on countable sequences. Bourbaki, however, rather than including these latter concepts in the treatise, encouraged Henri Cartan to publish them, while elaborating on their relation to topological concepts, under his own name.[20]

Over the next years alternative approaches to questions of continuity and convergence were developed by other mathematicians, based on concepts such as directed systems and nets. The equivalence of the various alternative systems and those of Bourbaki was proven in the USA by Robert G. Bartle in 1955.[21] Thus, the history of the development of topology, at least from 1935 to 1955, cannot be told without considering in detail the role played in it by both Bourbaki as a group and its individual members.

The significance of Bourbaki's work for the development of algebra has less to do with the redefinition of basic concepts than with the refinement and promotion of the conception of this discipline as a hierarchy of structures.[22] The following quotation of Dieudonné seems to reproduce faithfully Bourbaki's images of algebra during the group's early years of activity:

> The development of "abstract" algebra begins around 1900 when it is recognized that the notion of *algebraic structure* (such as the structure of group, ring, field module, etc.) is the fundamental notion in algebra, putting the nature of the mathematical objects on which the structure is defined in the background, whereas, up to then, the majority of the algebraic theories dealt with calculations principally over the real or complex numbers. (Dieudonné 1985, 64-65)

---

18. Cf. Mac Lane 1987a, 166: "The recognition of the importance of compactness and of its description by coverings is a major step in the understanding of topological spaces. It developed only slowly—and was not really codified until Bourbaki, in his influential 1940 volume on topology, insisted." See also Mac Lane 1988, 337.

19. In his own research on topological groups André Weil showed that the metric plays only a secondary role in defining the main topological concepts. He thus developed an alternative approach based on uniform spaces as the adequate conceptual framework for this purpose (Weil 1937). See also Weil 1978, 538.

20. They appeared as Cartan 1937, 1937a. The debates leading to the publication of Bourbaki's book on topology are described in some detail in Beaulieu 1990, 39-41.

21. In Bartle 1955.

22. To a considerable extent what Bourbaki did for topology in terms of a unified presentation, van der Waerden had already done for algebra. See below § 7.3.3.

This image of algebra provided the conceptual framework within which Bourbaki members produced important contributions to the body of knowledge in their own research. This was particularly the case with algebraic geometry which, as Dieudonné said, was "one of the principal beneficiaries" of this conception.[23] In fact, a natural outcome of the above conception was the transformation of the disciplinary aims and scope of algebraic geometry through a redefinition of its classical concepts, by replacing the field of complex numbers by an arbitrary field. In fact, since the mid-forties algebraic geometry underwent a deep transformation which has often been characterized as a reformulation of the discipline as a part of commutative algebra, based on the concepts of "sheaf" and "scheme."[24] Not surprisingly perhaps, among the central figures who brought about this transformation one finds several leading French mathematicians, who were also members of Bourbaki: André Weil, Jean-Pierre Serre and Alexandre Grothendieck.[25] The body of algebraic knowledge, i.e., the concepts, techniques and methods, needed as the common background for understanding these works was precisely that developed in Bourbaki's treatise.[26] And no less important than that: the image of algebra within which their research was produced was the image of algebra first introduced by van der Waerden, and then so eagerly promoted by Bourbaki.

But there was also a second main source for the transformation of algebraic geometry coming from the work of Oscar Zariski (1899-1986) and his followers.[27] Zariski, like his French colleagues (and especially Weil), though independently of them, also set out to redefine the main problems and the conceptual foundation of algebraic geometry in terms of the newly consolidated,

---

23. See Dieudonné 1985, 59. However, Dieudonné's many accounts of the historical development of mathematics, and in particular of the role played by Bourbaki within it, deserve to be addressed with a critical attitude. (This is done in some detail below, especially in §§ 7.4 and 8.4.) In particular one should point out here, that Bourbaki's own activity did not begin with the structural image of algebra in mind. Rather, this image was absorbed during the first years of the project, as the preparation of the first volumes advanced. See Beaulieu 1994, 247-248.

24. Grothendieck & Dieudonné 1971, 2; Israel 1977, 68-69; Zariski 1950, esp. 77.

25. For their respective seminal works in this area see Grothendieck 1958; Serre 1955; Weil 1946. See also Weil 1954.

26. Cf. Weil 1946, *xiii*.

27. A detailed biography of Zariski, including technical appendices summarizing the significance of his work, appears in Parikh 1991. For Zariski's work on the "algebraization" of the foundations of algebraic geometry see Mumford 1991; Parikh 1991, 87-89.

structural view of algebra (though sometimes laying stress on different points of emphasis).[28] During the sixties several of Zariski's students (e.g., Michael Artin, David Mumford, Heisuke Hironaka) undertook to merge the perspective developed by Zariski with the techniques recently introduced by Grothendieck and Serre.[29] In the introduction to a book on Geometric Invariant Theory, Mumford described the aims of his work in terms that could have been taken from any chapter of Bourbaki, as follows:

> It seems to me that algebraic geometry fulfills only in the language of schemes that essential requirement of all contemporary mathematics: to state its definitions and theorems in their natural abstract and formal settings in which they can be considered independent of geometric intuition. (Mumford 1965, iv)[30]

The influence of Bourbaki, both as a group and through the works of its individual members, thus played a central role in shaping the development of algebraic geometry since the mid-1940s, but it is clear as well, that Bourbaki was not alone at that. Moreover, one should also note, that in spite of the strong structural orientation that has been characteristic of this field of research since that time, in the last decades algebraic geometers have returned to old motivations and classical questions.[31]

But if topology and algebra are among the disciplines in which Bourbaki's influence has been felt most strongly, at the other end of the spectrum disciplines like logic and most fields of applied mathematics seem not to have been aware of or influenced by Bourbaki at all. This last generalization, however, must be qualified, because Bourbaki has directly influenced mainstream trends in mathematical economics since the 1960s, through the work of Nobel laureate Gérard Debreu.[32]

---

28. For the mathematical divergences and the collaboration between Weil and Zariski (in particular concerning their respective views on algebraic geometry), see Parikh 1991, 84-85 & 90-95. A third author whose contribution to the "algebraization" of the foundations of algebraic geometry must be mentioned is van der Waerden. He has described his own contribution in van der Waerden 1970, esp. 172-176.

29. See Artin 1991; Mumford 1991; Parikh 1991, 147-161.

30. A similar assessment appears in Safarevich 1971, v.

31. As Zariski (1972, xiii) wrote: "There are signs at the present moment of the pendulum swinging back from 'schemes', 'motives', and so on towards concrete but difficult unsolved questions concerning the old pedestrian concept of a projective variety (and even of algebraic surfaces)." Cf. also Israel 1977, 69.

32. Cf. Ingrao & Israel 1990, 280-288; Weintraub & Mirowski 1994.

Beyond the objective difficulties encountered in analyzing Bourbaki's influence there are additional, subjective problems connected with the mythological status of the group. Ever since the name Bourbaki first appeared in public, the group became the focus of much attention and curiosity among mathematicians, and a full-fledged mythology came to surround it. The essence of this mythology is condensed in the following quotation taken from an article published by Paul Halmos in 1957:

> His name is Greek, his nationality is French and his history is curious. He is one of the most influential mathematicians of the 20th century. The legends about him are many, and they are growing every day. Almost every mathematician knows a few stories about him and is likely to have made up a couple more. His works are read and extensively quoted all over the world. There are young men in Rio de Janeiro almost all of whose mathematical education was obtained from his works, and there are famous mathematicians in Berkeley and in Göttingen who think that his influence is pernicious. He has emotional partisans and vociferous detractors wherever groups of mathematicians congregate. The strangest fact about him, however, is that he doesn't exist. (Halmos 1957, 88)[33]

The legend surrounding the group and the professional stature of the researchers who composed its membership have occasionally impaired the objectivity of appraisals of Bourbaki's scientific contributions. The following is an inspired description of these difficulties:

> Confronted with the task of appraising a book by Nicolas Bourbaki, this reviewer feels as if he were required to climb the Nordwand of the Eiger. The presentation is austere and monolithic. The route is beset by scores of definitions, many of them apparently unmotivated. Always there are hordes of exercises to be worked painfully. One must be prepared to make constant cross references to the author's many other works. When the way grows treacherous and a nasty fall seems evident, we think of the enormous learning and prestige of the author. One feels that Bourbaki *must* be right, and one can only press onward, clinging to whatever minute rugosities the author provides and hoping to avoid a plunge into the abyss. Nevertheless, even a quite ordinary one-headed mortal may have notions of his own, and candor requires that they be set forth. (Hewitt 1956, 507. Italics in the original)

---

33. Among the many additional articles and books that deal with the myth of Bourbaki the following may be mentioned: Boas 1986; Dieudonné 1970, 1982; Fang 1970; Guedj 1985; Israel 1977; Queneau 1962; Toth 1980. According to Mehrtens 1990, 320, already in the seventies Bourbaki had transformed from myth into history. This assessment may be debated as to its exact dating, but probably not as to its essence.

Among the many laudatory commentaries of Bourbaki's work one also finds articles intended for a popular audience that bolster the myth or borrow from it. One also finds several reviews in which well-known specialists in particular disciplines of mathematics present detailed analyses of specific books within the treatise, pointing out their virtues in approach, clarity of presentation, or in the excellent choice of the exercises. These reviews, however, are sometimes so untypically effusive in extolling the merits of Bourbaki's books that their credibility becomes questionable. Take, for example, the following review by a leading mathematician (himself a member of Bourbaki) of a new edition of one of Bourbaki's books:

> If the preceding editions [of the book] were meant to represent an almost perfect account of the bases for present day mathematics, this is now the perfect basis; the author is sufficiently representative of the mathematical community to make such a claim quite close to the truth. Furthermore, in a time in which indiscriminate use of science and technology threatens the future of the human race, or at least the future of what we now call civilization, it is surely essential that a well integrated report about our mathematical endeavors be written and kept for the use of a later day "Renaissance." As Thucydides said about his "History of the Peloponesian War", this is ... a treasure valuable for all times. (Samuel 1972)

Or as Emil Artin wrote in his review of Bourbaki's book on algebra: "Our time is witnessing the creation of a monumental work."[34]

But obviously not all technical reviews of the *Eléments* are as laudatory as those quoted above. Several reviews, written by specialists in their respective fields, have pointed out the shortcomings (mainly in notation and approach), of this or that specific chapter within Bourbaki's treatise. Thus, for instance, in a review of Bourbaki's book on set theory, B. Jonsson claimed that: "due to the extreme generality, the definitions are cumbersome, and all the results derived are of a very trivial nature."[35] Likewise Paul Halmos, reviewing Bourbaki's book on integration wrote: "I am inclined to doubt whether their point of view will have a lasting influence."[36]

---

34. Artin 1953. As further instances of reviews praising Bourbaki's work see Gauthier 1972; Mac Lane 1948; Rosenberg 1960. Remarkably, most reviewers of Bourbaki, favorable and critical alike, describe the choice of exercises as excellent; this choice is usually attributed to Jean Dieudonné. Cf. Eilenberg 1942, 1945; Kaplansky 1953, 1960; Kelley 1956; Thom 1971, 698. Examples of praise of Bourbaki's work couched in non-technical language can be found in Fang 1970, Queneau 1962, Toth 1980.

35. Jonsson 1959.

Beyond this kind of criticism, directed at particular issues within Bourbaki's output, one also finds critical attitudes concerning the more general
influence of Bourbaki on the overall picture of twentieth-century mathematics. The following quotation is an instance of the latter, expressed by Fields-
medalist Michael Atiyah, who, without explicitly mentioning the name of
Bourbaki, obviously alludes to it. In assessing the dangers of an unrestricted
reliance on the axiomatic approach, Atiyah said:

> Most books nowadays tend to be too formal most of the time. They give too much
> in the way of formal proofs, and not nearly enough in the way of motivations and
> ideas. Of course it is difficult to do that—to give motivations and ideas... French
> mathematics has been dominant and has led to a very formal school. I think it is
> very unfortunate that most books tend to be written in this overly abstract way
> and don't try to communicate understanding. (Minio 1984, 17)[37]

The opinions quoted above cover a period of nearly thirty years. This has
to be taken in account when evaluating them. The present discussion does not
aim at evaluating the overall import and influence of Bourbaki's approach on
contemporary mathematics. Neither does it analyze the significance of any of
Bourbaki's specific contributions to the mainstream disciplines addressed in
Bourbaki's treatise. Rather, the present discussion focuses on a restricted, and
usually overlooked, aspect of Bourbaki's work, namely, Bourbaki's reflexive
attempt to produce a formal concept, the concept of *structure*, meant to elucidate the non-formal idea of a mathematical structure. This particular aspect of
Bourbaki's work, however, is central to the group's images of mathematics
and its influence is more clearly manifest at that level. In particular, the concept of *structure* also plays an important role in Bourbaki's own historiography. Moreover, Bourbaki, or at least some of its members had hoped, at a
relatively early stage of their work, that *structures* would also play a central
role in the body of knowledge considered in the *Eléments* (and thus in mathematics at large). This hope, however, was not fulfilled. Thus, understanding
the role that *structures* play in Bourbaki's work provides further insights

---

36. See Halmos 1953. Further critical reviews appear in Bagemihl 1958; Gandy 1959 ("It is
possible, then, that this book may itself soon have only historical interest"); Hewitt 1956, 1966;
Mathias 1992; Michael 1963; Munroe 1958.

37. For a much harsher attack, both on the significance of the Bourbaki project for mathematics
as a whole and on the motivations behind it, see Mandelbrot 1989, 11-12. Further criticism of the
view, that accords such a central place in mathematics to the axiomatic method (by Bourbaki and by
others) appear in Browder 1975; Israel 1981; Spohn 1961.

regarding both the overall import and influence of that work and the development of the structural approach in mathematics.

7.2 Structures and Mathematics

The concept of structure has been often associated with Bourbaki's mathematics as well as with Bourbaki's putative philosophy of mathematics. Thus, together with the widespread identification of the structural trend in mathematics with the name of Bourbaki, a "structural philosophy of mathematics" has been attributed to the group. What is meant by this? This issue has rarely been discussed in detail. Despite the centrality that many authors claim for Bourbaki's program in contemporary mathematics, very little has been done to elucidate that purported philosophy. In general, when the term "structuralist" is invoked in connection with Bourbaki it is seldom followed by a detailed explanation of the exact meaning of the term.[38] In fact, it seems unlikely that Bourbaki's views can be accurately captured in either a single formula or a fully articulated philosophical picture of mathematics. What was then Bourbaki's conception of mathematics?

It is not unusual to come across pronouncements of Bourbaki members, who insistently characterize Bourbaki's approach as that of the "working mathematician" whose professional interest focuses variously on problem solving, research and exposition of theorems and theories, and which has no direct interest in philosophical or foundational issues. Thus Bourbaki formulated no explicit philosophy of mathematics and in retrospect individual members of the group even denied any interest whatsoever in philosophy or even in foundational research of any kind.[39]

---

38. An isolated example of a more articulated attempt to analyze Bourbaki's putative structuralist philosophy of mathematics appears in Fang 1970. Fang's book is a lengthy exegesis of Bourbaki's contribution to the vitality of contemporary mathematics. At the same time it is a harsh attack on all those who would dare criticize Bourbaki's views. His account of Bourbaki's philosophy of mathematics, however, adds up to no more than declaring the philosophical and metamathematical formula "mathematics = logic" to be false. It would be exaggerated to claim that Fang's account is supported by sound philosophical arguments. Instead, Fang claims that Bourbaki's own mathematical work is the ultimate representative of contemporary mathematics and the best example that the essence of mathematical thinking cannot be subsumed under the narrow equation proposed by the logicists. Although one could accept this latter claim, it is far from being an explanation of what Bourbaki's "structuralist" program is. See also Kline 1980, 25, for an unsatisfactory account of Bourbaki's "structuralist" program.

Yet, even if it was true that the group steadfastly avoided pronouncements on issues other than pure theorem-proving and problem-solving, Bourbaki's work, like that of any other scientist or group of scientists, proceeded within a particular framework of images of mathematics. Moreover, like any other scientist's system of images of knowledge, Bourbaki's own system has been subject to criticism, it has evolved through the years, and, occasionally, it has included ideas that are in opposition to the actual work whose setting the images provide. Since Bourbaki gathered together various leading mathematicians, it has also been the case that members of the group professed changing beliefs, often conflicting with one another at the level of the images of knowledge. This point will be developed in what follows.

Therefore, when trying to understand Bourbaki's "structuralist" conception of mathematics, it seems more convenient to speak of Bourbaki's images of mathematics rather than of Bourbaki's philosophy of mathematics.[40] Bourbaki's images of mathematics can be reconstructed by directly examining the mathematical work and the historical accounts of the development of mathematics published by the group, by examining pronouncements of different members of the group, and from several other sources as well. Particular caution must be exercised in this regard concerning the status of pronouncements by different members of the group. Jean Dieudonné has no doubt been the most outspoken Bourbaki member, and—more than anyone else—he has spread Bourbaki's name along with what he saw as Bourbaki's conception. André Weil has been the second most active spokesman in this regard. The views of the majority of the group's members—in particular, those views concerning the structural conception of mathematics and the role of the concept of *structure* in the work of Bourbaki—have been usually much less documented or not documented at all. The present account of Bourbaki's images of mathematics will be based mainly on an analysis of the actual contents of the group's work. At the same time, and without attempting to portray Dieudonné

---

39. Jean Dieudonné (1982, 619) once summarized Bourbaki's avowed position regarding these kinds of questions "as total indifference. What Bourbaki considers important is *communication* between mathematicians. Personal philosophical conceptions are irrelevant for him." (Italics in the original)

40. Giorgio Israel's articles on Bourbaki have described in detail central elements of the group's images of mathematical knowledge. In Israel 1977 he characterizes Bourbaki's views as an "ideology" rather than a philosophy. The term "ideology", however, is far from unequivocal, and is in need of further clarification.

or Weil as official spokesmen for the group, their pronouncements will help provide a fuller picture of those images.

Bourbaki began its work amidst a multitude of newly obtained results, some of them belonging to as yet unconsolidated branches of mathematics; the early years of Bourbaki's activity witnessed a boom of unprecedented scope in mathematical research.[41] In 1948 Dieudonné (signing with the name of Bourbaki) published a now famous article (already quoted above), that was later translated into several languages and which has ever since come to be considered the group's programmatic manifesto: "The Architecture of Mathematics."[42] According to the picture of mathematics described in that article, the boom in mathematical research at the time of its writing raised the pressing question, whether it could still be legitimate to talk about a single discipline called "mathematics", or:

> ... whether the domain of mathematics is not becoming a tower of Babel, in which autonomous disciplines are being more and more widely separated from one another, not only in their aims, but also in their methods and even in their language. (Bourbaki 1950, 221)

In fact, this same question had occupied Hilbert's thoughts several decades before. Hilbert's 1900 list of twenty-three problems called attention to the diversity of problems facing contemporary mathematics. Apparently Hilbert himself was able to embrace all this variety, but he closed his address by raising the question of the unity of contemporary mathematics in terms very similar to those later used by Dieudonné. Hilbert said:

> The question is urged upon us whether mathematics is doomed to the fate of those other sciences that have split up into separate branches, whose representatives scarcely understand one another and whose connections become ever more

---

41. In order to illustrate the boom of mathematical knowledge during the twentieth-century, Davis & Hersh 1981, 29, have pointed out that in 1868 the *Jahrbuch über die Fortschritte der Mathematik* divided mathematics into 12 disciplines and 30 sub-disciplines, while in 1979 the *Mathematical Reviews* registered 61 disciplines and 3400 sub-disciplines. As for the initial years of Bourbaki's activity, the index of the *Zentralblatt für Mathematik und ihre Grenzgebiete* of 1934 registers 68 disciplines and 197 sub-disciplines.

42. Mehrtens 1990, 318, writes, following a report of Liliane Beaulieu, that Dieudonné published the article without first discussing it as usual in the framework of the group's meetings. Nevertheless, its contents were never contradicted by other Bourbaki members, at least not publicly. Beaulieu's well-documented account of Bourbaki's first years of activity shows how the views expressed by Dieudonné in this manifesto were a consequence of Bourbaki's early stages of activity, rather than a motivation for it. See Beaulieu 1994.

loose. I do not believe this nor wish it. Mathematical science is in my opinion an indivisible whole, an organism whose vitality is conditioned upon the connection of its parts. For with all the variety of mathematical knowledge, we are still clearly conscious of the similarity of the logical devices, the relationship of the ideas in mathematical theory and the numerous analogies in its different departments. We also notice that, the farther a mathematical theory is developed, the more harmoniously and uniformly does its construction proceed, and unsuspected relations are disclosed between hitherto separate branches of the science. ... Every real advance [in mathematical science] goes hand in hand with the invention of sharper tools and simpler methods which at the same time assist in understanding earlier theories and cast aside older, more complicated developments.[43]

Obviously Hilbert was a main source of inspiration for Bourbaki, and the further the group developed the original plan, the more pressing became the question of the unity of mathematics. In the "Architecture" manifesto, Dieudonné also echoed Hilbert's belief in the unity of mathematics, based both on its unified methodology and in the discovery of striking analogies between apparently far-removed mathematical disciplines. For Dieudonné it was the axiomatic method that accounted for these two unifying tendencies in mathematics. Dieudonné wrote:

Today, we believe however that the internal evolution of mathematical science has, in spite of appearance, brought about a closer unity among its different parts, so as to create something like a central nucleus that is more coherent than it has ever been. The essential part of this evolution has been the systematic study of the relations existing between different mathematical theories, and which has led to what is generally known as the "axiomatic method." ... Where the superficial observer sees only two, or several, quite distinct theories, lending one another "unexpected support" through the intervention of mathematical genius, the axiomatic method teaches us to look for the deep-lying reasons for such a discovery. (Bourbaki 1950, 222-223)[44]

But here Dieudonné went beyond Hilbert and proposed a further idea, directly connected with the axiomatic method and central to Bourbaki's own

---

43. Quoted from the English translation: Hilbert 1902, 478-479. The Tower-of-Babel metaphor is also used, and a similar warning against the dangers of the situation described appear also in Klein's lectures on the history of nineteenth century mathematics (Klein 1926-7 Vol. 1, 327.)

44. But Bourbaki was in fact hesitant, in the early meetings of 1935, concerning the wholehearted adoption of the axiomatic approach that has come to characterize the group's work so markedly. These qualms are described in Beaulieu 1994, 246.

unified view of mathematics, namely, the idea of mathematical structures. Dieudonné described the unifying role of the structures as follows:

> Each structure carries with it its own language, freighted with special intuitive references derived from the theories from which the axiomatic analysis ... has derived the structure. And, for the research worker who suddenly discovers this structure in the phenomena which he is studying, it is like a sudden modulation which orients at once the stroke in an unexpected direction in the intuitive course of his thought and which illumines with a new light the mathematical landscape in which he is moving about.... Mathematics has less than ever been reduced to a purely mechanical game of isolated formulas; more than ever does intuition dominate in the genesis of discoveries. But henceforth, it possesses the powerful tools furnished by the theory of the great types of structures; in a single view, it sweeps over immense domains, now unified by the axiomatic method, but which were formerly in a completely chaotic state. (Bourbaki 1950, 227-228)

Thus in the "Architecture" manifesto Dieudonné attributed to the structures—and especially to "the theory of the great types of structures" as well as to the tools provided by it—a central role in the unified picture of mathematics. In order to understand exactly what he meant by this, we must take a closer look at Bourbaki's work.

As the Bourbaki project evolved from writing a course in analysis into an up-to-date, comprehensive, account of the central branches of mathematics, Bourbaki laid greater stress on presenting the whole of that mathematical knowledge in a systematic and unified fashion, and within a standard system of notation. As we saw in Part One, a similar task had been successfully undertaken some years before by van der Waerden, albeit for the more limited context of algebra alone. In fact van der Waerden's book was a main source of inspiration for Bourbaki in the group's early stages of activity.[45] At that time, the images of classical, nineteenth-century algebra were still the dominant ones in France, especially due to the lasting influence of Serret's textbook of algebra (§ 2.1 above). The new methods and the recent achievements of Emmy Noether, Emil Artin and their followers in Germany, were still foreign to the general mathematical audience in France. Commenting on the strong impact

---

45. Cf. Dieudonné 1970, 136-137. However, at the very initial stages of the project, there was no consensus among members of the group as to the convenience of following van der Waerden's approach. As a matter of fact, in the early meetings there was substantial disagreement as to the extent and specific content of the "abstract package" (Bourbaki's expression) that should be included in the book. See Beaulieu 1993, 30; Beaulieu 1994, 244-246.

caused by the first reading of this book, and its influence on the Bourbaki project Dieudonné wrote: "I had graduated from the Ecole Normale, and I did not know what an ideal was, and barely knew what a group was."[46] In retrospect, the main thrust of Bourbaki's initial motivation may be seen as an attempt to reorient French mathematics away from its traditionally dominant conceptions and into the new perspectives lately developed in Germany. In particular, Bourbaki's treatise, as it gradually came to be conceived and worked out, may actually be seen as an extension of van der Waerden's achievement to the whole of mathematics; that is, much the same as van der Waerden had succeeded in presenting the whole of algebra as a hierarchy of structures, so did Bourbaki present much larger portions of mathematics in a similar way.[47]

There is, however, a noteworthy difference between van der Waerden's systematic exposition of modern algebra and Bourbaki's own, more ambitious attempt. Van der Waerden's unification of algebra consisted in a successful restructuring of a whole discipline, which was attained through a redefinition of the images of knowledge of that particular discipline. While his innovation remained mainly at the level of the images of mathematics, it provided a convenient framework within which much important progress in the body of knowledge was later attained. As was seen in § 1.3 above, van der Waerden felt no need of providing an explicit explanation, either formal or non-formal, of what is to be understood by an "algebraic structure" or by "structural research in algebra." Bourbaki, unlike van der Waerden in this respect, not only attempted on various opportunities to explain what the structural approach is and why it is so novel and important for mathematics, but, moreover, in putting forward the theory of *structures* as part of the treatise, they manifested an eagerness to endorse those explanations, and in fact Bourbaki's whole system of images of mathematics, by means of an allegedly unifying, mathematical theory. This eagerness is to be understood, in the first place, as

---

46. Dieudonné 1970, 137. But perhaps not only French mathematicians were unaware of recent developments in algebra in Germany. According to Zariski's testimony, for instance, having got most of his mathematical instruction in Italy, it was only while writing his important treatise on Algebraic Surfaces (Zariski 1935) that he began to study the works of Emmy Noether as well as Krull's and van der Waerden's books. See Parikh 1991, 68; Zariski 1972, *xi*. For a more detailed account of algebraic research in Italy in the early twentieth century see Brigaglia 1984.

47. Also the extent of the more modern subjects of mathematics that Bourbaki intended to cover in the treatise changed over the first years of activity. See Beaulieu 1994, 246-247.

an instance of that image of knowledge characteristic of twentieth-century mathematics, according to which meta-issues in mathematics are considered to have been meaningfully elucidated only insofar as they have been articulated in *formal* mathematical theories.

A reflexive, formal-axiomatic elucidation of the idea of mathematical structure could prove useful not only as a general frame of reference but also in addressing some central open questions concerning the role of structures in mathematics. One such central issue was the issue of selection. The issue of selection is a central question in science in general, at the level of the images of knowledge. What an individual scientist selects as his discipline of research, and the particular problems he selects to deal with in that particular discipline will largely determine, or at least condition, the scope and potentialities of his own research. What a community of scientists establishes as main open problems and main active subdisciplines will substantially influence the future development of the discipline as a whole. Clearly, the contents of the body of knowledge directly delimits the potential selections of scientists. On the other hand, these contents *alone* cannot provide clear-cut answers to the issue of selection. Criteria of selection are open to debate and, obviously, there are several possible factors that will determine a particular scientist's choice, when confronted with a given body of knowledge.

Bourbaki was very conscious of the centrality of the issue of selection and, from the very beginning of the group's activities, considerable effort was invested in debating it.[48] In the early meetings, that eventually led to the creation of the core Bourbaki group, an important criterion for the selection of issues to be treated in the projected treatise on analysis was their external applicability and their usefulness for physicists and engineers. Over the first years of activities, however, given the more abstract inclinations of certain members and the way in which the writing of the chapters evolved, gradual changes affected the criteria of selection guiding the group's work.[49]

---

48. Evidence for this can be gathered by browsing through the issues of "La Tribu." Below in § 9.4, one particular debate on selection is discussed in detail, namely, the one concerning the possible inclusion of categories and homological algebra in the treatise. See also Beaulieu 1990, 39-41, concerning discussions on the most suitable axiomatic definitions for topology. It should also be stressed here, that, according to Dieudonné's retrospective appraisal (Dieudonné 1970, 22), selection "is the crucial part in Bourbaki's evolution. I think that we can attribute much of the hostility that has been shown towards Bourbaki ... to this strict selection."

49. Cf. Beaulieu 1993, 30-31.

As the axiomatic approach increasingly became a dominant concern for Bourbaki, the problem of selecting, and especially that of justifying, the most interesting theories to be included in the treatise became a pressing one.[50] As Henri Cartan wrote in retrospect, on the face of it the choice of axioms could seem to be completely arbitrary; in practice, however, a very limited number of such systems constitute active mathematical research disciplines, since theories "built upon different axiomatic systems have varying degrees of interest."[51] Moreover, certain axiomatic systems on which mathematicians may decide to invest their research efforts were variously dismissed by others as "axiomatic trash."[52] Thus, Bourbaki's image of mathematics as it became consolidated around the 1940s and as it was expressed in Dieudonné's "Architecture"—an image reportedly centered around the view of mathematics as the science of axiomatic systems—implied the necessity of formulating criteria to explain how the chaff of "axiomatic trash" is winnowed from the grain of the mathematically significant axiomatic systems.

Given the success attained by reflexive theories in mathematical research in the decades preceding Bourbaki's activity, and, more specifically, given the pervasive influence of Hilbert on Bourbaki's image of mathematics, it would be natural, or at least plausible, to expect that an answer to the above-posed question be given by means of a reflexive mathematical theory, within which the correctness of the choice could be endorsed by mathematical proof. Bourbaki's formulation of the theory of *structures* could be seen as an actual response to that expectation.

But on the other hand, considering the intellectual inclinations of the mathematicians involved in the Bourbaki project, one is justified in thinking that each member of the group had strongly conceived opinions of what should be considered as mathematically interesting and what should not, independently of the elaboration of a formal theory meant to "classify the fundamental disciplines of mathematics."[53] The reliance on each mathematician's

---

50. A general discussion of this issue appears in Spalt 1987. Spalt, following Lakatos, contrasts two different kinds of mathematics: Structure-mathematics vs. informal mathematics (*inhaltlichen Mathematik*). While in the latter, according to Spalt, the problem of justification (*Rechtfertigung*) of concepts and methods is more naturally solved, in the former it appears in all of its acuteness and remains unsolved.

51. Cartan 1980, 177.

52. A term also used by Dieudonné (e.g., in Dieudonné 1982, 620).

intuition in order to decide on this question was asserted by Henri Cartan as follows:

> There is no general rule in mathematics by which one can judge what is interesting and what is not. Only a thorough understanding of existing theories, a critical evaluation of the problems at hand or a sudden, unexpected flash of intuition can enable the researcher to choose the appropriate axiom system. (Cartan 1980, 177)

One might thus expect, that in many of the issues discussed in the Bourbaki meetings, the intuitions of the various members concerning the correct selections often differed and perhaps even clashed, until common agreement was finally reached.[54] Thus, the publication of a formal theory of *structures* and its connection with the issue of selection bring to the fore interesting dualities and tensions, which are characteristic of much of the Bourbaki undertaking: standardized techniques vs. ingenuity in problem-solving, collective work vs. individual genius, formalized generalization vs. intuitive grasping of peculiar mathematical situations.

At any rate, one can see how the thorough adoption of the axiomatic approach as the main tool for the exposition of mathematical theories, together with the images of knowledge associated with that approach, create a direct connection between the issue of selection and Bourbaki's formulation of the theory of *structures*.[55] As it happened, however, and as will be seen in the following sections, this theory did not effectively provide answers to this, or to any other reflexive issue. Nevertheless, Bourbaki's images of mathematics, and in particular the group's actual choices, though obviously not derived

---

53. As claimed in Cartan 1980, 177: "The concept of *structure* ... allowed a definition of the concept of isomorphism and with it a classification of the fundamental disciplines within mathematics."

54. To quote Cartan once more (1980, 179): "That [a final product] can be obtained at all [in Bourbaki's meetings] is a kind of miracle that none of us can explain."

55. As with other issues, Dieudonné retrospectively attempted to rationalize Bourbaki's choices, explaining them with arguments other than the professional authority of the group's members. Cf. Dieudonné 1982, 620. It is worth stressing, however, that in spite of formulating Bourbaki's alleged objective criteria of selection, Dieudonné wrote that (1982, 623): "No one can understand or criticize the choices made by Bourbaki unless he has a solid and extended background in many mathematical theories, both classical and more recent." For a criticism of the views expressed here by Dieudonné, see Hermann 1986. For a more general criticism of Bourbaki's choices see Benoit Mandelbrot's comments in Albers & Alexanderson (ed.) 1985, 222. In Mandelbrot 1989, 11, he also wrote: "For Bourbaki the fields to encourage were few, and the fields to discourage were many."

from a formal theory, proved to be enormously fruitful in certain quarters of mathematics. Still more interesting, Bourbaki's criteria of selection have very often been accepted *as if* they were actually backed by such a reflexive theory. This point will be further elaborated below.

A second clue to understanding Bourbaki's images of mathematics and the attempt to include a formal theory of *structures* in Bourbaki's treatise is connected with the group's view of mathematics as the science dealing with axiomatic formal systems, as well as its self-proclaimed status as "legitimate heir" to Hilbert's alleged "formalist" doctrine. Whether or not Bourbaki's outlook, or those of its individual members, faithfully reflects Hilbert's own views is debatable, although this is not the place to do so. What is relevant for our discussion here is to recognize that a main reason why Bourbaki adopted a formalist position was to avoid philosophical difficulties. As already stated, members of Bourbaki consistently declared themselves first and foremost to be "working mathematicians", and their actual views concerning philosophical or foundational issues is perhaps most frankly expressed in the following quotation of Dieudonné:

> On foundations we believe in the reality of mathematics, but of course when philosophers attack us with their paradoxes, we run to hide behind formalism and say: "Mathematics is just a combination of meaningless symbols" and then we bring out chapters 1 and 2 [of the *Eléments*] on Set Theory. Finally we are left in peace to go back to our mathematics and do it as we have always done, working in something real. (Dieudonné 1970, 145)

This position of "Platonism on weekdays and formalism on Sundays", which is so widespread among working mathematicians, becomes especially worthy of attention in the case of Bourbaki. It has been claimed elsewhere that such a position is untenable as a consistent philosophical account of mathematics, since it involves both logical inconsistency and a distorted description of the actual doings of the mathematician.[56] Nevertheless, it is an accepted image of mathematics, that has at least helped many a twentieth-century mathematician confer some meaning to his own scientific work. This seems to be the case as well for Bourbaki.

Yet since Bourbaki does not represent rank-and-file mathematicians, but rather those who assumed a role of leadership and intended to come forward

---

56. See Hersh 1971 (1985), 11 ff.

with an innovative, meaningful interpretation of what mathematics essentially is, it may come as a surprise that, while raising the banner of rigor and parsimony in mathematics, Bourbaki was willing to adopt the above-mentioned philosophical position without any reluctance. It is not a criticism of Bourbaki's philosophical sophistication, or lack of it, which concerns us here but rather the question, how is the elaboration of the theory of *structures* connected with Bourbaki's images of mathematics. The above-described mixture of a declared formalist philosophy with a heavy dose of actual Platonic belief is illuminating in this regard. The formalist imperative, derived from that ambiguous position, provides the necessary background against which Bourbaki's drive to define the formal concept of *structure* and to develop some immediate results connected with it can be conceived. The Platonic stand, on the other hand, which reflects Bourbaki's true working habits and beliefs, has led the very members of the group to consider this kind of conventional, formal effort as superfluous. Indeed, of all the apparatus developed in the first book of the treatise following that formalist imperative, only feeble echoes appear in the other volumes, where Bourbaki's real fields of interest are developed. This desire to avoid philosophical issues by adopting an inconsistent position, and the image of mathematics associated with that position, prepare us to understand Bourbaki's need to elaborate a formal theory of *structures* and the futility of this undertaking.

This general picture of Bourbaki's images of mathematics and of their connections with the elaboration of a theory of *structures* allegedly standing at the basis of the building of mathematics provides an adequate background against which to examine in greater detail the contents of Bourbaki's treatise and the actual role played by *structures* within it.

7.3 *Structures* and the Body of Mathematics

The present form of the *Eléments*, composed of ten books, was nearly attained in the early seventies. The ten books are: I. Theory of Sets; II. Algebra; III. General Topology; IV. Functions of a Real Variable; V. Topological Vector Spaces; VI. Integration; Lie Groups and Lie Algebras; Commutative Algebra; Spectral Theories; Differential and Analytic Manifolds.[57] The first six books of the treatise were intended to be more or less self-contained. The other four presuppose knowledge of the first six volumes and, for that reason, Bourbaki gave them no number. The French edition of the *Elé-*

ments bears the subtitle "The Fundamental Structures of Analysis", and the headings of the various volumes reflect one of Bourbaki's most immediate innovations, namely the departure from the classical view according to which the main branches of mathematics were taken to be geometry, arithmetic and algebra, and analysis. Each book in the treatise is composed of chapters that were published successively, though not necessarily in the order indicated by the index. The theory of *structures* appears in Book I, which deals with the set theory. We will now examine the concept of *structure* and the role played by it in this and in the other books of the *Eléments*.

7.3.1 Set Theory

As with other subjects included by Bourbaki among the "Fundamental Structures of Analysis", such as algebra and topology, the group's initial plans did not envisage a systematic, axiomatic elaboration of the theory of sets as an independent subject. Rather, the original idea was to use only elementary set-theoretical notions, introduced from a naive perspective, such as the direct needs of a treatise on analysis would require. This approach reflected a long-standing tradition with respect to set theory in France. During the first three years of actual work on the treatise, however, attention shifted gradually away from the classical issues of analysis and most of the effort was in fact directed toward more basic and "abstract" issues.[58] The chapters of Bourbaki's volume on set- theory were published only during the 1950s, but a summary of results appeared as early as 1939.

Book I of the *Eléments, Theory of Sets*, is composed of four chapters and a "Summary of Results": 1. Description of Formal Mathematics; 2. Theory of Sets (first French edition of both chapters: 1954); 3. Ordered Sets, Cardinals, Integers (1956); 4. Structures (1957). The "summary" was first published in French in 1939.[59] The book is preceded by an introduction on formalized languages and the axiomatic method. No mathematician, it is said, actually works in a fully formalized language, but rather in natural language. However, "his

---

57. From the first meetings onwards, several different suggestions were put forward as to the desired organization of the treatise into separate volumes. Several subjects that had been contemplated in the plans drafted about 1941, and ranging from algebraic topology to partial differential equations, to numerical analysis, were left out, first because of the impossibility of collective work during the war, and later because of sensible changes in the main interest of the group following the admission of younger members. See Beaulieu 1994, 249-251.

58. See Beaulieu 1994, 246-247.

experience and mathematical flair tell him that translation into formal language would be no more than an exercise of patience (though doubtless a very tedious one)." The first aim of the book then is to present one such formalized language. This language should be general enough to allow the formulation of all the axiomatic systems of mathematics. The existence of such a language is warranted by the fact that:

> ... whereas in the past it was thought that every branch of mathematics depended on its own particular intuitions which provided its concepts and primary truths, nowadays it is known to be possible, logically speaking, to derive practically the whole of mathematics from a single source, the theory of sets. (Bourbaki 1968, 9)

However, since even the complete formalization of set theory alone turns out soon to be impracticable, strings of signs that are meant to appear repeatedly throughout the book are replaced from the beginning by symbolic abbreviations, and condensed deductive criteria are introduced, so that for every proof in the book it will not be necessary to explain every particular application of the inference rules. The final product is a book which, like any other mathematical book, is partially written in natural language and partially in formulae but which, like any partial formalization, is supposed in principle to be completely formalizable. At any rate, the claim is made that the book on set theory lays out the foundations on which the whole treatise may be developed with perfect rigor.[60]

Naturally, when dealing with formal systems the problem of consistency immediately arises. Bourbaki did not attempt to address the problem of consistency through any kind of formalistic device. Rather, Bourbaki deviated here from the professed formalist position and, without further ado, reverted to empiricist considerations. Thus, Bourbaki stated that a contradiction is not expected to appear in set theory because it has not appeared after so many years of fruitful research.[61] But this is not the only sense in which the formalistic apparatus introduced by Bourbaki fulfills no real foundationalist role.

Take for example the discussion on formalized languages, following the introduction. A proof is defined as a series of relations of terms formed according to specified rules. A theorem is defined as any relation appearing in a

---

59. There have been several reprints (with some minor changes) and translations of this volume into other languages. Unless otherwise stated, all quotations of *Theory of Sets* below, are taken from the English translation (Bourbaki 1968). There are minor differences between the successive editions, but these are referred to and commented separately in this section.

60. Cf. Bourbaki 1968, 11.

proof. All these steps and the stress laid upon the claim that the formal expressions are devoid of meaning are standard for a formalistically-oriented foundation of mathematics. However Bourbaki added a rather unusual commentary:

> This notion [of a theorem] is therefore essentially dependent on the state of the theory under consideration, at the time when it is being described. A relation in a theory τ, becomes a theorem in τ when one succeeds in inserting it into a proof in τ. To say that a relation in τ "is not a theorem in τ" cannot have any meaning without reference to the stage of development of the theory τ. (Bourbaki 1968, 25)

In like manner the concept of a contradictory theory was defined here by Bourbaki as time-dependent: as long as a contradiction has not been proven to exist within the theory, that theory is free from contradiction. Time, however, could prove it to be otherwise. Now the notion of a meta-mathematical concept (and a central one, at that) as being time-dependent, although debatable in itself, could be accepted or overlooked if appearing in an introduction to a book in any standard mathematical discipline. It would likewise not be surprising were it found, e.g., in an intuitionistic book. But it seems strangely out of place in the framework of a book allegedly intended to provide the sound basis and the framework on which the whole picture of the basic branches of mathematics is to be developed from a declared *formalistic* perspective.

Bourbaki's style is often described as one of uncompromising rigor with no heuristic or didactic concessions to the reader.[62] This characterization fits perhaps the bulk of the treatise, but not *Theory of Sets*. In fact, the further one

---

61. Cf. Bourbaki 1968, 13. Cf. also one of Bourbaki's earlier publications (1949, p. 3): "... absence of contradiction, in mathematics as a whole or in any given branch of it, thus appears as an empirical fact rather than as a metaphysical principle ... We cannot hope to prove that every definition ... does not bring about the possibility of a contradiction." Imre Lakatos claimed, in an article entitled "A Renaissance of Empiricism in Recent Philosophy of Mathematics?" (Lakatos 1978 Vol. 2, 24-42), that foundationalist philosophers of mathematics, from Russell onwards, when confronted with serious problems in their attempts to prove the consistency of arithmetic, have not hesitated to revert to empirical considerations as the ultimate justification for it. Although Bourbaki is not mentioned among the profusely documented quotations selected by Lakatos to justify his own claim, it seems that these passages of Bourbaki could easily fit into his argument. Bourbaki's empiricist solution to the issue of consistency is also discussed in Israel & Radice 1976, 175-176.

62. As Queneau 1962, 9 puts it: "La seule concession que fasse Bourbaki aux considèrations heuristiques, ce sont ces notes historiques." Of course, there may be utterly different opinions about what constitutes "heuristic considerations" when it comes to the thinking of a layman and that of a mathematician.

advances through the chapters of *Theory of Sets*, encountering ever-new symbols and results, the more one finds additional heuristic explanations of the meaning of the statements, even when they are not especially difficult. Typical is the following example towards the end of the chapter, in the section dealing with quantified theories:

> C36. Let A and R be relations in τ, and let x be a letter. Let τ' be the theory obtained by adjoining A to the axioms of τ. If x is not a constant of τ, and if R is a theorem in τ', then $(\forall_A x)R$ is a theorem in τ. (Bourbaki 1968, 42)

After a two-line proof of C36 is given, the following comment appears:

> In practice we indicate that we are going to use this rule by a phrase such as 'Let x be any element such that A'. In the theory τ', so defined, we seek to prove R. Of course, we cannot assert that R itself is a theorem in τ. (*ibid.*)

In this way, the formal language that was introduced step by step is almost abandoned and quickly replaced by the natural language. The recourse to extra-formalistic considerations in the exposition of results within a text-book is, of course, perfectly legitimate. What should be noticed here, however, is the departure from Bourbaki's behavior in other books of the *Eléments* and the divergence between Bourbaki's pronouncements and what is really done in *Theory of Sets*.

There is, in fact, written evidence of Bourbaki's uncertainty about how to address these problems. The kinds of problems addressed in the first chapters of *Theory of Sets* were not a major concern of the entire group; as a matter of fact, it overlapped the research fields of very few of its members. Nevertheless, those issues had to be addressed if the desired formal coherence of the treatise was to be achieved. Several issues of "La Tribu" record different proposals regarding the desired contents of *Theory of Sets*, as well as several technical problems encountered while developing them in detail. We will consider this point below; yet it is pertinent to quote here a report on the progress in the work on *Theory of Sets* by 1949, in which the above issues of formalized languages and inference rules were dealt with. This report reads:

> Since the first session, Chevalley raised objections concerning the notion of a formalized text, which threaten to hinder the whole publication. After a night of contrition, Chevalley turned to more conciliatory opinions and it was agreed that there are serious difficulties to it, which he was assigned to mask as unhypocritically as possible in the general introduction. A formalized text is in fact an ideal notion, since one has seldom seen any such a text and in any case Bourbaki has

none. One should therefore speak with discretion about those texts in chapter 1 and indicate clearly in the introduction what separates us from them.[63]

Set theory and the foundational problems of mathematics were close to Chevalley's own research, and this may be the reason why he insisted, perhaps more than any other member of the group, that *Theory of Sets* be published soon.[64] The publication, however, was constantly postponed since problems surrounding the formalized language continued to appear. The book in its final form, rather than being the outcome of a coherently envisaged foundation for mathematics, is a compromise between the desire to bring formalization to its most extreme manifestation, as demanded by Chevalley, and the need to produce a readable book that would fit the overall style of the treatise and the standard reader's interest. This is also probably one of the reasons for the ad-hoc character of *Theory of Sets* when regarded as part of the treatise.

In Chapter 2 the axioms for sets are introduced and some immediate results are proven. Many concepts are treated here using a rather idiosyncratic notation. Bourbaki's overall influence is very often manifest in the extended use, in several fields of mathematics, of new nomenclature and notation introduced in Bourbaki's treatise. But as Paul Halmos has remarked, many of the concepts and notations introduced here are used no more than once and therefore could have been dispensed with. Moreover, as Halmos claimed:

> It is generally admitted that strict adherence to rigorously correct terminology is likely to end in being pedantic and unreadable. This is especially true of Bourbaki, because their terminology and symbolism are frequently at variance with commonly accepted usage. The amusing fact is that often the "abuse of language" which they employ as an informal replacement for a technical name is actually conventional usage: weary of trying to remember their own innovation,

---

63. "La Tribu" - April 13-25: 1949: "Dès la première séance de discussion, Chevalley soulève des objections relatives à la notion de texte formalisé; celles ci menacent d'empêcher toute publication. Après une nuit de remords, Chevalley revient à des opinions plus conciliantes, et on lui accorde qu'il y a là une sérieuse difficulté qu'on le charge de masquer le moins hypocritement possible dans l'introduction générale. Un texte formalisé est en effet une notion idéale, car on a rarement vu de tels textes et en tous cas Bourbaki n'en est pas un; il faudra donc ne parler dans le chap. I qu'avec beaucoup de discrétion de ces textes, et bien indiquer dans l'introduction ce qui nous en sépare."

64. In fact, in his student years at the Ecole Normale Chevalley developed a strong friendship with the early-deceased Jaques Herbrand, with whom he shared his early interest in logic. Later on Chevalley worked mainly on class field theory, group theory, algebraic geometry and the theory of Lie algebras. See Dieudonné & Tits 1987.

the authors slip comfortably into the terminology of the rest of the mathematical world. (Halmos 1957, 90)[65]

Chapter 3 deals with ordered sets, cardinals and integers. Ordered structures, as will be seen below, are among the so called "mother structures" of mathematics, to which Bourbaki accords a central role in its picture of mathematics; it is surprising, then, that they are considered in no other place in the entire treatise. Lattices, for example, are only briefly mentioned (p. 146), although it should be noticed that a relatively large number of exercises concerning them are included at the end of the chapter.[66]

Finally, Chapter 4 develops the concept of *structure*, Bourbaki's formalized notion of structure. Before defining *structures* Bourbaki introduced some preliminary concepts. The basic ideas behind those concepts can be formulated as follows: take a finite number of sets $E_1, E_2, ..., E_n$, and consider them as the building blocks of an inductive procedure, each step of which consists either of taking the Cartesian product ($E \times F$) of two sets obtained in former steps or of taking their power set $B(E)$. For example, beginning with the sets E, F, G the outcome of one such procedure could be: $B(E)$; $B(E) \times F$; $B(G)$; $B(B(E) \times F)$; $B(B(E) \times F) \times B(G)$ and so forth. Bourbaki introduces a formal device for defining and characterizing every possible construction of the kind described above. The last term obtained through a given construction of this kind for n sets $E_1, E_2, ..., E_n$ is called an "echelon construction scheme S on n base sets" and it is denoted by $S(E_1, E_2, ..., E_n)$. Given one such scheme and n additional sets E_i', and n mappings $f_i : E_i \rightarrow E_i'$, a further formal straightforward procedure enables one to define a function from $S(E_1, E_2, ..., E_n)$ to $S(E_1', E_2', ..., E_n')$ (i.e., to the corresponding system built over the sets $E_1', E_2', ..., E_n'$ instead of $E_1, E_2, ..., E_n$). This function is called the "canonical extension with scheme S of the mappings $f_1, ..., f_n$" and it is denoted by $<f_1, ..., f_n>^s$. This function is injective (resp. surjective, bijective) when each of the f_i's are. To define a "species of structure" Σ, one takes:

(1) n sets $x_1, x_2, .., x_n$, as "principal base sets."

(2) m sets $A_1, A_2, .., A_m$, the "auxiliary base sets", and finally

(3) a specific echelon construction scheme $S(x_1, ..., x_n, A_1, ..., A_m)$.

---

65. For a detailed review of Chapters 1-2 of *Theory of Sets*, see Halmos 1955. A technical criticism of Bourbaki's system of axioms for the theory of sets is developed in Mathias 1992.

66. Only half a page of the 17 pages in Section 1 of Chapter 3 is devoted to lattices (section 1.11). In contrast, out of 24 exercises to this section, 7 deal with lattices. In fact, lattices are treated in some detail in Chapter 7 of Bourbaki 1972, especially pp. 512-529.

This scheme will be called the "typical characterization of the species of structure Σ." Such a scheme is obviously a set. A *structure* is now defined by characterizing some of the members of this set by means of an axiom of the species of structure. This particular axiom is a relation which the specific member $s \in S(x_1,...,x_n, A_1,...,A_m)$ together with the sets $x_1,...,x_n,A_1,...,A_m$ must satisfy. The relation in question is constrained to satisfy the conditions of what Bourbaki calls a "transportable relation", which means roughly that the definition of the relation does not depend upon any specific property of S and the sets in themselves but only refers to the way in which they enter in the relation through the axiom. The next example introduced by Bourbaki makes things clearer.

An internal law of composition on a set A is a function from $A \times A$ into A. Accordingly, given any set A, form the scheme $B((A \times A) \times A)$ and then choose from all the sub-sets of $(A \times A) \times A$ those satisfying the conditions of a "functional graph" with domain $A \times A$ and range A. The axiom defining this choice is a special case of what we call algebraic *structures*.[67]

Together with this example Bourbaki also showed, using the previously introduced concepts, how ordered-*structures* or topological-*structures* may be defined. That these are Bourbaki's first examples is by no means coincidental. These three types of *structures* constitute what Bourbaki calls the *mother structures*, a central part of Bourbaki's images of mathematics which we shall discuss below.

After defining *structures* Bourbaki introduced further concepts connected with that definition. However, in the remainder of the chapter, recurring reference to n principal base sets and m auxiliary base sets is avoided by giving all definitions and propositions for a single principal base set (and for no auxiliary set) while stating that "the reader will have no difficulty in extending the definitions and results to the general case" (p. 271). This is a further instance of how Bourbaki ignored in *Theory of Sets* their own self-imposed strict rigor of the other books in the treatise.

These concepts deserve closer inspection since they reveal the ad-hoc character of the notions set forth in *Theory of Sets*. Bourbaki's purported aim in introducing such concepts is to expand the conceptual apparatus upon which the unified development of mathematical theories will be developed. However, all this work turns out to be rather superfluous, since, as will be

---

67. This example appears in Bourbaki 1968, 263.

seen, these concepts are used in a very limited—and certainly not especially illuminating or unifying—fashion in the remainder of the treatise.

* Isomorphism: Let U, U' be two *structures* of the same type Σ on n principal base sets, $E_1, E_2, ..., E_n$ and $E_1', E_2', ..., E_n'$ respectively, and let n bijections $f_i \colon E_i {\rightarrow} E_i'$, be given. If S is the echelon construction scheme of Σ, then $<f_1, ..., f_n>^{s}$ is defined as an isomorphism if

$$<f_1, ..., f_n, Id_1, ..., Id_m>^{s}(U) = U'$$

where Id_i denotes the identity mapping of an auxiliary set A_i into itself. This definition uses the concept of canonical extension introduced above to express in a precise fashion the desirable fact that the isomorphism 'preserves' the structure.

* Deduction of *structures*: Bourbaki defines a formal procedure for deducing a new species of structures from a given one. For instance, if the species of topological group structures is defined on a single set A by a generic *structure* (s_1, s_2), where s_1 is the graph of the composition law and s_2 the set of the open sets of A, then each of the terms s_1 and s_2 is a procedure of deduction and respectively provides the *group* and the *topology* underlying the topological group *structure* (s_1, s_2). Likewise, a commutative group structure can be deduced from either a vector space, or from a ring or from a field.

* Poorer-Richer *structures*: Among the examples introduced in order to clarify the mechanism of deduction of *structures* defined above, a criterion is defined which enables one to order *structures* with the same base sets and the same typical characterization as *poorer* or *richer*, according to whether the axiom defining the latter can be "deduced" from the former. For example, the species of a commutative group is *richer* than the species of groups.

* Equivalent species of *structures*: This definition enables one to identify the same *structure* when it is defined in different ways (e.g. commutative groups and Z-Modules).

* Finer-Coarser *structures*: This is a further relation of order defined between *structures* of the same species. Roughly, a given species of *structures* will be finer the more morphisms it contains with E as source and the fewer morphisms it contains with E as target.

These concepts will be discussed below, and their limited generalizing value within the treatise will be examined.

The subsequent sections of *Theory of Sets* are devoted to special constructions which can be made within the framework of the *structures*: inverse image of a *structure,* induced *structure,* product *structure,* direct image and quotient *structure.* The last section of the chapter deals with Universal Mappings. These are defined for an arbitrary *structure,* and the conditions are stated for the existence of a solution to the universal mapping problem in a given *structure.* It is proven that when this case holds its solution is essentially unique. The unwieldiness of the *structure*-related concepts is here perhaps more apparent than in any other place, since, for this specific problem, a fully developed and highly succinct version of the categorical formulation of the Universal Mapping Problem is available.[68] This point will be further developed in § 8.3 below.

After all this painstaking work, the book closes with a "Summary of Results" ("Fascicule de résultats") containing all the results of set theory which will be of some use in the remainder of the treatise. However, the term "Summary" does not accurately describe the contents of this last section. "Fascicule de résultats" seems a more precise name, because what one finds is neither all the results nor a presentation of them exactly as they appeared in the book but rather "all the definitions and all the results needed for the remainder of the series." If the book's stated aim was to show that a sound, formal basis for mathematics can be given, the Fascicule's purpose was to provide the lexicon needed for what follows and to explain the non-formal meaning of the terms within it. This sudden change of approach, from a strict formal style to a completely informal one, is clearly stated and justified by Bourbaki in the opening lines of the Summary:

> As for the notions and terms introduced below without definitions, the reader may safely take them with their usual meanings. This will not cause any difficulties as far as the remainder of the series is concerned, and renders almost trivial the majority of the propositions. (Bourbaki 1968, 347)

---

68. See also Mac Lane 1971, Chpt. 3.

Thus, for example, the huge effort invested in Chapters 2-3 is reduced to the laconic statement: "A set consists of *elements* which are capable of possessing certain *properties* and of having relations between themselves or with elements of other sets" (p. 347. Italics in the original). A footnote explains further:

> The reader will not fail to observe that the "naive" point of view taken here is in direct opposition to the "formalist" point of view taken in chapters I to IV. Of course, this contrast is deliberate, and corresponds to the different purposes of this Summary and the rest of the volume.

The purpose of the summary, then, is to provide, in completely non-formal terms, the common basis upon which the specific theories will later be developed. It is only in this non-formal fashion that Book I is related to the rest of the treatise and, in particular, that the concept of *structure* appears as a unifying concept.

As for *structures,* the whole formal development is reduced in the Fascicule to a short, intuitive explanation of the concepts (even shorter than the one given in the present account) in which the main ideas are explained. The only important concept associated with *structure* which is mentioned, is that of isomorphism. No mention at all is made of derived-, initial-, quotient-, coarser- and finer-, and other *structures* defined in Chapter IV. This summary of results is essentially different from its counterparts in the other books of the series (for example that of "Topological Vectorial Spaces"),[69] both because of its variance from the actual contents of what it allegedly summarizes and because of the striking and total absence of technicalities.

As already noted, the "Fascicule" first appeared in French in 1939, whereas the first edition of the four chapters of *Theory of Sets* appeared (in French) only between 1954 and 1957. This interval saw many important developments in mathematics and, in particular, the emergence of category theory. It is likely that these developments stimulated Bourbaki's own thinking and that this contributed to the gap between the contents of the "Fascicule" and that of the book itself. These developments will be discussed again in § 8.5 below.

---

69. It is important to remark, however, that the kind of Summary appearing in *Topological Vectorial Spaces* is not itself free of problems. One reviewer (Hewitt 1956, 508) wrote: "The 'Fascicule de Résultats' is of doubtful value. It would seem difficult to appreciate or use this brief summary without first having studied the main text; and when this has been done, the summary is not needed."

Yet beyond the gap between the content of the Fascicule and that of the chapters, a shift in Bourbaki's conception of the role of *structures* within the treatise, and therefore within the whole picture of mathematics, is detectable already within the fascicule itself. In fact, a small but notable difference between the first and the third edition of the "Fascicule" exists, namely the addition of a footnote in the third edition. This footnote states that: "The reader may have observed that the indications given here are left rather vague; they are not intended to be other than heuristic, and indeed it seems scarcely possible to state general and precise definitions for structures outside the framework of formal mathematics." (p. 384)

By "outside the framework of formal mathematics", one should understand here "outside the conceptual framework proposed by Bourbaki in *Theory of Sets*." Thus, in spite of declarations to the contrary elsewhere, Bourbaki here implicitly admitted (concealing this confession, as it were, in a footnote) that the link between the formal apparatus introduced in *Theory of Sets* and the activities of the "working mathematician" (Bourbaki's declared main addressee) is tenuous, and, at best, of purely heuristic value.

After this account of the way in which *Theory of Sets* was constructed to enable the final definition of a structure and its related concepts, it is time to inspect more closely the use to which these concepts are put in the different books of the treatise.

7.3.2 Algebra

Bourbaki's book on algebra comprises nine chapters, the first editions of which appeared in print between 1942 and 1959 and which later underwent several re-editions.[70] The image of algebra dominating Bourbaki's book on algebra is essentially the same as that of *Moderne Algebra* in the sense that different algebraic structures are presented in a somewhat hierarchical manner. Thus, for instance, vector spaces are presented as a special case of groups and, therefore, all the results proven for groups hold for vector spaces as well. However, this hierarchy is absolutely non-formal since it is not anchored in terms of the concepts defined in the four chapter of *Theory of Sets*.

Neither commutative groups nor rings are presented as *structures* from which a group can be "deduced", nor is it proven that $\mathbf{Z}$-modules and commutative groups are "equivalent" *structures,* to take but two concepts. Some of

---

70. All quotations below are taken from the English version Bourbaki 1973.

the *structure*-related concepts do appear in the opening sections of the book, but the rather artificial use to which they are put and their absence from the rest of the book suggests that this initial usage was an *ad-hoc* recourse to demonstrate the alleged subordination of algebraic concepts to the more general ones introduced within the framework of *structures*. For example, readers are told that the definition of an "isomorphism of magmas" (§ 1.1), namely a bijection between two sets endowed with internal laws of composition which "preserves" the laws of composition, conforms to the "general definitions." However, the formal verification of this trivial fact is actually much more tedious than it may appear at first sight. In fact, according to the definitions, one must first specify the echelon construction scheme of a "magma" (this is done as an example in Bourbaki 1973, § 1.4); then one should show, as explained above, that the defining axiom (namely the relation "$F \in B((A \times A) \times A)$ as a functional graph whose domain is $A \times A$") is a "transportable relation" for the given scheme, and finally, that

$$<f>^{S}(U) = U'$$

where $<f>^{S}$ is the canonical extension with scheme S and the function f, and U the *structure* in question.

All this exacting verification is neither accomplished not even alluded to in the book, nor is any similar assertion thoroughly verified in what follows. For example, the reader is reminded that the central theorem for a monoid of fractions of a commutative monoid can be expressed in the terminology introduced in *Theory of Sets* by saying that the problem in question "is the solution of the universal mapping problem for E, relative to monoids, monoid homomorphisms and homomorphisms of E into monoids which map the elements of S to invertible elements." It follows, from a theorem proven in *Theory of Sets* for universal mappings, that the solution given here is essentially unique. This is one of the very few results of *Algebra* which can be pointed out as being obtained as a consequence of the general results obtained in *Theory of Sets*. However, due to the unwieldiness of the concepts, the formal verification of the conditions under which the particular case in question can be treated by using the general one is itself an elaborate process that is not carried out in the book, rendering doubtful, once again, the advantages of having invested so much effort in the general concepts.

The only theorems proven in terms of *structures* are the most immediate ones, such as the first and second theorems of isomorphism (§ 1.6, prop. 6&7),

and even they receive a special reformulation for groups later on in the same book (§ 4.5, Theorem 3). No new theorem is obtained through the *structural* approach and standard theorems are treated in the standard way. The Jordan-Hölder theorem (§ 4.7) aptly illustrates this situation, especially because elsewhere it had been proven within a wider conceptual framework of which group theory is a particular case,[71] while Bourbaki's proof was rather more restricted.

These remarks are not intended to imply that there is one best way to prove this, or any other, theorem. The point is merely to stress the fact that *structure*-related concepts, even within the framework of Bourbaki's own treatise, do not actually stand behind any generalization that is operationally important.

These are the only, feeble connections between algebraic structures and *structures* in Bourbaki's *Algebra*. As the book advances further into the subsequent theories in the hierarchy of algebraic structures, the connection with *structures* is only scarcely mentioned, if at all. Ironically, the need for a stronger unification framework was indeed felt in later sections. Such was the case, for instance, in Chapter 3 where three types of algebras defined over a given commutative ring are successively discussed: tensor-, symmetric- and exterior- algebras. Although a separate treatment is accorded to each type of algebra, this treatment nearly repeats itself in its details three times, one after the other. Thus Bourbaki defines each kind of algebra and then discusses, for each case separately: "functorial properties", "extension of the ring of scalars", "direct limits", "Free modules", "direct sums", etc.[72] This is worth mentioning not only because a unified presentation of the three could have been more economic and direct but especially because all the above mentioned issues lend themselves naturally to a categorical treatment and this possibility is not even mentioned here. The "functorial properties" of the algebras are explained through the use of the standard categorical device of "commutative diagrams",[73] but without mentioning the concepts of functor or category.[74] A further interesting point in this context is that for all the three cases a side

---

71. See for example Ore 1937 or George 1939. See Birkhoff 1948, 88 for a survey of different proofs of this theorem.

72. Bourbaki 1973, 484-522.

73. See § 8.1 below.

74. A similar situation, where categories and functors could have made the presentation more concise and more general, but their use was avoided, is found in Bourbaki's book on Commutative Algebra. This is discussed in greater detail below in § 7.3.4.

comment was added to the effect that a certain elementary fact, proven for the three cases "is a solution of the *universal mapping problem*",[75] for which the reader is referred to Chapter IV of *Theory of Set*. This result, however, is not formally proven, and what is more significant, it is not used for any purpose in the rest of the section or of the book.

7.3.3 General Topology

Bourbaki's book on general topology comprises ten chapters and a Fascicule de résultats, whose first editions appeared successively between 1940 and 1953, and then underwent several re-editions.[76] In this book one finds the single most outstanding example of a theory presented through Bourbaki's model of the hierarchy of *structures,* starting from one of the "mother structures" and descending to a particular *structure,* namely that of the real numbers.[77] According to the plan in the introduction of the book, the theory of topological spaces is presented in the opposite way to that in which it historically originated. The approach is characterized by the introduction of topological structure independent of any notion of real numbers or any kind of metric.

However, as with *Algebra,* the hierarchy itself is in no sense introduced in terms of the *structure*-related concepts. Thus for instance, topological groups are not characterized as a *structure* from which the *structure* of groups can be "deduced." *Structure*-related concepts appear in this book more than in any other place in the treatise but, instead of reinforcing the purported generality of such concepts, a close inspection of their use immediately reveals their *ad-hoc* character.

As a first example, take the concept of homeomorphism, which is defined (as was "isomorphism" defined in *Algebra)* as a bijection preserving the *structure* of the topology. This definition is claimed to be "in accordance with the general definition."[78] Again, the verification of this simple fact (which is nei-

75. Bourbaki 1973 485, 497, 507.

76. All quotations are taken here from the English version Bourbaki 1966.

77. However, it took considerable work and discussion within Bourbaki to arrive at this conception. Liliane Beaulieu (1990, 39) has described the first report on topology prepared by André Weil and presented for discussion in a Bourbaki congress of 1936 as follows: "One striking feature in Weil's report is that he first introduced the most familiar examples of topological concepts and spaces as a motivation to admit progressively more general ones. This is opposite of what later became Bourbaki's principle 'proceeding from the general to the particular'." Beaulieu also documented the subsequent transformation of the initial report into its definite formulation.

ther done nor suggested in the book) is a long and tedious (though certainly straightforward) formal exercise.

A more elaborate example appears in Chapter 3, dealing with topological groups, and more explicitly, in section 7.1 on "Inverse limits of algebraic structures." Since these notions mix algebraic and topological ideas, one would expect to find *structural* ideas applied here in order to analyze the relationship between these two separate mathematical domains. And in fact, one does find them, only to discover that their mention is somewhat deceptive. Thus Bourbaki relates the idea of "inverse limits" to *structural* ideas in the following words:

> Let Σ be a species of algebraic structures, and let Σ_0 be the *impoverished* structure corresponding to Σ. Whenever we speak of an inverse system of sets $(X_\alpha f_{\alpha\beta})$ endowed with structures of species Σ, we shall always suppose that the $f_{\alpha\beta}$ are *homomorphisms* for these structures. If we endow $X = limX_\alpha$ with the internal and external laws of the X_α, then X carries an algebraic structure of *species* Σ_0. Naturally it remains to be seen in each particular case whether or not this structure is of species Σ. (Bourbaki 1966 Vol. 1, 285. Italics in the original)

On the face of it, this could provide a felicitous instance of the application of *structural* concepts in order to elucidate an interesting mathematical situation. Yet, beyond the declaration of what should be done in *structural* terms, nothing of the sort is actually done. Instead, the above quotation is followed by the reformulation of the general setting for the particular cases of groups and rings, in which cases the question in the last sentence of the quotation may be answered in the affirmative.

Thus, the failure of *structures* to play a significant role as a generalizing concept is illustrated in *General Topology* not only by the infrequency of its applications, but *precisely* through the uses to which the concept is actually put. Far from being general concepts used in apparently *different* situations (as claimed by Bourbaki), many *structure*-related concepts appear *only* in a few instances of *Topology*.[79] Such concepts seem, therefore, to have been defined in *Theory of Sets* just to be handy for *General Topology*, but no other use was found for them in the whole treatise. Naturally, this is perfectly legitimate from the formal point of view, but it is much to the detriment of any claim

---

78. Bourbaki 1966, 18.

79. Such as in section 4.2., where a partial ordering of topologies is defined. The topologies are ordered from *coarser* to *finer*.

about the generalizing value of *structures*. Moreover, it certainly contradicts a leading principle of Bourbaki concerning the axiomatic treatment of concepts, namely that "a general concept is useful only if it is applicable to a number of more special problems and really saves time and effort."[80] Following that principle, Bourbaki did not hesitate to qualify other theories, as "insignificant and uninteresting." By now, it should be clear that Bourbaki's own theory of *structures* does not satisfy that principle.

7.3.4 Commutative Algebra

The remaining books of Bourbaki's treatise rely mainly on concepts taken from *Algebra* and *General Topology* and the concept of *structure* is totally absent from them.[81] In Bourbaki's *Commutative Algebra*, consisting of seven chapters whose first editions appeared between 1961 and 1965,[82] one finds a remarkable departure from the group's self-imposed methodological rules. In this book the limitations of *structures* as a generalizing framework are interestingly manifest and, in fact, they are explicitly acknowledged.

Consider the discussion on "flat modules." As it happens, this is a concept which is better understood in terms of concepts taken from homological algebra, a mathematical discipline which was not dealt with in the treatise until 1980. While it is often the case that when formally introducing concepts in a book of the treatise, Bourbaki illustrates those concepts by referring to an example which had not been yet introduced in that specific book, if the example is not a logical requisite for a full understanding of the concept itself and it appears in another place of the treatise, Bourbaki presents the example between asterisks and gives the corresponding cross-reference. This policy is explained in the "Mode d'emploi" that serves as preface to each of the books in the treatise.

In the case of flat modules, a whole section (§ 4) was included "for the benefit of the readers conversant with homological algebra", in which Bourbaki showed "how the theory of flat modules is related to that of the Tor functors."[83] The concept of functor and the particular case of the Tor functor are

---

80. Cartan 1980, 180.

81. As a matter of fact, the term "structure" is used once, but with a completely different meaning. See Book X on "Differential and Analytic Manifolds": F. § 6.2.1, p.62.

82. All quotations are taken here from the English version Bourbaki 1972.

83. Bourbaki 1972, 37.

not developed in the treatise, but Bourbaki thought it important to present the parallels between the two approaches: Bourbaki's own approach and the functorial approach to homological algebra. In order to do this, Bourbaki freely used concepts and notations foreign to the treatise. This is one of the few instances in the treatise where, instead of sticking to the usual notation between asterisks, Bourbaki gave a reference to a book or article outside it. Thus, the reader is referred to a forthcoming volume of the treatise where categories, and in particular, Abelian categories were eventually to be developed. Until then, however, one could also consult Cartan & Eilenberg 1956 or Godement 1958.

A cursory examination of issues of "La Tribu" during the fifties uncovers recurring attempts to write chapters on homological algebra and categories for the *Eléments*. Eilenberg himself, who had initiated together with Saunders Mac Lane the study of categories (§ 8.2 below), was commissioned several times to prepare drafts on homology theories and on categories, while a Fascicule de résultats on categories and functors was assigned successively to Grothendieck and Cartier.[84] However, the promised chapter on categories never appeared as part of the treatise. The publication of such a chapter could have proved somewhat problematic when coupled with Bourbaki's insistence on the centrality of *structures*; the task of merging both concepts, i.e. categories and *structures*, in a sensible way, would have been arduous and not very illuminating, and the adoption of categorical ideas would probably have necessitated the rewriting of several chapters of the treatise.[85] In this regard, it is interesting to notice that when the chapter on homological algebra was finally issued (1980) the categorical approach was not adopted. Although the conceptual framework provided by categories had become the standard one for treating homological concepts since the publication of the above mentioned textbook of Cartan and Eilenberg, in Bourbaki's own presentation these concepts are defined within the narrower framework of modules. And naturally, the concept of *structure* was not even mentioned there.

---

84. Cf. for example "La Tribu" #28 (June 25 - July 8; 1952); # 38 (March, 11-17; 1956); # 39 (June 4-July 7; 1956); #40 (October 7-14; 1956).

85. This point is elaborated in detail below in § 8.4.

7.4 *Structures* and the Structural Image of Mathematics

Although the Bourbaki project started with a relatively limited aim in mind, namely, the writing of an up-to-date treatise in analysis, in the early 1940s, after several years of activity, a much more ambitious program was consolidated. The *Eléments* eventually assumed the form of a unified, comprehensive presentation of *the whole picture* of the essentials of mathematics from a single, best point of view.[86] This conception, however, eventually proved overly sanguine and Bourbaki soon realized that they must limit themselves to only a more reduced, if still highly significant portion of mathematics.[87] Moreover, it became clear that the accelerating pace of developments in research would make it impossible to bring the *Eléments* fully up-to-date. Nevertheless, Bourbaki continued to regard each volume as a definitive survey containing all the basic knowledge needed for understanding and pursuing research in the particular disciplines considered.[88] The *Eléments* was intended to provide the basis for the "classical" component of mathematical knowledge, which it was assumed would remain basically unchanged during the foreseeable future. Thus, it was supposed to provide all the tools needed for developing the second component of mathematics, the living one, as made manifest in current mathematical research.[89]

The evidence presented above suggests that *Theory of Sets*, and particularly the concept of *structure* defined in it, are not essential to the contents of

---

86. Diuedonné (1970, 145) expressed this conception in retrospect as follows: "Bourbaki sets off from a basic belief, an unprovable metaphysical belief we willingly admit. It is that mathematics is fundamentally simple and that for each mathematical question there is, among all possible ways of dealing with it, a best way, an optimal way." As with other issues, the opinions of other members on this point are less documented, if at all. However, the spirit of the whole project and of the specific discussions on each chapter of the treatise, as documented in the various issues of "La Tribu", indicate in this case that Dieudonné's report expresses faithfully an idea shared by other members of the group.

87. Dieudonné 1970, 136: "Little by little, as we became rather more competent and more aware, we realized the enormity of the job that had been taken on, and that there was no hope of finishing it as quickly as [planned]." See also Fang 1970, 43: "The grand plan notwithstanding ... [Bourbaki's] work will remain unfinished because modern mathematics will never be completed." See also Israel 1977, 67.

88. Cf. Boas 1970, 351.

89. Dieudonné opened his book *A Panorama of Pure Mathematics - As seen by Nicolas Bourbaki* (1982a), with an account of mathematics as composed of two different parts, in the above terms, namely a 'classical' and a 'living' one. What Dieudonné includes under the 'classical' part of mathematics equals the contents of the *Eléments*.

the *Eléments*. One can read and understand any book of Bourbaki's treatise without first learning the theory of *structures*. *Theory of Sets* could in principle be omitted from the series, for it has neither heuristic value nor logical import for any particular theory discussed in the other volumes of the treatise, which form the heart of Bourbaki's real interests. More than any other section of *Theory of Sets*, the theory of *structures* could safely be skipped by any potential reader. Seen from the vantage point of what Bourbaki envisaged for the treatise, namely, to provide the necessary, basic toolkit for the working mathematician, the concept of *structure* seems to be forced and unnatural.

Yet it is not only within Bourbaki's own work that the concept of *structure* plays no mathematically meaningful role. While the various books of the *Eléments* generally turned into widely quoted and even classic references for the topics covered therein, and a considerable portion of the concepts, techniques, notation and nomenclature, introduced by Bourbaki were readily adopted by the practitioners of those branches, this was not the case for *Theory of Sets* and the *structure*-related concepts.[90]

This conclusion can be easily confirmed by examining any scientific review journal. Consider for instance the *Index of Scientific Citations*, during the period 1962 to 1966, the apogee of Bourbaki's influence. The index during these five years includes over 435 quotations of the *Eléments*, but only three of them refer to the chapter on *structures*. Of these three quotations, one appears in a theoretical biology article.[91] In general, the ideas of *Theory of Sets* seem to have inspired organizational schemes for non-mathematical disciplines more than they directly influenced mathematical research.[92]

But if the description of Bourbaki's work presented here is correct, why—it may be asked—have "mathematical structures" come to be generally identified with Bourbaki? The answer is quite simple, and it has to do with the distinction, stressed all along in the present book, among the various formal and non-formal meanings of the term "structure." This distinction has been often

---

90. Nevertheless, it should be stressed, that a renewed interest in Bourbaki's concept of *structure* has arisen lately in the framework of current research in model theory, incidentally in connection with the work of José Sebastiao e Silva (mentioned in footnote 3 in the introduction to Part Two above). See, e.g., Da Costa 1987, 144-145; Da Costa & Chuaqui 1988. According to Da Costa (1986, 143 ff.), the limitations of Bourbaki's theory is a consequence of its focusing on syntactic issues. Sebastiao e Silva's ideas, in Da Costa's view, if properly elaborated (using also techniques developed by Alfred Tarski (1901-1983)), could provide the semantical dimension lacking in Bourbaki's concepts, thus providing the conceptual basis for new avenues of research in model theory.

91. Cf. Gillois 1965.

left vague in the historical writings of Bourbaki and of some Bourbaki members. In order to discuss this point properly, it is first necessary to elaborate briefly on the issue of Bourbakian historiography.

Each of the books of Bourbaki's treatise is accompanied by an account of the historical evolution of the discipline considered in it. These accounts were later collected and published in a volume entitled *Eléments d'histoire des mathématiques* (1969), which was widely read among mathematicians and often praised by them.[93] Dieudonné and Weil have been among the Bourbaki members who have expressed a clear and sustained interest in the history of mathematics. Besides having taken active part in the writing of the *Eléments d'histoire*, they independently published abundantly on the history of mathematics.[94]

Bourbaki's historiography, as manifest in the *Eléments d'histoire* as well as in the individual writings of Dieudonné and of Weil, has been strongly connected with their overall conception of mathematics. In particular, they have applied similar criteria to differentiate important from unimportant ideas in both *present* mathematical research and *past* mathematical theories. Naturally enough, this has also been the case regarding the centrality of structures in mathematics.

Bourbakian historiography has been criticized in the past for its "Whiggish" approach. As a matter of fact, this is perhaps the domain in which Bourbaki's writings have been more harshly criticized.[95] Yet most of the criticism

---

92. A typical example of this is provided by the so-called "structuralist" trend in contemporary philosophy of science. (For an account of the development of the trend see Diederich 1989). Wolfgang Stegmüller (1979), for instance, described the structuralist approach as "A Possible Analogue to the Bourbaki Programme in Physical Sciences." In attempting to clarify the internal structure of physical theories he applied an axiomatization procedure which allegedly follows "Bourbaki's programme." According to Stegmüller Bourbaki performed a formalization of all mathematics using "set-theoretical rather than metamathematical methods." By following a similar approach in order to formalize physical theories, Stegmüller intends to enable a "realistic" formal treatment of them and of their semantics. Bourbaki's approach, claims Stegmüller, represented an improvement over Russell's "impractical" foundational work in mathematics; thus it is bound as well to represent a parallel improvement over Carnap's "'impractical' foundational work" in the philosophy of empirical science. Stegmüller was pursuing here an approach initially proposed by Suppes 1969. See also Da Costa 1987; Moulines & Sneed 1979.

93. For typical examples of enthusiastic appraisals of Bourbaki's *Eléments d'histoire*, see Mac Lane 1986 and Scriba 1961.

94. The bibliography at the end of this book include some, but not all, of the historical writings of Dieudonné and of Weil.

directed towards Bourbaki's historiography has not addressed the problematic status of *structures* within it. This latter issue can be addressed here following the above analysis of the roles played by the formal and by the non-formal concepts of "structure" in Bourbaki's mathematics. In fact, one can observe in Bourbaki's historiography a noteworthy combination of the tendency to apply to historical research criteria of selection initially conceived for refashioning mathematics in terms of structures, on the one hand, and of the alleged central-ity of *structures* for mathematics, on the other hand. This combination has brought forward an historical narrative, according to which the idea of struc-ture (and even perhaps the concept of *structure*) not only is essential to the present overall picture of mathematics, but has even been instrumental in bringing about its historical ascendancy. The equivocal use of the term "struc-ture" in its various meanings has only complicated things more.

Consider, for example, the following quotation of Dieudonné that was already cited above:

> Today when we look at the evolution of mathematics for the last two centuries, we cannot help seeing that since about 1840 the study of specific mathematical objects has been replaced more and more by the study of mathematical *struc-tures*. (Dieudonné 1979, 9. Italics in the original)

Taken as a very general statement, this is a seemingly straightforward his-torical assessment, although its import cannot be understood without knowing the precise meaning of the term "mathematical structure." But as it is, one tends to accept such a general claim, and even more so after Dieudonné adds

---

95. Spalt 1987 is in fact a book-long criticism of Bourbaki's historiography. See especially pp. 2-4 & 24. See also Spalt 1985: "Strukturalistische Mathematikgeschichtsforschung ist kaum etwas anderes als ein kriminalistisches Aufsuchen jener Begriffe (oder deren struktureller Pendants) in der Vergangenheit, die aus heutiger—sprich: Bourbakis—Sicht die "wahre Natur" der mathematischen Theorien ausmachen. Lehrsätze oder Methoden früherer Zeit sind solcher Historie nur dann und insoweit von Bedeutung, als sie sich als "Spezialfälle" oder "Vorläufer" zeitgenössischer Verallge-meinerungen erfassen lassen."

Grattan-Guinness 1979 harshly criticizes the historiographical approach of Dieudonné (ed.) 1978, especially on grounds of its retrospectively applying present criteria of selection and of its fail-ing to refer to any existing secondary literature. Freudenthal 1981 praises that volume edited by Dieudonné, claiming that "in spite of its shortcomings this *is* a good history of mathematics." How-ever the shortcomings stressed by Freudenthal in some detail are very serious and can hardly be overseen. For further criticism of Bourbaki's historiography see also Israel 1977, 64-65; 1978, 63-69; 1981, 209-211; and Lakatos 1976, 135 & 151.

a dose of caution and states that, indeed, the concept of structure was foreign to mathematicians even as late as 1900. Thus, he further wrote:

... this evolution was not noticed at all by contemporary mathematicians until 1900, because not only was the general notion of mathematical structure foreign to them, but the basic notions of specific structures such as group or vector space were emerging very slowly and with a lot of difficulty. *(ibid.)*

The quotation suggests that after 1900, the general notion of "mathematical structure" became known or somewhat clearer to mathematicians. This is a questionable assertion in itself, but not one that requires an elaborate criticism. In fact, Dieudonné's claim is highly problematic because it *cannot* be taken as an assertion merely about the general emergence of structures. To be sure, his claim is followed by a footnote specifying that the term "mathematical structure" is to be taken *in the specific technical sense defined by Bourbaki in the fourth chapter of the first book of the Eléments*! The quotation should, then, read as the claim that since about 1840, and more explicitly after 1900, mathematics has increasingly become the study of *structures*! This statement is quite different from the general one suggested above and, as the preceding chapters clearly show, it can in no sense be accepted as historically correct.

Of course, not all articles by or about Bourbaki assume the identity of the non-formal and formal sense of the term "structure" as explicitly as Dieudonné did in the above-quoted passage.[96] But even when this identification appears in more ambiguous terms,[97] it supports a pervasive assumption that seems to underlie many accepted accounts of the structuralist approach to mathematics and of the central role played by Bourbaki in its consolidation

---

96. Cf., e.g., Dieudonné 1982, 619: "[T]he connecting link [between the diverse theories within the treatise] was provided by the notion of *structure*." (Italics appear here in the original, but not following the convention adopted in the present book to denote the formal term. It is therefore not clear, in Dieudonné's text, whether he means the formal or the non-formal sense.

97. Cf., for instance, the following account of Bourbaki's early years of activity, in which both senses of the term are subtly intermingled (Weil 1992, 114): "In establishing the tasks to be undertaken by Bourbaki, significant progress was made with the adoption of the concept of structure, and of the related notion of isomorphism. Retrospectively these two concepts seem ordinary and rather short on mathematical content, unless the notions of morphism and category theory are added. At the time of our early work these notions cast light upon subjects which were still shrouded in confusion: even the meaning of the term 'isomorphism' varied from one theory to another. That there were simple structures of group, topological space, etc., and then also more complex structures, from rings to fields, had not to my knowledge been said by anyone before Bourbaki, and it was something that needed to be said." A similar statement appears also in Weil 1978, 537.

and expansion. This is particularly the case when it comes to the putative link between the hierarchy of structures and the alleged centrality of the so-called "mother structures."[98] The mother structures appear in Bourbaki's *Architecture* manifesto as follows:

> At the center of our universe are found the great types of structures, ... they might be called the mother structures ... Beyond this first nucleus, appear the structures which might be called multiple structures. They involve two or more of the great mother-structures not in simple juxtaposition (which would not produce anything new) but combined organically by one or more axioms which set up a connection between them... Farther along we come finally to the theories properly called particular. In these the elements of the sets under consideration, which in the general structures have remained entirely indeterminate, obtain a more definitely characterized individuality. (Bourbaki 1950, 228-29)

It should be stressed again that this description of the mother structures is *not* an integral part of the formal, axiomatic, theory of *structures* developed by Bourbaki. The classification of structures according to this scheme is mentioned several times in *Theory of Sets*, but only as an illustration appearing in scattered examples.[99] Many assertions that were suggested either explicitly or implicitly by Bourbaki or by its individual members—i.e., that all of mathematical research can be understood as research on structures, that there are mother structures bearing a special significance for mathematics, that there are exactly three, and that these three mother structures are precisely the algebraic-, order- and topological-structures (or *structures*)—all this is by no means a logical consequence of the axioms defining a *structure*. The notion of mother structures and the picture of mathematics as a hierarchy of *structures* are not results obtained within a mathematical theory of any kind. Rather, they belong strictly to Bourbaki's non-formal images of mathematics; they appear in non-technical, popular, articles, such as in the above quoted passage,[100] or in the myth that arose around Bourbaki.[101]

---

98. Which is incidentally hinted at in Weil's passage quoted in the preceding footnote.

99. For example in *Theory of Sets* pp. 266, 272, 276, 277, 279.

100. And also, in greater detail, in Cartan 1980, 177.

101. It should be stressed, however, that in Bourbaki's "Architecture" manifesto, one also reads (Bourbaki 1950, 229), that the picture of mathematics as a hierarchy of *structures* is nothing but a convenient schematic sketch, since "it is far from true that in all fields of mathematics, the role of each of the structures is clearly recognized and marked off." Furthermore, "the structures are not immutable, neither in number nor in their essential contents."

The equivocal blurring of the formal and non-formal meanings of the term "mathematical structure" and the related issues surrounding the mother structures allows one to identify the influence of Bourbaki's images of mathematics in places where the name Bourbaki is not even mentioned. In fact, the centrality accorded by Bourbaki to the idea of structure, while associating it with the theory of *structures*, has been implicitly taken for granted even by historians who consciously attempt to adopt a historiographical approach opposed to that of Bourbaki. Thus, whenever an author classifies mathematical structures as algebraic, topological and ordered structures, one may assume that he has taken Bourbaki's scheme for granted and has accepted that the mother structures are a meaningful mathematical idea.

A noteworthy instance of this implicit acceptance of Bourbaki schemes appears in Wussing's book on the rise of the abstract concept of group. In this book, contrary to Bourbaki's historiographical tendency, much effort is invested in order to avoid hindsight in the exposition of the development of mathematical ideas. However, in its epilogue, when the author explains the connection between the rise of the abstract concept of group and the rise of the structural trend in mathematics in general, he wrote:

> Within the limits of my study, and to the extent to which these limits bear on the history of the structural concept of "group", the connections between structural thinking and classical mathematics are relatively clear. The very advanced systematization of algebraic structures within contemporary mathematics, that is, the existence of "universal algebra", suggests analogous studies of the genesis of other algebraic structures. But in view of the absence of methodological models and the state of modern mathematics, one is likely to encounter far greater difficulties in the study of ordered and topological structures. (Wussing 1984, 258)

In this quotation, Wussing wholly adopts not only Bourbaki's mother structures scheme, but also the assumption that the possibility of properly elucidating the idea of structure depends on the existence of a formal concept of "structure" for a specific domain. Thus Wussing claims that it was the existence of a concept of "universal algebra" which should encourage the research into the rise of other particular algebraic structures. On the contrary, the absence of a general formal concept of topological- or order-structure has, in his view, hindered the undertaking of historic studies in those areas. But as the present book intends to show, the situation is precisely the opposite. The existence of formal concepts of structures may lead to incorrect historical interpretation, since it induces overlooking the important non-formal aspects of the actual historical process. It should be added, however, that in Wussing's book, as in several other similar works, the issue of the rise of the structural method is just an offshoot of the main argument, an afterthought or a programmatic statement for future works. This is probably the reason why he and other authors are much less critical on this issue than in the general tone of their works.[102]

It is remarkable that in some of Bourbaki's writings one finds an ambiguous attitude towards the validity of the notion of the hierarchy of *structures*. On the one hand, Bourbaki has cautioned that the picture of mathematics as a hierarchy of *structures* is nothing but a convenient scheme since "it is far from true that in all field of mathematics the role of each of the structures is clearly recognized and marked off." Furthermore, "the structures are not immutable, neither in number nor in their essential contents."[103] On the other hand, the inclusion of those examples in *Theory of Sets*, amidst Bourbaki's formal treatment of a theory of structures has had the effect, intentionally or unintentionally, of conferring upon them, metonymically as it were, that special kind of truth-status usually accorded to deductively obtained propositions.

This ambiguous link between the idea of mother structures and a formal mathematical theory has been manifest not only in works on the history of mathematics but also in works that address questions concerning the nature of mathematics and its applicability to other disciplines. Perhaps the best known example of this kind is reflected in Piaget's manifest enthusiasm for Bourbaki's work, already mentioned above. Hans Freudenthal has commented on this wrong-headed view as follows:

> The most spectacular example of organizing mathematics is, of course, Bourbaki. How convincing this organization of mathematics is! So convincing that Piaget could rediscover Bourbaki's system in developmental psychology. ... Piaget is not a mathematician, so he could not know how unreliable mathematical system builders are. (Freudenthal 1973, 46)[104]

---

102. For additional accounts of the rise of the structural approach in mathematics in which the mother structures are similarly alluded to see Behnke 1956, 29-31; Birkhoff 1974, 336; Novy 1973, 223; Purkert 1971, 23. Even historians who have critically approached Bourbaki's pronouncements on mathematics in general, seem to have admitted Bourbaki's claims on the centrality of mother *structures* at face value. Cf. Mac Lane 1987a, 33 ff. (and § 9.2 below); Israel 1978, 60-61; Mehrtens 1990, 139.

103. Bourbaki 1950, 229. And of course the term "structure" is equivocally used in this passage, so that it is not completely clear whether it should be taken in its formal or in its non-formal meaning.

But Piaget was not alone in failing to realize how unreliable mathematical system builders can be. Some mathematicians seem to have been equally gullible.[105] Bourbaki' members, especially in the first years, hardly saw themselves as "unreliable system builders" nor did they see their formulations as provisional. Doubts about the certainty of Bourbaki's program arose only later on, whereas the image of mathematics as revolving around the concept of *structure* persisted long after that. This change in attitude is shrouded in the ambiguity of claims advanced by Bourbaki members, especially Dieudonné, indicating that "the connecting link [between the diverse theories within the treatise] was provided by the notion of *structure*."[106] If "structure" is taken to mean *structure* then Dieudonné's claim reflects Bourbaki's initial confidence. In that case, however, they are imprecise. If, on the contrary, "structure" is given its non-formal meaning, then Dieudonné's claims may be sound, but they say something different, and indeed significantly less, than they were meant to assert.

The present chapter has analyzed the role played by the concept of *structure* within Bourbaki's *Eléments de mathématique*. The influence of Bourbaki's treatise on twentieth-century mathematics presents significant parallels to that of van der Waerden's book on algebra. As textbooks that compiled much important, previously existing research work, their central contribution consisted in the restructuring of the disciplines studied in them. This restructuring implies a redefinition of the subdisciplines involved, and of their boundaries and interrelations. It also implies a selection of basic concepts, basic tools and basic problems in each subdiscipline. In the case of Bourbaki, given the ambitious scope of the enterprise, this restructuring had implications for mathematics as a whole. One should also bear in mind the fact that both works exerted a strong influence mainly on the images of mathematics, rather

---

104. For a further criticism of Piaget's reliance on Bourbaki's schemes see Lurcat 1976, esp. 278-280.

105. Consider for example the following assessment of Bourbaki's contributions (Gauthier 1972, 624): "Le Chap. IV sur les 'Structures' est sans doute le plus novateur: on sait que ce thème de structure est proprement bourbakiste. C'est le groupe Bourbaki qui a, en effet, thématisé les structures en mathématiques et les a catégorisées selon les trois grandes espèces de structures-mères." Cartan, on the other hand, explicitly declared that, after twenty years of activity, "there may be some concepts among the fundamentals in Bourbaki's textbook which have already become outdated." Cf. Cartan 1980, 180.

106. Dieudonné 1982, 619. Italics in the original.

than directly through the body of mathematics. This is not intended to belittle the scope or importance of this influence, but rather the opposite. In framing the main open problems, the accepted tools, the aims of mathematical education, etc., in the decades that followed its publication, the *Eléments*, and the picture of mathematics implied by it were to become a central force in the development of mathematical research, and in the growth of mathematical knowledge.

The concept of *structure*, allegedly a central pillar in Bourbaki's building of mathematics, plays no actual role in the presentation of theories within the treatise. Nevertheless this concept and the associated hierarchy based on the three mother structures have been often considered as if they in fact provide a solid, reflexive foundation of Bourbaki's images of mathematics and, in a certain sense, endow them with a kind of justification that is usually absent from alternative schemes.

Bourbaki's work and the way it influenced the subsequent development of mathematics interestingly illuminate the interplay between images and body of knowledge. On the one hand, Bourbaki's creation of a new system of images of knowledge together with the impressive body of knowledge—which was produced under the influence of this image—were to direct and condition a considerable portion of mathematical research for several decades. On the other hand, the inherent force of these images of knowledge was partially bolstered by a reflexive body of knowledge, the theory of *structures*, which in fact did *not* provide the kind of foundational support attributed to it. The mathematical significance of Bourbaki's overall contribution remains the same if one deletes or ignores the existence of the theory of *structures*. Moreover, it is arguable, although by no means certain, that Bourbaki's actual influence on mathematics would have remained the same had not the entire book, *Theory of Sets*, and the theory of *structures* been published at all. Nevertheless, one may wonder how the history of the idea of structure (both in and outside mathematics), as well as its historiography, would have looked like had not Bourbaki formulated the theory of *structures* and had not the distinction between the formal and non-formal meanings of the term "mathematical structure" been blurred in the writings of various individual members of the group.

Chapter 8

Category Theory: Early Stages

The concepts of "Category" and "Functor" were first defined by Samuel Eilenberg and Saunders Mac Lane in 1942. This initial presentation did not elicit any special interest among contemporary mathematicians. Many years were still needed until the theory became an autonomous mathematical discipline. Since the 1960s category theory has afforded a unifying conceptual framework enabling a fairly effective manipulation of concepts belonging to branches like algebra, topology, logic and others, from a single, unified perspective. The generalizing possibilities afforded by category theory also led to several attempts at providing an abstract foundation for all of mathematics in terms of categorical concepts, in the hope of overcoming the difficulties encountered when this task was attempted from the set-theoretical point of view.

The subject-matter of category theory is the various mathematical disciplines, abstractly considered. Thus, the theory of categories provides tools and a convenient perspective from which to elucidate, from a formal mathematical point of view, the non-formal notion of a mathematical structure. Because of their very nature, categories concern a wider spectrum of mathematical disciplines than Ore's **structures**. Categories and *structures,* on the other hand, could in principle be considered as suitable for dealing with similar mathematical situations. As a matter of fact, however, the historical interrelations between the two notions is far from straightforward. Although it will not be possible to discuss here in all detail the complex motivations that underlay category theory, it will be seen that this theory provided a richer language and more meaningful results than the theory of *structures* did. Thus, the present chapter describes the initial stages in the development of the theory, until it began to consolidate into an autonomous discipline with a particular methodology, a set of open problems, a stable community of researchers and a sense of identity. It also considers some significant conceptual and historical connections between the theory of categories, on the one hand, and both the theory of *structures* and the theory of **structures**, on the other. But first, for the benefit of the uninitiated reader, it is convenient to present some of the basic ideas of category theory.

8.1   Category Theory: Basic Concepts

The concept of category formalizes the idea of a "mathematical domain", while the concept of functor formalizes the idea of the connections between two different mathematical domains. The central idea of category theory is that a mathematical domain may be abstractly characterized, not by describing the common features intrinsic to all the entities included in that domain, but rather by examining the *connections* among those entities. The category composed of all groups will be thus characterized, for instance, not by the intrinsic properties of the internal operation which defines the groups, but rather by the abstract properties of the homomorphisms defined among groups. It is assumed that by examining those connections, it will be possible to grasp the peculiarity of each mathematical domain. Adopting the abstract approach to mathematics implied overlooking the *nature* of the elements involved in the domains studied. The categorical approach goes one step further, and proposes to overlook not only the nature of the elements, but also, like in the theory of **structures**, their very *existence*. Thus, category theory proposes to formalize the idea of a mathematical structure through the properties of its morphisms, namely, through the properties of the interconnections among the different individual representatives of a given structure. As a matter of fact, the very idea of a category arose as subsidiary to the study of certain mathematical situations which necessitated the simultaneous consideration of *all morphisms* of a given mathematical domain.

A category Γ contains *objects* $A,B,C,...$ and *arrows* (or morphisms) $f,g,h,...$ Every arrow corresponds to two objects A,B called the domain and the codomain respectively. This situation is denoted by the following diagram:

$$f: A \rightarrow B$$

The set of all arrows from A to B in the category Γ is denoted $\text{Hom}_\Gamma(A,B)$. To every object A in a category there exists a unique identity arrow $1_A: A \rightarrow A$.

If the domain of an arrow $g: B \rightarrow C$ equals the codomain of an arrow $f: A \rightarrow B$, there exists always a *composition* arrow

$$(g{\circ}f): A \rightarrow C$$

In addition, two simple axioms hold in every category. First, the associative law for the composition states that given three arrows

$$f: A \rightarrow B, \, g: B \rightarrow C, \, h: C \rightarrow D$$

one can obtain two identical triple compositions

$$h{\circ}(g{\circ}f) = (h{\circ}g){\circ}f: A \rightarrow D$$

The second axiom states that for every arrow $f: A \rightarrow B$, the following identity holds:

$$f{\circ}1_A = f = 1_B{\circ}f: A \rightarrow B.$$

Immediate examples of categories are the category *Grp* of all groups and homomorphisms of groups; the category *Set* of all sets and functions among them; *Ab* of all Abelian groups and homomorphisms of Abelian groups; *Top* of all topological spaces and continuous functions; and so forth.[1]

It is natural to move one step further and consider the categories as forming a category in themselves. Obviously the arrows of this category are the functors, namely arrows sending an object in a given domain category to an object in the codomain category, and simultaneously sending each morphism of the first category into a morphism in the second. Thus, given two categories Γ, Ω, F is a functor from Γ to Ω if and only if for every object A in Γ there exists an object $F(A)$ in Ω, and for every morphism $f: A \rightarrow B$ in Γ, there exists another morphism $F(f): F(A) \rightarrow F(B)$ in Ω. Moreover, it is also demanded that identities and compositions be preserved by the functors, namely that

$$F(1_C) = 1_{F(C)} \text{ and } F(f{\circ}g) = F(f){\circ}F(g).$$

An immediate example is the forgetful functor from *Grp* to *Set*. The image of any groups under this functor is its underlying set, while the image of a morphism between two given groups is the corresponding underlying function between the sets. A less trivial example could be the functor which sends an

---

1. Basic expositions of the elements of category theory, in the lines described here, appear in Jacobson 1980, esp. Chpt. 1; Mac Lane 1970, esp. Chpts. 1 & 2. They do not differ in any essential sense from the initial definitions given by Eilenberg and Mac Lane.

integral domain into its field of fractions. More interesting examples of functors appear in algebraic topology, associating algebraic domains (e.g., an Abelian group) to each topological space, in order to obtain from the former new information about the latter.

It is possible to consider also functors in which the direction of the arrows are reversed, i.e., in which the image of the morphism $f: A \rightarrow B$ in Γ is $F(f): F(B) \rightarrow F(A)$ in Ω. Such functors are called *contravariant*, as opposed to the *covariant* functors defined above.

Category theory, then, studies the properties of the arrows connecting diverse objects; morphisms connect objects within a given category, functors connect objects of different categories. One can advance a step further still and ask about the arrows connecting two functors. This question leads to another central concept of the theory: *natural transformations*.

Given two categories Γ, Ω and two functors F, G between them, a natural transformation $\theta: F \rightarrow G$, from F to G, is defined as a law assigning to every object A in Γ, an arrow $\theta_A: F(A) \rightarrow G(A)$ in Ω, such that the following commutative diagram is satisfied:

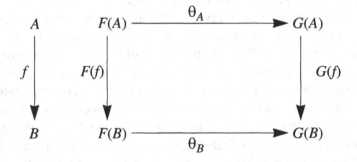

To say that the above diagram commutes means that given an object in $F(A)$, going from $F(A)$ to $G(B)$ by any of the possible paths afforded by the diagram, one is always led to the same object in $G(B)$. Commutative diagrams became one of the central tools of category theory. In this particular case, the demand of commutativity for the diagram implies that for any arrow $f: A \rightarrow B$

$$G(f) \circ \theta_A = \theta_B \circ F(f).$$

We will see below that it was the attempt to elucidate this kind of situations which led to the initial definitions of category and functor.

In order to illustrate the possibilities afforded by these categorical concepts, two examples will be now considered which are simply formulated in categorical terms. The first example is "one-to-one morphisms." A morphism $m: A \rightarrow B$ in a category Γ is called *monic* if given any two morphisms f_1, f_2 in the same category $f_1, f_2: C \rightarrow A$, then from the equation $m \circ f_1 = m \circ f_2$ one can deduce that $f_1 = f_2$. In other words a morphism is monic if it is left-cancellable. Clearly this definition is equivalent to the usual definition of a one-to-one function, only that here nothing is said about the specific action of the function on the individual elements of the objects involved. A surjective mapping may be characterized in similar terms as a right-cancellable function. Such a mapping will be called an *epic* in category theory. Naturally, an isomorphism is a morphism which is both epic and monic.

The second example will be useful for the short historical account of the rise of the theory given in the following sections. Given two sets X, Y define their Cartesian product $X \times Y$, as usual, as the collection of ordered pairs (x, y) with x belonging to X and y belonging to Y, and define two "projections" $p: X \times Y \rightarrow X$, $q: X \times Y \rightarrow Y$ in the usual way. Let two functions f, g be given, having the same domain:

$$f: Z \rightarrow X \quad g: Z \rightarrow Y$$

Define a new function $h: Z \rightarrow X \times Y$, sending every element in Z to the pair $(f(z), g(z))$. It can be shown that there exists only one function h, satisfying this property. Now the function h can also be defined in purely categorical terms without explicitly mentioning the particular form in which it is defined on every element of Z: h is the only function from Z to $X \times Y$ for which $p \circ h = f$ and $q \circ h = g$. One can start from this arrow-defined property and use it to define the Cartesian product in categorical terms, without making reference to the elements of the objects involved.

Given a category Γ and two objects X, Y in that category a "product" of X and Y in the category is an object P together with two arrows p, q

$$X \xleftarrow{\quad p \quad} P \xrightarrow{\quad q \quad} Y$$

such that given any object Z in Γ, and two arrows $g: Z \to Y, f: Z \to X$, there exists a unique arrow $h: Z \to P$ such that $p{\circ}h = f$ and $q{\circ}h = g$. In other words, the following diagram is required to be commutative:

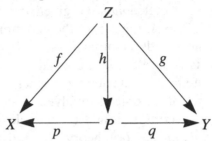

This particular kind of diagram appears very often in category theory. It is possible to formulate the condition more succinctly by saying that the first diagram is required to be *universal,* namely that for any two arrows sending objects of Z to X and Y simultaneously, it must be possible to "factorize those arrows through P" as described in the second diagram above. The commutativity of the diagram implies that any other product P' would be isomorphic to P. An important early problem of category theory was to determine the existence or non-existence of universals for particular situations. This example will be retaken below.

This short account will suffice for the task of describing the initial stages of category theory and their connection with the attempts to elucidate the idea of a "mathematical structure" within an elaborate, axiomatic mathematical theory.

8.2   Category Theory: A Theory of Structures

The decade preceding Eilenberg and Mac Lane's introduction of categories was extremely rich in developments in both algebra and topology. The basic notions of category theory had their origins in a wide variety of recurrent mathematical situations that originally appeared in many mathematical fields, but especially in the latter two. These situations were eventually isolated and given a name within category theory, while pointing out the important mathematical consequences that can be drawn simultaneously in various domains, from a systematic reliance on them. Thus, for instance, the notion of an exact sequence. In categorical terms, given three objects A,B,C in a category, and two arrows connecting them,

$$f: A \rightarrow B, \; g: B \rightarrow C,$$

the diagram

$$A \xrightarrow{f} B \xrightarrow{g} C$$

is called an exact sequence if the image of f equals the kernel of g. The formulation of the concept appeared in print in this way only in 1952, in a textbook on algebraic topology jointly written by Eilenberg and Norman Steenrod (see below § 8.3). The significance of this situation for understanding topological problems, however, had been realized as early as 1941 by Witold Hurewicz (1904-1956) in an article on homotopy theory.[2] Later, in 1945, Henri Cartan wrote a review article in which the centrality of the concept for algebraic topology was explicitly pointed out. At this stage, Cartan used neither the categorical formulation of algebraic topology nor a separate name for exact sequences.[3]

A detailed account of the rise of category theory should thus describe systematically the connections between the formulation of the central concepts of the theory and situations of the kind mentioned in the above parragraph. This complex process, however, is beyond the scope of our more specific concern here. We limit ourselves in the present section to a brief account of Mac Lane and Eilenberg's first article on categories, and of its immediate background.

One recurrent, easily identifiable, mathematical situation which Eilenberg and Mac Lane had come across in their joint work of many years in algebra and topology involved the notion of "naturality." This notion, however common and apparently clear, lacked a precise, formal definition. Eilenberg and Mac Lane attempted to elucidate this notion in the first article in which categories and functors were introduced. The opening passage of the article addresses the issue as follows:

---

2. See Hurewicz 1941, 562. Hurewicz was also among the first to use the notation $f: A \rightarrow B$ to denote simultaneously continuous functions and group homomorphisms. See Hurewicz 1936, 220.

3. See Cartan 1946, 6.This article (submitted in May 1945) was explicitly intended by Cartan as an attempt to systematically introduce Bourbaki's terminology into current research in the field. Thus, it suggested the use of the term "compact", instead of the then more usual "bicompact" (as in Alexandrov & Hopf 1935).

Frequently in modern mathematics there occur phenomena of "naturality": a "natural" isomorphism between two groups or two complexes, a "natural" homeomorphism of two spaces and the like. We here propose a precise definition of the "naturality" of such correspondences, as a basis for a general theory. (Eilenberg & Mac Lane 1942, 537)

In this first article, naturality was defined and examined within the somewhat restricted framework of group theory.[4] A general axiomatic formulation of the concept remained as a task for a promised, future article. At this stage, however, it was already clear that such a general definition would necessitate considering collections of *all* groups and their morphisms, *all* topological spaces and their continuous functions, and so on.

The promised definition appeared in a second joint article of 1945 under the title of "General Theory of Natural Equivalences." The article presented an abstract formulation of the concepts of category and functor, but as the title has it, its main concern was with natural transformations. The categorical concepts introduced in this article were subsidiary to the main issue, that of elucidating the phenomena of naturality.

Eilenberg and Mac Lane's article of 1945 opened by presenting one of the best known examples of a natural mapping. Let L be a finite-dimensional vectorial space defined over the real number field. Let $T(L)$ be the dual space of L, namely, the space of all linear applications from L to R (in the article, the dual space is called "conjugate space"). L and $T(L)$ are two vectorial spaces with the same dimension, and therefore they are isomorphic. However, the isomorphism between L and $T(L)$ cannot be explicitly defined without specifying a basis for each of the spaces.

Take now the double dual space of L, $T(T(L))$. Again, it is clear that $T(T(L))$ is isomorphic to L as a vectorial space. But at variance with the case of $T(L)$, it is now possible to actually define the isomorphism without specifying the bases.[5] Eilenberg and Mac Lane took the existence of these two kinds of isomorphism as their starting point, as a phenomenon worthy of a more pre-

---

4. Mac Lane 1988, 333-334, explains in greater detail the specific calculation of cohomology groups considered in that article.

5. Given $l \in L$ define the function $E_l: T(l) \to R$, mapping $f \in T(L)$ to $E_l(f) = f(l)$. It is clear that $E_l \in T(T(L))$.

Define now the set $R_E = \{E_l / l \in L\}$. Clearly R_E is a sub-space of $T(T(L))$. On the other hand, it is clear that R_E is isomorphic to L. If we define now the mapping $E: L \to T(T(L))$ by $E(l) = E_l$, E is an isomorphism of vector spaces, for L and $T(T(L))$ have the same dimension.

cise mathematical elucidation. The first step in this elucidation was the definition of a natural transformation, in terms similar to those used above. As they
explained in their article, the reason for the "naturality" of the isomorphism
between L and $T(T(L))$ lies in the fact that its definition embraces simultaneously all possible finite dimensional vectorial spaces. Therefore:

> A discussion of the "simultaneous" or "natural" character of the isomorphism
> $L \cong T(T(L))$ clearly involves a simultaneous consideration of all spaces L and all
> transformations μ connecting them. (Eilenberg & Mac Lane 1945, 233)

The need to consider simultaneously both all the spaces and all the morphisms among them led the authors to define categories. Already in this article
of 1945, they stressed the potential advantages of introducing the concept into
broader mathematical contexts. In the years before 1945, Eilenberg and Mac
Lane had been working intensively on some homological problems within
which natural transformations appear quite often, some of them explicitly in
the form of commutative diagrams, as described above.[6] Broadly speaking,
homological procedures provide techniques for associating a suitably defined
group to every given topological space (or to every given algebraic system, in
the case of homological algebra), so that properties of the space (resp., of the
algebraic system) may be drawn from more easily deducible properties of the
homology groups. These procedures also involve associating a group homomorphism to every continuous function between the given topological spaces
(resp., between the given algebraic systems). Thus, it is not surprising that
while attempting to define naturality from the perspective of homological theories, the concept of functor also arose.

Although categorical concepts arose from a specific problem-setting
within the context of homological research in algebra and topology, Eilenberg
and Mac Lane were aware of the metamathematical projections of their newly
introduced concepts. They were also aware of the fact that theirs was not the
only attempt in that direction. Thus they wrote:

---

6. See, e.g., Eilenberg & Mac Lane 1942a, 815-816, in which natural transformations of complexes are dealt with. In the same article (pp. 778-779) Eilenberg and Mac Lane dealt also with the
group of characters of a topological space. Groups of characters is a further example of natural transformations introduced by the authors in Eilenberg & Mac Lane 1942, 540. Every group is isomorphic to its group of characters but not in a natural way. The group obtained by finding the characters
of the group of characters is "naturally" isomorphic to the original group. It is worth remarking that
in Eilenberg & Mac Lane 1942a, the naturality of the double-character group is not mentioned.

In a metamathematical sense our theory provides general concepts applicable to all branches of abstract mathematics, and so contributes to the current trend towards uniform treatment of different mathematical disciplines. In particular, it provides opportunities for the comparison of constructions and for the isomorphism occurring in different branches of mathematics; in this way it may occasionally suggest new results by analogy.

The theory also emphasizes that, whenever new abstract objects are constructed in a specified way out of given ones, it is advisable to regard the construction of the corresponding induced mappings on these new objects as an integral part of the definition ... This emphasis on the specification of the type of mappings employed gives more insight into the degree of invariance of the various concepts involved. (Eilenberg & Mac Lane 1945, 236)[7]

Thus in its very initial stages, the categorical program suggested—in terms similar to those used by Ore in introducing his own program ten years before—that the structural essence of mathematics may be better understood when overlooking the existence of elements in each of the structures considered. Ore had proposed to examine the links of the given algebraic system with certain of its subsystems in order to elucidate that essence. Category theory proposes a similar perspective, but within a much broader scope: to overlook the existence of elements and to examine the links of the given domain with *all* possible domains of the same type, and even the links of a structure of a given type with those of a different type. By examining those links, one is bound to uncover the essence of the structural character of mathematics. Moreover, the categorical perspective affords the means to clarify and to establish a hierarchy of structural properties, according to their degree of invariance. For instance, Eilenberg and Mac Lane claimed that, concerning the theory of groups, the theory of categories will establish that:

... the concept of commutator subgroup of a group is in a sense a more invariant one than that of a center, which in its turn is more invariant than the concept of the automorphism group of a group, even though in the classical sense all three concepts are invariant.

The invariant character of a mathematical discipline can be formulated in these terms. Thus, in group theory all the basic constructions can be regarded as the definitions of co- or contravariant functors, so we may formulate the dictum: The

---

7. According to Mac Lane's own account (Mac Lane 1988, 334), he and Eilenberg were also aware of the fact that the extreme abstraction characteristic of this paper might hinder its publication in a respectable journal. Mac Lane even expressed doubts regarding the possibility that the same article would have been published, had it been submitted by wholly unknown authors.

subject of group theory is essentially the study of those constructions of groups which behave in a covariant or contravariant manner under induced isomorphism. More precisely, group theory studies functors defined on well specified categories of groups, with values in another such category. (Eilenberg & Mac Lane 1945, 236-237)

It is easy to see that the relationship between a group and the lattice of its subgroups may be expressed in functorial terms. According to the above quotation, the degree of naturality and invariance of such a functor could also be defined. Thus, from the formal point of view, Ore's program might be seen in retrospect as an early manifestation of certain ideas that later appeared in category theory. Mac Lane had studied algebra in 1929-30 under Ore in Yale, and in 1933-34 he worked again with him as a post-doctoral fellow. Mac Lane was certainly aware of the leading ideas of Ore's program, and in particular of Ore's belief that the structural character of mathematics is to be elucidated by overlooking the existence of elements in the different domains. Mac Lane himself published a research announcement concerning an attempt to formulate a generalizing concept for algebra, along the lines suggested by Ore in his program.[8] This remark is not intended to suggest that Mac Lane formulated the central ideas of category theory as a continuation, or as an extension of Ore's program. Any categorical formulation of the leading ideas of Ore's program was not included by Mac Lane and Eilenberg in their articles, even as an example. However, it is worth pointing out that both category theory and Ore's theory of **structures** approach a similar metamathematical question from a similar perspective, and that at least one of the founders of category theory was well-aware of the basic conceptions of Ore's program, its limitations and its prospects.

Category theory and Ore's theory of **structures** provide two separate frameworks to formalize the non-formal notion that mathematics (or parts of it) deals with structures. It was seen that Ore took a single feature characteristic of the structural approach and attempted to build around it a full abstract foundation for algebra. Category theory proposed a broader perspective from the outset. The question arises, however, whether indeed category theory was able to account for all the aspects of the idea of structure, or in other words, whether category theory could provide a full formalization of the notion of

---

8. Mac Lane 1934. The research itself was never published. Mac Lane 1943, however, is a two-pages reply to an article of Ore on **structures** and the Jordan-Hölder theorem (Ore 1943b). See above § 6.3.

structure. Moreover, one may ask, how is this formalization related to Bour-baki's concept of *structure*. These questions are considered below in §§ 8.4-8.5.

Another issue which will be considered below (§ 8.4.), and which arises as soon as a category is defined, is the issue of the foundational problems related to the theory of sets. When defining the category of groups, one must consider the collection of all groups. Similarly, one considers the collection of all topological spaces, the collection of all sets, etc. The paradoxes of set theory appear in all these cases. Of course, this problem was mentioned by Mac Lane and Eilenberg already in their first article, and it continued to appear in several later articles. In their 1945 article, Mac Lane and Eilenberg suggested that this problem may be overcome by choosing reduced domains of discourse. One suggested alternative consisted of taking "*a* category of groups" (i.e., a certain legitimate collection of groups), rather than "*the* category of groups." A different suggestion was to adopt the unramified theory of types, and talking about a category of "groups of order *m*." A third suggestion was to adopt an axiomatic definition of sets, or simply to limit the cardinals of the objects involved in the categories. The authors enumerated the advantages and shortcomings of each of these suggestions, but postponed the definite solution of the problem to a future opportunity.

Before discussing some further milestones of the initial stages of the development of category theory in the next section, we conclude the present one by pointing out that the term "category" had been used, prior to Eilenberg and Mac Lane's 1945 article, in a different, though closely related mathematical context of which Eilenberg was well-aware. In fact, during the late 1920s and early 1930s the Russian mathematicians L.A. Lusternik and Lew Schnirelmann had applied topological methods to the calculus of variations, in order to investigate the existence of geodesics on an arbitrary surface of the topological type of the sphere.[9] In this context they introduced the term "category" to denote the cardinal of a certain partition of a space into closed, connected subspaces. This concept was later re-elaborated by the Polish mathematician Karol Borsuk (1905-1982), with whom Eilenberg was in close relationship, and with whom later he published several papers. One of Borsuk's articles on "categories"[10] appeared in the same issue of *Fundamenta Mathematicae* in

9. An account of their work appears in Lusternik & Schnirelmann 1934.
10. Borsuk 1936.

which Eilenberg's doctoral dissertation, submitted to the University of Warsaw, was published.[11] In 1941 the American topologist Ralph H. Fox still used the term in the same context and with the same meaning.[12]

8.3   Category Theory: Early Works

The first article of Eilenberg and Mac Lane on categories and functors was not followed by an intensive development of the theory. On the contrary, even the authors did not divert themselves from their current research interests in order to develop the ideas involved in that article; about five years were needed before categories and functors began to be considered as an object of mathematical interest deserving specific research. The present section deals with the first works in which categorical ideas were further developed. This discussion also brings to the fore the connections between the early development of the theory and the developments discussed above in Chapters 6 and 7.

In a conference in honor of Marshall Stone in 1970, Mac Lane delivered a lecture on the contributions of Stone to the origins of category theory.[13] Mac Lane pointed out that algebraic research conducted in the 1930s emphasized, above all, the subobjects and the quotient objects of the different domains investigated. Recently, according to Mac Lane, a particular concern with morphisms and their properties had increasingly arisen as an additional focus of study. This concern had developed only since 1939 and, in Mac Lane's view, the work of Marshal Stone on Boolean algebras—discussed above in § 6.5— was instrumental in bringing about this development.

In his historical account, Mac Lane claimed that Stone's ideas—emphasizing as they did the mathematical significance of the passage from one domain to a different one—provided a meaningful instance of one of the leading principles of category theory.[14] That such passages are possible, and that they provide important mathematical information, was well known before Stone. However, before Stone, such passages had been attempted in order to solve specific problems that had arisen in a particular domain. Moreover, they had relied on techniques provided by a second particular domain. Thus, for

---

11. Eilenberg 1936. Eilenberg also published an article (1936a) dealing with issues closely connected to those discussed in Borsuk 1936.

12. In Fox 1941.

13. Mac Lane 1970.

14. Mac Lane 1970, 229.

instance, early research in algebraic topology had been based on associating a specific kind of group to every topological space in order to gain, through our knowledge of the former, new information about the latter. The very correspondence between both domains—according to Mac Lane—had never before been considered for its own sake. Moreover, Mac Lane identified in the details of Stone's proofs particular cases of several central categorical concepts, such as functor and equivalence of categories. And, as a matter of fact, Stone had not limited himself to point out the equivalence between the elements of a Boolean algebra and those of its corresponding topological space, but he had also stressed the correspondence among functions in both domains. Thus he proved that to every homomorphism between Boolean algebras there corresponds a continuous function (in the opposite direction) between their assigned topological spaces. Stone was far from proposing a categorical line of inquiry, but he did provide an example of the implicit use of the concept of contravariant functor. Likewise the double passage, from the algebra to the topology and from the topology to the algebra was—in Mac Lane's view—a non-trivial instance of "functorial equivalence."

Without denying either the innovative character of Stone's ideas or the importance of his achievements, it may be useful to analyze the above assessment of Mac Lane within a broader context. What led Eilenberg and Mac Lane to define categories and functors was the study of the mathematical phenomenon of naturality. This phenomenon manifests itself mainly within a given category and not in the passage from one category to another. This was the case in the examples introduced by Eilenberg and Mac Lane themselves: double-dual vectorial spaces and double group of characters. Naturality had also appeared in their homological research, where very specific passages, from topology to algebra, were considered. It is doubtful whether in the early stages of category theory the connection between passages such as those considered by Eilenberg and Mac Lane, on the one hand, and those considered by Stone, on the other, was evident. Moreover, if Stone's work was accorded a very high mathematical value soon after its publication, it was due to its deep and comprehensive account of the theory of Boolean algebras. Mac Lane's appraisal of the importance of passages from one domain to another, such as appeared in Stone's work, and their identification as a peculiar, meaningful mathematical procedure worthy of research for its own sake was a result of hindsight that only the vantage point of the concept of functor could afford.

But it is also important to remark that the connections between algebra and topology, and the possibility of establishing a correspondence between two apparently distant mathematical theories, in the particular case of Stone's work, had originated in research on lattices. In fact, Stone had realized and stressed the connection of the theory of Boolean algebras with modern algebra at large, by focusing on the properties of those algebras when seen as lattices. In doing so, he was underscoring the role of lattices as a unifying concept in algebra. Stone had even defined the topology to which a Boolean algebra is equivalent, on the basis of properties satisfied by the algebra when considered as a lattice, rather than when considered as a Boolean ring. The relationship between Boolean rings and topology is built on the lattice-theoretical properties common to both domains. It is remarkable that Mac Lane did not mention this fact in his article on Stone, and limited himself to discussing the connection between topology and algebra in general terms. Lattice theory, and in particular the unifying perspective initiated by Ore, played an important role in the consolidation of the images of mathematics within which the theory of categories arose. Ore's program indicated that it is possible to attain meaningful results by abstracting from the mathematical domains considered. It indicated as well a specific direction for realizing that further abstraction (namely, the overlooking of the elements of the domains). Stone's work on Boolean algebras implied that Ore's program also provided a direct motivation for certain works in which proto-categorical ideas were successfully implemented. Thus, while Ore's program of **structures** virtually disappeared after 1942, one should not conclude that it never provided useful ideas which left their imprint on posterior, significant developments in twentieth-century mathematics. The program was abandoned at about the same time as the publication of the first articles on categories, and it is not unlikely that the latter event had some bearings on the former, though no direct evidence exists to confirm this conjecture.

Parallel to the relation between **structures** and categories, one may also wonder about the relationship between the rise of category theory and the development of Bourbaki's theory of *structures*. In this regard, plenty of unpublished evidence does exist indicating that an internal debate was held among Bourbaki's members on the issue of whether or not to adopt the categorical point of view, and how this could be done within the existing framework of the *Eléments*. This unpublished evidence is examined below in § 8.5. As a matter of fact, it is not even necessary to resort to unpublished evidence in order to document the existence of this internal debate. Echoes of the debate come to the fore in published material, in which one can detect an underlying attempt to exploit the advantages of categorical thinking without explicitly introducing categorical concepts and without giving up the *structural* framework. The following is a noteworthy example of this.

In the opening section of the present chapter the study of "Universal constructions" was pointed out as a mathematical situation suitable for categorical formulation. Individual examples of universal constructs were known before the rise of category theory, but it was only the categorical language which enabled their clear recognition and exact definition. At the same time, Bourbaki addressed the problem of universal mappings and even proved that this problem has essentially a single solution (§ 7.3.1). Therefore, universal constructions seem to offer a good example in which the connections between *structures* and categories may be analyzed.

The ideas contained in Bourbaki's section on universal constructions were first formulated in an article by Pierre Samuel,[15] a second-generation Bourbaki member. In his article Samuel attempted to present a single generalized formulation of a problem that had arisen in diverse branches of mathematics. This problem could therefore be seen as a real test for the usefulness of *structures* and related concepts. But what Samuel actually did in addressing the problem was to adopt a categorical perspective, albeit couched in Bourbakian language. Although the words category or functor do not appear in the article the emphasis throughout is on morphisms as the key concept in elucidating the problem. Samuel's article, then, underscores the limitations of *structures*, rather than showing any of its advantages.

The central focus of interest in Samuel's article are the free topological groups, a construct that had been considered previously by several mathematicians. Samuel believed that previous treatments of the issue had been excessively cumbersome, and that he would be able to provide a simpler formulation by considering their "universal properties."[16] Since he wanted to demonstrate the applicability of the concept of "universal construction" to situations in other fields of mathematics, Samuel tried to connect these ideas to Bourbaki's *structures*. Nevertheless, although Bourbaki's concepts are indeed

---

15. Samuel 1948.

16. Previous works on free groups which Samuel mentions are Kakutani 1944; Markoff 1942; and Nakayama 1943.

formally applicable to the situations considered by Samuel, they do not positively contribute to the generalization and solution of those problems.

Samuel first described the universal properties of the Cartesian product. He pointed out the existence of projections, which he defined in terms that strongly suggest categorical formulations, such as exposed in the opening section of the present chapter. Nevertheless, he did not explicitly mention categories or functors. Samuel defined a *T*-mapping as a mapping between two *structures* of the same type *T* which must also satisfy a number of conditions. These conditions happen to be exactly the same conditions that are imposed on morphisms in the axioms of category theory. An *ST*-mapping is a mapping sending an *S-structure* into a *T-structure,* and which satisfies some further conditions. Again, these conditions closely resemble the axioms defining functors in category theory. On the contrary, this whole emphasis on morphisms was in no sense characteristic of Bourbaki's theory of *structures*, neither as announced in the *Fascicule*, nor in its published form of 1956.

Samuel opened his article with a brief account of the classical examples of universal constructions: the field of fractions of an integral domain, free products, completion of uniform spaces, Cech compactification, etc. All these examples appeared again in the corresponding section of *Theory of Sets* (first published in 1957). In that section, one also finds additional examples of the universal mapping problem, which appear in greater detail in other books of the treatise. However, from the point of view of the internal organization of *Theory of Sets*, the natural connection between this section dealing with universal constructions and those preceding it is far from obvious. Although the problem is stated there in structural terms, as it happens with other issues in the treatise, one does not find even a cursory verification that the mentioned examples do, in fact, satisfy the formal conditions of the definition. Neither can one find such a verification in the corresponding sections of those books of the treatise in which the particular examples are developed. Thus, this section on universal mappings is but one further instance of the *ad-hoc* fashion in which the theory of *structures* is formulated and applied in the *Eléments*. In 1948, Samuel referred the reader to Bourbaki's forthcoming book on set theory as a reference for the precise definitions of some *structural* concepts that appeared in his article. These concepts do not play any actual role in the latter, but Samuel expected that they would eventually do so in Bourbaki's planned volume. It is very likely that these concepts had been discussed in meetings of Bourbaki that he had attended, and that he was therefore certain that the chapter on *structures* was soon to appear, thus providing the necessary concepts for his article. When that chapter finally appeared, the ideas developed in it were less useful than Samuel had envisaged in 1948. It is noteworthy that the section of the *Eléments* in which free topological groups are discussed happens to be one of the few places in which Bourbaki refers to a work outside the treatise. Bourbaki cited there Samuel's 1948 article and claimed that the reader may use the results appearing in the article in order to verify that free topological groups constitute an instance of a universal construction. On the other hand, Bourbaki's chapter of *General Topology,* in which topological groups are discussed does not mention free topological groups at all, even in the many exercises proposed at the end of the chapter.

All these considerations clearly attest for a many-sided conceptual interrelation between categorical and *structural* ideas within Samuel's work. As will be seen in the next few pages, Mac Lane was well-aware of these interrelations when he began to develop his own early contributions to the theory of categories.

At about the same time of publication of Samuel's article Mac Lane was working on related issues. Those issues appeared in a long article in 1950, which was preceded by a research announcement, and which constitute one of the first contributions to the development of category theory.[17] Mac Lane had also reviewed Samuel's article and had indicated the existence of a gap in Samuel's proof and the need to add some further assumptions on the ST-mappings in order to make the proof valid.[18] Mac Lane reviewed Samuel's article against the background of his own work on the same problem, relying fully on the categorical perspective.

Mac Lane's own article of 1950 bore the name "Duality in Groups" and it attempted to characterize duality phenomena in categorical terms. There are many situations in group theory which may be paired into mutually dual cases: G as the domain of a morphism and G as the codomain of a morphism; S as a subgroup of G and S as a quotient group of G; monomorphisms and epimorphisms; etc. Any statement of group theory yields a dual statement by interchanging dual concepts. However, the dual statement of a true statement is not necessarily true. Take for instance the following theorem: If S is a subgroup of

---

17. The article appears in Mac Lane 1950, while the announcement appears in Mac Lane 1948a.

18. See Mac Lane 1948b.

the quotient group G/N and M is a subgroup of G containing N as a normal subgroup, then S is in fact a quotient group M/N. The dual statement reads as follows: If S is a subgroup of G, and N is a normal subgroup of S, then the quotient S/N is a subgroup of the quotient G/N. This last assertion is, in general, not true, since the subgroup N is not necessarily normal in G. How can one characterize those statements which are true if and only if their duals are also true? This is the question addressed by Mac Lane in his article. He claimed that the required characterization should be given, not in terms of the elements of the groups, but rather in terms of their subgroups and their morphisms. In other words, Mac Lane proposed to address this question strictly in categorical terms. By doing that, Mac Lane was publishing the first article in which categories were used to solve a substantive mathematical problem.

But before describing the categorical contents of Mac Lane's article, it is necessary to point out how ideas that had earlier been part of Ore's program appear once again in the background of this development. The phenomena of duality, as was already seen, lie at the heart of the development of lattice theory from Dedekind's work on and, in particular, in Ore's program. Mac Lane himself saw the phenomena of duality not as one concerning the elements of the groups, but rather the interrelations among the subgroups of the group. The categorical perspective adopted by Mac Lane accounts for the stress laid in his article upon homomorphisms. But the combined interest in duality phenomena and in results which may be proven by overlooking the elements of the relevant domains had a forerunner in the works of Ore. Moreover, several specific phenomena considered in Mac Lane's article are directly connected with factorization properties and chain conditions, i.e., to those issues which constituted the main kinds of problems addressed by Ore. Thus, for instance, Mac Lane defined the dual of a composition series, "chief series", by taking a sequence of quotients instead of a sequence of subgroups. Mac Lane showed that a dual version of the Jordan-Hölder theorem holds for the chief series. As a matter of fact, in his research announcement of 1948, Mac Lane explicitly mentioned the fact that duality phenomena in group theory had become manifest, to a great extent, through lattice-theory. Nevertheless—he said—lattice theory can only partially account for this duality: there are several theorems of group theory whose lattice-dual formulations are not true. As an example Mac Lane mentioned a version of the Jordan-Hölder theorem which is valid under certain ascending chain conditions, but whose version for the corresponding descending chain fails to be true. In his article of 1950, Mac Lane no longer mentioned lattice-theory.

Let us return now to the categorical contents of Mac Lane's article. The example by means of which Mac Lane connected the phenomena of duality to categorical ideas is that of Cartesian and free products. In these examples one sees the connection between universal mappings, duality and morphisms, by means of the tool afforded by commutative diagrams. In such diagrams duality is expressed by changing the directions of the arrows involved.

Thus, the Cartesian product is characterized by the diagram:

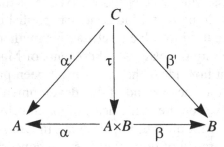

together with the demand that given a group C and morphisms α',β' as in the diagram, there exists a unique morphism $\tau: C \to A \times B$ such that the diagram commutes. The free product $A*B$ is defined as the dual of the Cartesian product, and is therefore defined by means of the dual diagram:

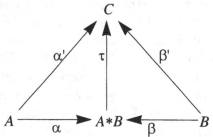

together with the demand that given a group C and morphisms α',β' as in the diagram, there exists a unique morphism $\tau: A*B \to C$ such that the diagram commutes.

Of course, there is nothing peculiar to groups in the above definitions, and therefore, the definition is general and may be used in an abstract category. Mac Lane thus raised the question: for which categories do the above constructs exist in the general case? Though he did not address here the issue in all its generality, Mac Lane proved that Cartesian and free products exist in the case of groups. The proof of the existence of those products involves the use of particular elements of the groups, and therefore Mac Lane claimed that this is not a proof of reciprocal duality. On the contrary, his proof of the uniqueness of the products was formulated strictly in categorical terms.[19]

As a further important example of a duality phenomenon, Mac Lane mentioned the axiomatic formulation of homology and cohomology which Samuel Eilenberg and Norman Steenrod were working on by that time, and which was eventually published in their joint book of 1952. Mac Lane was well-acquainted with the details of the manuscript being prepared by his colleagues, and in his own article he explained that in the formulations of Eilenberg and Steenrod, the axioms do not refer to the elements of the homology groups but only to certain homomorphisms among them. The axioms defining the homology, concluded Mac Lane, are dual to those defining the cohomology.[20] According to his own testimony, the concept of Abelian category, which was introduced in this article for the first time, originated at the time when Eilenberg and Steenrod were writing their book. The usual definition of homology could be reformulated in categorical terms, by defining the latter as a functor from a given category into that of Abelian groups. The question now arose whether an abstract categorical characterization may be given of those categories on which homologies can be defined. In Mac Lane's words:

> This study arose from a discussion with Sammy [Eilenberg] and my wish to do Eilenberg-Steenrod axiomatic homology theory with values not just abelian groups but objects of a more general abelian category.[21]

The aim of the article was the formulation of an axiomatic framework within which the phenomena of duality might be properly elucidated. Mac Lane formulated the axioms of a bicategory, namely, a category in which two particular, basic kinds of morphisms may be singled out: projections and injections. The axioms defining the bicategories are similar to those of the categories, with some additional special conditions imposed on the projections

---

19. Mac Lane 1950, 490.
20. Mac Lane 1950, 494.
21. Mac Lane 1976a, 136. See also Mac Lane 1971, 205. It should be noticed, however, that, as we shall see below, the concept defined in 1950 by Mac Lane was slightly different from what is known today as an Abelian category.

and the injections. Since the axioms for categories happen to be self-dual, the following main duality principle may be easily deduced:

> If any statement about a category is deducible from the axioms for a category, the dual statement is likewise deducible.[22]

The two additional primitive terms of bicategories, projections and injections, are mutually dual, and therefore, the dual of a statement about bicategories is obtained by interchanging those two terms.

But beyond a precise characterization of duality phenomena, Mac Lane expressed his belief that bicategories may provide a generalizing conceptual framework for mathematics, better than the existing ones:

> The bicategory language appears to be the appropriate vehicle for many of the theorems of universal algebra ... often giving simpler formulations, because the axiomatic formulation avoids the inevitable cumbersome explicit description of the general form of any algebraic or mathematical system. This is especially the case when universal algebra is extended to include those algebraic systems, which occur so frequently, in which several groups, homomorphisms, functions, and so forth, together constitute a single algebraic system. (Mac Lane 1950, 503)

Mac Lane cited here several works included under the heading of "universal algebra", and which the categorical language was supposed to supersede; these include Bourbaki's work, but also works of Birkhoff, Jonsson and Tarski.[23] Although this is not explicitly stated, it is clear that Mac Lane was comparing the categorical language, above all, with Bourbaki's theory, which had been alluded to in Samuel's article published in the same journal two years before, and was reviewed by Mac Lane himself.

Mac Lane dedicated a special section to show that the bicategorical language allows for a generalization of all those results of universal algebra previously obtained in the framework of lattice theory. In assessing the generalizing potential of categorical ideas Mac Lane compared it again to those previous attempts in algebra of which Ore's program was also a part. Moreover, Mac Lane even defined a concept, the *lattice-ordered category*, in which both conceptions, the categorical and the lattice-theoretical are mixed. In his view this concept represented:

> ... a simultaneous generalization of the notions of group and lattice; specifically, every group is a lattice ordered category, in which every mapping is an equiva-

---

22. Mac Lane 1950, 498.
23. Specifically Birkhoff 1935; 1948; Jonsson & Tarski 1947; and Tarski 1949.

lence and there is only one identity, and every lattice is a lattice ordered category in which every mapping is an injection. (Mac Lane 1950, 504)

After proving the basic results on bicategories Mac Lane introduced "Abelian bicategories", which are defined by a system of self-dual axioms. Mac Lane claimed, finally, that all standard properties of the homomorphisms among Abelian groups (as well as those of the linear mappings among vector spaces) are derivable in this system. As a matter of fact, this claim was an overstatement that was not fully justified on grounds of later developments, yet the idea of duality and of Abelian categories became central to the future evolution of category theory.[24] He believed that the ideas contained in this article were important enough to be presented in a one hour lecture delivered to the AMS, but there was almost no immediate reaction to it. A reviewer of the article described it simply as an "expanded version of an earlier note."[25]

The contribution of Eilenberg to the development of category theory in its initial stages was manifest in the publication of two books in which the unifying power of the categorical language was fully exploited for the first time. The two books quickly became classics, each in its own field: *Foundations of Algebraic Topology* (1952), written in collaboration with Norman Steenrod, and *Homological Algebra* (1956), in collaboration with Henri Cartan.[26]

Norman Steenrod (1910-1971) was among the first mathematicians to enthusiastically salute the appearance of the categorical language and he was eager to test its power in the precise formulation and solution of substantive open problems.[27] As early as 1945 Eilenberg and Steenrod published a research announcement on their planned axiomatic research of homological theories. The dominant approach to algebraic topology before Eilenberg and Steenrod's work required a fatiguing reliance on considerations drawn from analytic geometry.[28] Under the approach advanced by Eilenberg and Steenrod

---

24. See Mac Lane 1971, Chp. 2, for a more detailed technical exposition of these ideas.

25. See Mac Lane's own account in Mac Lane 1988, 339.

26. A detailed account of Eilenberg's overall contribution to the theory of categories appears in Mac Lane 1976a.

27. As a curiosity, I may mention here an e-mail message circulated among categorists in 1996, concerning Steenrod's axiomatization of homology, in which Eilenberg testifies that Steenrod had said to him, upon seeing Eilenberg and Mac Lane 1942, that "no paper had ever influenced his thinking more." At the same time, according to Eilenberg, the algebraic topologist P.A. Smith told him that "he had never read a more trivial paper in his life." I thank Colin Mclarty for communicating this message to me.

all this redundant work could be avoided. Using an axiomatic definition, it became possible to define the homological concepts directly, and to address the central problems of algebraic topology in a more straightforward fashion.

Given two topological spaces X, Y and two closed sets in them A, B respectively, it is possible to define the q-dimensional relative homology group of X mod A, $H_q(X, A)$, for every integer q, in such a way that for every continuous function

$$f: (X, A) \to (Y, B)$$

one defines—provided $f(A) \subseteq B$—a unique "induced group homomorphism"

$$f_*: H_q(X, A) \to H_q(Y, B)$$

satisfying certain simple conditions, similar to those imposed on functors as they had been defined in earlier articles on category theory. Although Eilenberg and Steenrod did not explicitly mention categorical concepts in their early announcement, their formulation of the homology groups of a given topological space closely followed the definition of a functor from *Top* (the category of topological spaces) to *Ab* (the category of Abelian groups), and could easily be translated into purely categorical terms.

The research announced by Eilenberg and Steenrod was never published as a finished article, but the ideas developed by them circulated among algebraic topologists and soon became widely used.[29] These ideas appeared finally in print in the form of a book in 1952, becoming the standard approach to the field. Besides the specific axiomatic definitions and techniques introduced into algebraic topology through the categorical approach, the use of arrow diagrams as a central mathematical tool was adopted here for the first time in the exposition of a full-fledged theory. The two most useful properties to be checked in a diagram are *commutativity*, (i.e., that in a given diagram, any two paths connecting different vertices yield one and the same morphism) and *exactness* (i.e., that the image of a morphism entering a vertex equals the kernel[30] of the morphism leaving that vertex). On the use of diagrams the authors wrote:

> The diagrams incorporate a large amount of information. Their use provides extensive savings in space and in mental effort ... in the case of many theorems,

28. A standard pre-categorical formulation of algebraic topology appears in Lefschetz 1942, especially chapter 3.

29. See, e.g., Kelley & Pitcher 1947, 687.

the setting up of the correct diagram is the major part of the proofs. (Eilenberg & Steenrod 1952, *xi*)

The whole book is actually built on proofs, at the heart of which lie commutative and exact diagrams. The authors suggested that the reader's degree of mastery of the new techniques introduced in the book, may be measured according to his mastery of the technique of diagrams.

Also the ideas presented in Cartan and Eilenberg's book on homological algebra were known to the practitioners of the discipline well before the book was actually published in 1956. Homological algebra, a mathematical discipline which began in the early 1940s, deals with the application of homological techniques to different branches of algebra (e.g., groups, associative algebras, Lie algebras), on the one hand, and with the algebraic elucidation of homological concepts, on the other. Cartan and Eilenberg's book provided a comprehensive synthesis of much work that had been done over the preceding decades on the two above-mentioned aspects of the discipline. This synthesis was made possible by the systematic application of a handful of central concepts, such as complex and projective resolution. These concepts had originally been formulated, in a rather intricate fashion, in terms of modules and morphisms of modules. However, they happen to be much more simply defined in categorical terms, using the notion of a derived functor. In their book, categories and functors were already taken for granted and they were therefore given no separate consideration on their own. The reader in need of further information about categories and functors was referred directly to Eilenberg and Mac Lane's original article of 1945. Some additional results concerning derived functors, which did not appear in Cartan-Eilenberg's book, appeared, shortly before its publication, in an article by the Japanese mathematician Nobuo Yoneda (1930-1996) including an important lemma that has become associated with his name.[31] Yoneda had worked for a while with Eilenberg, and he was among the first to realize the importance of the

---

30. In any algebraic theory the kernel of a morphism $f: A \rightarrow B$ is the set of all elements $a \in A$ such that $f(a) = 0$, where 0 denotes the zero element of B. Since in the categorical language there is no reference to the specific elements of the objects, it is necessary to find an alternative definition for kernel. Abelian categories were introduced in order to allow for the definition, within the categorical framework, of concepts such as kernels, exact sequences, etc.

31. Yoneda 1954. For a later formulation of the lemma and its importance, see, e.g., Mac Lane 1971, 66. See also Kinoshita 1998; Mac Lane 1998.s

new perspective afforded by Cartan-Eilenberg's book and to apply it for his own research.

In a detailed review of Cartan and Eilenberg's book written in 1956, Mac Lane discussed the constant tension between the desire to produce generalizing conceptual frameworks in modern mathematics and the impossibility of doing this successfully. Thus Mac Lane wrote:

> The authors' approach in this book can best be described in philosophical terms as monistic: everything is unified... Perhaps mathematics now moves so fast— and in part because of vigorous unifying contributions such as that of this book— that no unification of mathematics can be up-to-date. (Mac Lane 1956, 622)

The joint works of Eilenberg and Mac Lane, as well as their separate contributions, are among the most salient examples of these opposed tendencies of mathematics: many important concepts developed by them provided unifying ideas with important consequences for mathematics. On the other hand, the richness they recurrently found in their newly introduced theories underlined the impossibility of pushing the whole of mathematics into a unique narrow conceptual framework.

8.4   Category Theory: Some Contributions

After the contributions of Mac Lane and Eilenberg to the first stages of category theory, the next milestone in the development of the theory was the publication of a handful of articles which transformed the theory into an autonomous research field by the end of the fifties. First, the definition and study of Abelian categories in separate articles by Alexander Grothendieck and David Buchsbaum. Second, the introduction of adjoint functors by Daniel Kan. Third, and somewhat later, the first foundational attempts based on categorical ideas. These contributions are briefly discussed in the present section.

In 1950 Mac Lane had defined bicategories in an attempt to find a natural framework in which to discuss duality phenomena. Aware of the advances of Eilenberg and his collaborators in defining general homological theories, Mac Lane believed that bicategories might also contribute to that task. However, ideas introduced by Mac Lane in his study of bicategories were first used to provide general definitions of homology only in the works of Grothendieck and Buchsbaum.[32]

Buchsbaum's ideas appeared in an article of 1955 and also as an appendix (without proofs) to the book of Cartan and Eilenberg.[33] They were thus read by a relatively broad audience. Buchsbaum saw in his definition of "exact categories" the basis for the conceptual foundation that Mac Lane had previously sought for Cartan-Eilenberg's general theory of homology. The Cartan-Eilenberg axioms defined homologies on functors which sent categories of modules over a given ring into categories of modules over a different ring. The dual of these axioms, obtained by inverting the direction of all arrows in the diagrams, yield the axioms of cohomology. Given a category A, they denoted by A^* the dual category of A, similar to A in most respects but whose diagrams are obtained by inverting the arrows in the diagrams of A. An exact category is a category A for which all statements in A are true if and only if their duals are true in A^*. If one translates the axioms defining the homology on A into their duals one then obtains the cohomologies for A^*. Thus the concept of exact category provides a more general framework to study homological theories. Rather than having to restrict attention to categories of modules, exact categories allow for the treatment of homological concepts—or, more generally, for the treatment of additive functors—over abstract categories satisfying certain conditions.

Ideas similar to those of Buchsbaum were developed, in a more comprehensive fashion, by Alexander Grothendieck. Grothendieck had started his research career in functional analysis by studying duality phenomena in vectorial topological spaces. Already the publication of his thesis on tensorial topological products in the early fifties brought him international recognition,[34] and he continued to publish many more important works in this field between 1950 and 1955. Besides, since 1954 Grothendieck began to develop ideas of homological algebra that had arisen within his previous work on algebraic geometry.[35]

In 1957 Grothendieck published an article which is considered one of the most important stages in the early development of category theory.[36] The first part of the article coincides, to a large extent, with Buchsbaum's work,[37] and it builds on the central concepts of an "additive category" and of an "Abelian category" (which is equivalent to Buchsbaum's exact categories); Grothend-

---

32. Mac Lane himself (1971, 205) has described his ideas of 1950 as a "clumsy" prelude to the development of Abelian categories. See also Hilton 1982, 81.
33. Cartan & Eilenberg 1956.
34. See, e.g., Grothendieck 1952.
35. For an account of Grothendieck's early works see Grothendieck & Dieudonné 1968.
36. See Grothendieck 1957.

ieck showed the importance of these concepts by proving several meaningful theorems. Without going too much into technicalities, an additive category may be characterized as one in which the set of morphisms between any two objects of the category may be endowed with the structure of an Abelian group, whereas an Abelian category is an additive one, in which all morphisms satisfy some additional properties of factorization into monics and epics. The best known result proven by Grothendieck in his article establishes the necessary conditions for the application of homological techniques in an Abelian category. One of these conditions includes the existence of a certain distinguished object within the category, capable of "separating the points" of it. The second condition accounts for the existence of "enough" injective objects in the category.

Grothendieck's article provided an important application of categorical ideas in the then incipient domain of sheaf theory, since sheaves of modules happen to constitute an Abelian category. Grothendieck's result, then, allowed the definition of very general cohomologies on a sheaf. Many mathematicians saw in this result a definite argument for the adoption of the categorical approach, and especially of the concept of derived functor afforded by it, to the study of homological algebra.[38]

It is worth pointing out that Grothendieck was one of the most prominent mathematicians of his generation in France, and was also a member of Bourbaki. His above mentioned work exemplifies the odd situation in which several Bourbaki members have sometimes found themselves: in their individual research they used and developed categorical ideas, while in the framework of their collective work in Bourbaki, they totally refrained from the use of such concepts. Bourbaki's reaction to the beginnings of category theory is examined in greater detail in the next section.[39]

---

37. According to Mac Lane's testimony (1988, 339), when he first heard Grothendieck in 1955 lecturing on the issue, it became clear to him, that Grothendieck had worked out his ideas independently, while being totally unaware of Buchsbaum's work, and in fact of all earlier work on Abelian categories.

38. A full exposition of the theory of Abelian categories appears in Mac Lane 1971, Chp. 8. For a cursory account of the development of sheaf theory, see Gray 1979. See especially pp. 60-61, in connection with Grothendieck's early research.

39. For Grothendieck's and Serre's use of categories in their work on algebraic geometry, and the significance of these works on the further development of category theory see Mac Lane 1988, 339 ff.

The concept of adjoint functor was developed by Daniel Kan, apparently working in isolation from the mathematical community, while in his duty service in the Israeli army. In 1956 Samuel Eilenberg visited Israel and Kan was sent to meet him in order to show Eilenberg his results. Eilenberg was the ideal reader for this work and among the few who could appreciate its value, and he recommended to the mathematical institute of the Hebrew University in Jerusalem to confer upon Kan a Ph.D. degree.[40]

Two functors, $F: \Gamma \to \Omega$ and $G: \Omega \to \Gamma$, are called adjoint if and only if for every pair of objects C,D in Γ and Ω respectively, there exists a bijective mapping yielding the equivalence

$$\text{Hom}_\Gamma(C,G(D)) \cong \text{Hom}_\Omega(F(C),D)$$

That is, if the class of morphisms between C and $G(D)$ in the category Γ is in a one-to-one relation with the class of morphisms between $F(C)$ and D in the category Ω. Moreover, it is required that the above correspondence be a natural transformation.

In his accounts of the development of category theory, Mac Lane has claimed that several particular cases of adjoint functors were known well before Kan's definition of the general concept in categorical terms. That is the case, for instance, in Stone's work on Hilbert Spaces.[41] However, many years elapsed between the particular work on these examples and Kan's general definition, although categories were formulated already in 1945. Can this delay be sensibly explained? Mac Lane's own explanation is worth quoting here:

> Ideas about Hilbert space or universal constructions in general topology might have suggested adjoints, but they did not; perhaps the 1939-1945 war interrupted this development. During the next decade 1945-55 there were very few studies of categories, category theory was just a language, and possible workers may have been discouraged by the widespread pragmatic distrust of "general abstract nonsense" (category theory). Bourbaki just missed. (Mac Lane 1971, 103)

According to E.T. Bell's statistics, already quoted above, in 1938 a relatively large number of mathematicians were indeed interested in generalizing concepts, such as would be later provided by category theory. Bourbaki's attempts at formulating a unified conceptual framework for mathematics began at about the same time. Mac Lane believed that the particular examples

---

40. This information was communicated to me by Professor Eilenberg in a personal conversation. Kan's work was first published in Kan 1958.

41. See Mac Lane 1970, 230-232; 1971, 103; 1988, 330. See also Stone 1970.

studied by Bourbaki should have allowed the identification of the general concept of adjoint functor, but the unwieldiness of *structures* prevented this from coming about. Thus, the above quotation is followed by a reference to supplement III of Bourbaki's book on multilinear algebra and the following comment:

> [Bourbaki's] definition of universal construction was clumsy, because it avoided categorical language ... Bourbaki's idea of universal construction was devised to be so general as to include more—and in particular to include the ideas of multilinear algebra which were important to French Mathematical traditions. In retrospect, this added generality seems mistaken; Bourbaki's construction problem ... missed a basic discovery; this discovery was left to a younger man, perhaps one less beholden to tradition or fashion. Put differently, good general theory does not search for the maximum generality, but for the right generality. (*ibid.*)

Mac Lane added here some technical details to explain more precisely the mathematical nature of the connection between Bourbaki and categorical concepts. His conclusion is clear, even if it is not explicitly stated: in the attempt to provide a universal language able to address successfully problems arising in diverse branches of mathematics, category theory proved to be richer, more flexible, and more illuminating than Bourbaki's *structures*. Bourbaki's attempt was guided by a desire to cover *all* aspects of mathematical reality, while category theory chose to deal with what its creators believed to be the more relevant aspects of mathematics. However, even this more limited attempt proved to be only partially successful; as Mac Lane himself claimed elsewhere, the mathematical reality is much more varied than even a good generalizing theory like category theory can exhaust.[42]

In any event, it was the case that the clear and, as it turned out, very useful formulation of the idea of adjoints in categorical terms provided a further instance of the power and flexibility of the categorical language. When Peter Freyd—also a student of Eilenberg—proved the existence theorem for adjoint functors in 1963, they became a central motive of the categorical discourse.[43]

A further aspect of the early development of category theory was the categorical research on the foundations of mathematics, a direction initiated by F. William Lawvere. As mentioned above in § 8.3, Mac Lane and Eilenberg

---

42. Cf. Mac Lane 1987a, 406-407.
43. Freyd 1964.

had already pointed out in their first paper the difficulties brought about by the need to consider very large collections of objects when defining a category. Mac Lane and Eilenberg suggested several possible directions for the solution of this problem, but the issue was left open and continued to appear as unsolved in subsequent articles on category theory.[44] In 1959 Mac Lane published a further attempt to solving this problem, which was not really succesful. He reviewed the possible solutions that had been suggested by that time, by him and by other mathematicians. Grothendieck, for instance, had proposed the use of inaccessible cardinals in order to build "Universes" in which the categories of Abelian groups could be safely defined.[45]

The problematic connection between category theory and the foundations of mathematics was addressed from a strikingly new perspective in the early sixties in the work of Lawvere. Instead of searching for a proper set-theoretical environment within which the usual paradoxes would be avoided when defining categories, Lawvere proposed to take the categorical ideas as the most basic ones. Thus, he started from the axiomatically defined concept of the category of all categories as a foundation for mathematics at large. True to the essence of categorical thinking, such a concept would emphasize the interrelations among categories over the internal composition of the categories themselves. In particular, it would provide an account of the category of sets without relying on concepts such as element and membership. The article in which Lawvere proposed his program opened with the following words:

> In the mathematical development of recent decades one sees clearly the rise of the conviction that the relevant properties of mathematical objects are those which can be stated in terms of their abstract structure rather than in terms of the elements which the objects were thought to be made of. The question thus naturally arises whether one can give a foundation for mathematics which expresses wholeheartedly this conviction concerning what mathematics is about and in particular in which classes and membership do not play any role. (Lawvere 1966, 1)

---

44. Even in their book on algebraic topology, although Eilenberg and Steenrod did not directly develop the issue of categories in itself, the set-theoretic problem raised by category theory was mentioned as an unsolved problem. See Eilenberg & Steenrod 1952, 120-121.

45. Bourbaki (1972a) dealt with the concept of "Univers." It constitutes an attempt (not followed by any serious work in the same direction) to bridge the gap between Bourbakian and categorical concepts, while at the same time providing the appropriate framework to define categories and overcoming foundational problems.

In other words, Lawvere explicitly suggested to attempt a reflexive clarification of an image of mathematical knowledge. In order to pursue this reflexive task Lawvere formulated a first order axiomatic system defining the category of all categories.[46] In his opinion, it would be possible to derive from it all the objects of mathematics and their known properties. Lawvere's article was but a first step in a series of foundational researches, in which his original ideas were further pursued.[47] Perhaps the most important development following this train of ideas was the study of "elementary toposes." This concept had originated from the application of Grothendieck's ideas to algebraic geometry, but later it also became the main tool for all applications of categorical ideas to foundational and logical issues in mathematics, and it has come to constitute a main focus of interest of present research in categories.[48] However, the possibility to develop the whole of mathematics based on the concept of the category of all categories, or on any other similar categorical idea, remains a question still open to debate and in need of further elucidation.

From what was said above, one can now summarize the initial stages in the development of category theory and its relation to the attempts to provide a generalizing conceptual framework and language for all of mathematics. The formulation of the basic concepts of the theory was not directly motivated by the desire to formalize the non-formal idea of mathematical structure, as was the case with Bourbaki's *structures*, or to give a general foundation for abstract algebra, as was the case with Ore's **structures**. In their joint research, especially in homological issues, Mac Lane and Eilenberg faced a very specific question, namely, how to characterize in formal terms a recurrent mathematical phenomenon usually called "natural." This attempt necessitated a conceptual framework allowing the simultaneous consideration of *all morphisms* of a given mathematical domain. This led to the definition of categories and functors as an auxiliary tool for examining the phenomenon of "naturality." The metamathematical potentialities afforded by the categorical ideas were gradually realized through Mac Lane's own inquiry into the duality

---

46. See the review of Lawvere's article by Isbell (1966). Isbell also detected a minor mistake of Lawvere.

47. See Mac Lane 1988, 339 ff.

48. Mclarty 1990 presents a detailed account of the origins of topos theory. In particular, he argues strongly against the view that topos theory arose as an attempt "to find an axiomatic characterization of the usual category of sets." See also Mac Lane 1988, 352 ff.

phenomena in group theory, and through the attempt of Eilenberg and his associates to provide axiomatic formulations for homological theories. Thus, in a roundabout way the theory became an alternative metamathematical attempt to understand the structural nature of mathematics.

Ore's **structure** theory had exhausted itself already by the mid-forties and it was no longer an actual alternative in this issue. Nevertheless the theory remained, to a large extent, a model of a reflexive theory of structures, and a standard against which alternative theories could be confronted. Category theory shared with Ore's program the leading principle according to which a formalization of the idea of structure may be attained by overlooking the existence of elements in the mathematical domains which one is trying to axiomatize. Moreover, through its influence on Stone's work on Boolean algebra, it provided an instance of the advantages to be expected from inquiries of this kind into the connection of seemingly distant mathematical theories.

When category theory became an advantageous alternative to mathematicians looking for a unifying language for mathematics, Bourbaki members were faced with the need to reevaluate their own concepts vis-à-vis the categorical ones. The reactions of Bourbaki members oscillated between serious attempts to combine the *structural* and the categorical approaches and utter disregard.

Category theory made a substantial metamathematical contribution in providing a language of considerable unifying power for many branches of mathematics. However, since it necessitated the simultaneous consideration of large collections of mathematical objects of a similar kind, the theory raised metamathematical problems from a different direction, namely, the usual foundational problems of set theory. This confrontation led to the attempt to transform category theory into a viable foundational alternative for mathematics, an attempt initiated with Lawvere's work. This attempt turned out to be far from unproblematic, but nevertheless it was adopted by many researchers who saw in it a possibility of merging two central trends of twentieth century mathematics: foundationalism and structuralism. Foundationalists tried to present mathematics as a body of knowledge which may be established deductively, and with full certainty, from a reduced collection of initial assumptions. It was tempting to suppose that if the concept of set had not succeeded in providing that reduced set of assumptions, then it would be possible to find it in categories, which formalize the concept of structure and therefore does away with the problematic relation of membership.

8.5   Category Theory and Bourbaki

From the account presented in the preceding sections it seems safe to advance the conjecture that, in the mid-fifties, Bourbaki must have addressed the question of whether or not to adopt some or all of the tools afforded by the increasingly expanding language of category theory, in order to use them for his own purposes. This was clearly manifest in the above quoted promise issued by Bourbaki—as early as 1961, in the first edition of *Commutative Algebra*—to put together a volume on Abelian categories. That volume never appeared, however. That promise and the failure to fulfill it together suggest Bourbaki's ambivalence about the value of the language of category theory to his project. Categorical ideas also pervaded Samuel's work on Universal constructs, added as the concluding section of *Theory of Sets*. Finally, several members of the group, in their individual research-work, used and developed concepts of category theory.

In reeditions of certain volumes of the *Eléments* it is also possible to detect the increasing tendency to adopt categorical ideas without explicitly adopting the categorical language. This is particularly evident in the volume on algebra. Even a cursory examination of the tables of contents of the second edition (1964) and of the English translation (1973) of the third edition (1971) suffices to learn about significant changes in the issues addressed and in the organization of the material. When one checks the actual contents of the 1971 edition one finds a strong emphasis on properties of morphisms, which was absent from earlier editions of the book. One finds commutative diagrams and exact sequences throughout (especially in the chapter on linear algebra). One also finds the concept of *Hom(E,F)*, used in the same important problem-settings in which category theory had previously exploited its advantages. Moreover, when dealing with tensors (another important basic concept of categorical discourse), Bourbaki devoted a special section to describe the "functorial properties of the tensor algebra."[49] Although the terms category and functor are not mentioned here, the reader with even the most elementary background readily recognizes the categorical notions underlying all of this discussion.

Bourbaki explicitly mentioned category theory in the 1973 edition of his *Algebra*, but, remarkably enough, he did so only in a footnote. This appears in section 7 of Chapter 1, which deals with free monoids and free groups, an issue not considered in the 1964 edition. Free objects provide a typical example of

---

49. Bourbaki 1973, 484-489.

an issue which is more easily dealt with in the categorical framework, since they appear in several different mathematical contexts, and since they involve universal properties of certain sets of morphisms.[50] Bourbaki analyzed those universal properties in this section, giving also sporadic references to *Theory of Sets*. As an important example Bourbaki mentioned the free product of a family of groups, which in the case of two groups is denoted as G_1*G_2. In a footnote Bourbaki wrote:

> Note that G_1*G_2 is not the "product" of G_1 and G_2 in the sense of *Set Theory* IV, § 2, no. 4 (nor in the sense of the "theory of categories"; in the context of this theory, G_1*G_2 is the "sum" of G_1 and G_2). (Bourbaki 1973, 88)

Since category theory is not mentioned in any other part of the book, one wonders why Bourbaki found it necessary to underscore the relationship between his definition and that of category theory precisely at this point. At any rate, this footnote is a stylistic anomaly reflecting a peculiar reaction of Bourbaki, when confronted with admittedly categorical concepts.

There are thus several published instances which indicate that the question of the adoption of category theory remained open for Bourbaki. But, in fact, there is more evidence of this ongoing debate than just these few clues. Bourbaki's internal discussions about the convenience of adopting the categorical point of view are explicitly documented in several issues of "La Tribu"; in particular, one finds them encapsulated in a brief commentary to the effect that in one of the meetings of Bourbaki:

> L'on a remarqué: une violente offensive de virue [sic] functoriel... ("La Tribu" # 40 [October 7-14, 1956])

Structures were first mentioned in the 'Fascicule' of *Theory of Sets*, which was published in 1939. Yet the first full version of Chapter IV, including the first exposition of the theory of structures appeared only in 1957. By that time category theory had developed considerably and had reached the status of an independent discipline that allowed for generalized formulations of certain mathematical situations. The early developments of the categorical formulation—more flexible and effective than the one provided by *structures*—made Bourbaki's early hope of finding *the* single best formulation for all of mathematics questionable, and cast strong doubt on the initially intended universal-

---

50. Cf. Jacobson 1980, 43-44 & 87-89.

ity of Bourbaki's enterprise. Let us now see how these questions and doubts manifested themselves in Bourbaki's meetings.

In various issues of "La Tribu" one can find echoes of technical problems that arose in connection with the detailed treatment of *structural* concepts. In the following quotation, for instance, the connection of such technical problems with categorical ideas is suggested by the allusion to "covariance" and "contravariance", concepts that had recently arisen within the first works on category theory:

> Our main undertaking has been to give a sense to the notion of "representation" of a structure, as well as to the notions of induced, product, and quotient structure. Chevalley has read a secret paper defining (whenever it works) the notion of induced structure, and Sammy [Eilenberg] has tried to extend the procedure to the definition of quotient structure. Besides, Sammy has proposed to define a structure as a subset of the echelon set (rather than as an element), which seems to provide a good definition of the representations (modulo certain small complications of "variance", co- or contra-). If this works one thus easily gets the notion of induced, product and quotient structures (cf. Samuel's paper on "Universal mappings"). It was decided to try Sammy's system.[51]

Publication of a final version of the results announced in 1939 concerning *structures* was delayed, among other reasons, because of unsolved questions suggested by category theory, on the one hand, and by ideas such as developed by Samuel in his article, on the other. This is indicated by the following report of 1956 on the tentative contents of *Theory of Sets*:

> Chap. IV (Structures) A paper by Cartier shows that Samuel's results on inductive limits are particular cases of ultra-general trinkets on the commutation of universal problems. These trinkets are well-stated only within the framework of categories and functors. Cartier proposes a metamathematical method to introduce the latter without modifying our logical system. But this system is vomited [sic] since it turns its back on the extensional point of view ... It was therefore

---

51. "On s'est principalement occupé de donner un sens à la notion de "répresentation" d'une structure, ainsi qu'aux notions de structures induite, produit et quotient. Chevalley a lu un papier secret définissant (lorsque çà marche) la notion de structure induite, et Sammy [Eilenberg] a essayé d'étendre le procédé à la définition de la structure quotient. D'autre part Sammy a proposé de définir une structure comme partie d'un ensemble de l'échelle (et non plus come élément), ce qui a l'air de donner une bonne définition des représentations (modulo certaines petites complications de "variance" co ou contra); si çà marche, on a alors facilement les notions de structures induite, produit et quotient (cf. papier Samuel des "Universal mappings"). On décide d'essayer le système Sammy." (La Tribu", #28 [June 25 - July 8; 1952] pp. 7-8)

decided that it will be better to enlarge the system in order to make room for the categories. At first sight, Godel's [sic] system seems to be convenient. In order ... not to delay the publication of a chapter on which we have worked much, it is decided ... to send chapter IV to press without modifying the inductive limits, and introducing the slight modifications concerning strict solutions of universal problems.

Regarding categories and functors, we are finally convinced of their importance. Therefore: Chap. V (Categories and functors): To begin with, Grothendieck will write down a kind of Summary of Results in a naive style, so that Bourbaki may realize what can be usefully done with it. Later on, it will be formalized.[52]

Further evidence of the interplay between structural and categorical concepts is provided by the following letter of Weil to Chevalley, dated October 15, 1950, and distributed among the members of Bourbaki as an appendix to one of the issues of "La Tribu." Thus Weil wrote:

I have just received chapters 2 and 3 of Set theory ... Should the word "function" be reserved for mappings sending a set to the "universe", as you have done (in which case, with your axioms, the values of the function constitute themselves a set properly understood)? Or is it perhaps convenient to name "function" anything to which we attach a functional symbol, e.g., $P(E)$, $A{\times}B$, $A{\circ}B$ (tens. prod.) etc.? Obviously, "function" in the second sense will not be a mathematical object, but rather a metamathematical expression. This is undoubtedly the reason why there are people (without giving names ...) who use the word "functor." Should we accept this term? It seems that a word is needed for this notion....

Regarding the theory of structures, your chapter does illuminate the issue. However, we can hardly avoid going much farther than you have, and find out

---

52. "Chap. IV (Structures) - Un papier de Cartier montre que les résultats de Samuel sur les limites inductives sont des cas particuliers de fourbis ultra-généraux sur la commutation des problèmes universels. Ces fourbis ne s'énoncent bien que dans le cadre des catégories et foncteurs. Cartier propose une méthode métamathématique d'introduire ces dernieres sans modifier notre système logique. Mais ce système est vomi car il tourne résolument le dos au point de vue de l'extension, ... On décide donc qu'il vaut mieux élargir le système pour y faire rentrer les catégories; à première vue le système Godel semble convenir. Afin de n'être pas le cul entre deux chaises, et aussi afin de ne pas retarder la publication d'un chapitre sur lequel on a beaucoup travaillé, on décide (malgré le véto de Dixmier, retiré in extremis) d'envoyer le chap. IV à l'impression sans modifier les limites inductives, et en ajoutant les petites modifications relatives aux solutions strictes des problèmes universels. Quant aux catégories et foncteurs, on est finalement convaincus que c'est très important. D'où:

Chap. V (Catégories et foncteurs) - Pour commencer Grothendieck rédigera une espèce de Fascicule de Résultats en style naïf, afin que Bourbaki se rende compte de ce qu'il est utile de pouvoir faire. On formalisera ensuite." ("La Tribu", #39 [4 June - 7 July 1956])

whether or not it is possible to give some generality to the notions of induced structure, product structure, homomorphism. As you know, my honorable colleague Mac Lane claims that every notion of structure necessarily implies a notion of homomorphism, which consists in indicating for each data constituting the structure, those which behave covariantly and those which behave contravariantly ... What do you think can be gained from this kind of considerations?[53]

These quotations, and many other statements scattered around several issues of "La Tribu", indicate that by the mid-1950s Bourbaki was reconsidering options concerning the possibility of using a unified conceptual framework for the treatise. This reinforces the impression created by a direct reading of published material, namely that Chapter IV of *Theory of Sets*, in which Bourbaki's theory of structures is developed, appeared at a stage when it was already clear that the concept of *structure* could not fulfil the expectations initially attached to it, and that the alternative provided by categories as a generalizing mathematical notion, could already be seen as one at least as comprehensive as that of *structure* and probably more satisfactory.[54]

A further aspect of the interrelations between category theory and Bourbaki's theory of *structures* regards the peculiar contributions of Charles Ehres-

---

53. "Je viens de recevoir les chap. II-III des Ensembles. ... D'abord: faut-il réserver le mot "fonction" à une application d'un *ensemble* dans l'univers, comme tu as fait (auquel cas, avec tes axiomes, les valeurs de la fonction forment aussi un ensemble, bien entendu)? ou bien convient-il de nommer "fonction" tout ce à quoi on attache un symbole fonctionnel, e.g. $P(E)$, $A \times B$, $A \circ B$ (prod. tens.) etc.? Evidemment "fonction" dans le second sens ne serait pas un objet mathématique, mais un vocable métamathématique; c'est sans doute pourquoi il existe (je ne veux nommer personne ...) des gens qui disent "foncteur"; devons-nous accepter ce terme? Il semble qu'on ait besoin d'un mot pour cette notion. "Fonction" dans les deux sens aurait peut-être plus d'avantages que d'inconvénients. ...

Quant à celle-ci [la théorie des structures], ton chapitre débrouille bien la question; mais nous ne pouvons guère ne pas aller plus loin que tu n'as fait, et chercher s'il est possible de donner quelques généralité aux notions de structure induite, structure produit, homomorphisme. Comme tu sais, mon honorable collègue Mac Lane soutient que toute notion de structure comporte necessairement une notion d'homomorphisme, consistant à indiquer, pour chacune des données constituant la structure, celles qui se comportent de manière covariante et celles qui se comportent de manière contravariante ... Que penses-tu qu'il y ait à tirer de ce genre de considerations?"

54. See also Mac Lane's own testimony regarding this issue (Mac Lane 1988, 337): "Early in the 1950s some members of Bourbaki, seeing the promise of category theory, may have considered the possibility of using it as a context for the description of mathematical structure. It was about this time (1954) that I was invited to attend one of Bourbaki's private meetings ... Bourbaki did not then or later admit categories to their volumes; perhaps my command of the French language was inadequate to the task of persuasion."

mann (1905-1979). Ehresmann was one of the founding members of Bourbaki, and his main research fields included differential topology and differential geometry. In the early fifties Ehresmann became increasingly interested in category theory as a unifying framework for mathematics and began to develop his own idiosyncratic ideas on this issue. In the sixties, Ehresmann drew himself away from the leaders of the French mathematical community (which included, naturally, the members of Bourbaki). This estrangement seems to have been motivated, among other things, by his deviating from the mainstream of mathematical research in France and, in particular, by his departing from Bourbaki's conception and the increasing stress laid by him on the categorical approach.

Ehresmann's original ideas are based on a combination of *structural* and categorical concepts. At variance with Bourbaki's formulation, however, Ehresmann did obtain new substantive results from his use of these concepts. Ehresmann's theory was not just a language allowing a better reformulation of existing results, but also an effective research tool leading to the discovery and proof of new results. His collected works include more than 2500 pages, in eight volumes, and a significant portion of it is composed of the results of his categorical/*structural* theory.[55] A systematic exposition of Ehresmann's theory appeared as a textbook on categories.[56]

Although Ehresmann constantly gathered around him a group of students who actively learned and developed his ideas, these ideas seem to have been somewhat too abstruse to become very popular, and Ehresmann's book was far from reaching the wide audience attained by Bourbaki's works on the one hand, or by the mainstream developments of category theory, on the other.[57] It would be beyond the scope of the present book to analyze in detail Ehresmann's work and its relation to Bourbaki's theory of *structures* or to category

---

55. Most of the work on this field appear in Volume II, part 1, of his collected works (Ehresmann 1981). The first representative article is Ehresmann 1957.

56. Ehresmann 1965.

57. As a possible explanation for the reduced audience that showed any interest in these works, Andrée Ch. Ehresmann—the late Ehresmann's widow and a mathematician in her own right (who edited and published her late husband's work)—described them as "visionary." In the introductory comments to the third part of the collected works (Partie III-1, p. XIV) she wrote: "Ces écrits, nettement en avance sur leur temps (ce qui, joint à des notations peu orthodoxes, a sans doute un peu diminué leur audience) annoncent les développements ultérieurs de la Théorie des Catégories et surtout la théorie des structures internes à une catégorie, dont la richesse s'est révélée avec l'etude des topos."

theory. Relevant to the present discussion, however, are Mac Lane's opinions on these issues. According to Mac Lane, Ehresmann understood very soon that Bourbaki's conceptual framework was not flexible enough as to allow a proper elucidation of the idea of mathematical structure, and he therefore preferred the categorical approach, which afforded a better tool for dealing with a greater variety of mathematical fields. Bourbaki's concepts, claimed Mac Lane, define "mathematical structures" by taking an abstract set and appending to it an additional construct; in category theory there is no subordination of "mathematical structures" to sets, and this is the source of the supremacy of this theory over that of Bourbaki.[58] Irrespective of one's assessment of Mac Lane's claim as correct or incorrect, the very existence of Ehresmann's theory and the special link that it built with category theory, without completely discarding *structural* ideas, indicate an additional path which Bourbaki could have chosen in facing the issue of categories. Eventually, Bourbaki decided not to include categorical ideas in their work and to stick to his original formulation. Yet the alternative presented by Ehresmann and the kind of results attained in its framework underscores the limitations of the *structural* formulation as an effective metamathematical tool.

In view of the above, it is somewhat surprising to read Dieudonné's pronouncement in 1982 concerning Bourbaki and categories:

> One often hears people wondering why Bourbaki has not undertaken to publish a chapter on categories and functors. I think one of the reasons is the following: the parts of mathematics where those concepts are extremely useful, such as algebraic geometry and algebraic and differential topology, are among those which Bourbaki cannot contemplate including in the treatise ... For many other parts of mathematics, it is certainly possible to use the language of categories and functors, but they do not bring any simplification to the proofs, and even in homological algebra, one can entirely do without their use, which would amount to introducing extra terminology. (Dieudonné 1982, 662)[59]

This comment calls for qualifications on several accounts. First, contrary to Dieudonné's claim above, in *Commutative Algebra* Bourbaki explicitly acknowledged the gains that may be expected from the use of categorical

---

58. See Mac Lane 1980.

59. Similarly in Samuel's review of Bourbaki's *Algebra* (Samuel 1972) we read: "From the way in which ... the sections ... on tensor products and Universal Algebras are written, one may guess that Bourbaki is not planning, for the time being, to write about categories or functors: one gets the feeling that, for him, functoriality is more a way of thinking than a way of writing."

ideas. Bourbaki made this point even at the cost of deviating from a restriction jealously preserved throughout most of the treatise, and referring the readers to works outside the *Eléments*. Second, after the publication of Cartan and Eilenberg's *Homological Algebra*, and especially after Grothendieck's famous article of 1957, the functorial approach was widely adopted in the field, mainly because of its simplicity (§ 8.3, above). Third, as the above analysis of the various books of Bourbaki's treatise shows, it is the *structural* approach that fails to simplify proofs and that ultimately does little more than introduce extraneous terminology. This claim of Dieudonné is a further example of his historical accounts in which mathematical structures are unrestrictedly identified with *structures*.

The present chapter has analyzed the birth and early stages of category theory. It has briefly discussed the kind of questions addressed in the first research works in which categorical concepts were used, and in particular some of the foundational issues to which the theory was expected to contribute later on. Likewise the conceptual and historical relationship between categorical and *structural* ideas was analyzed, as well as the ambiguous attitude of Bourbaki's members to categories and functors. The issues considered here, together with those analyzed in Chapters 6 and 7 illustrate the kind of reflexive mathematical research that followed the consolidation of the structural image of mathematics. The closing chapter of the present book will consider the subsequent changes in the images of knowledge brought about by this reflexive activity.

Chapter 9   **Categories and Images of Mathematics**

The first part of this book described the development of ideal theory from Dedekind to Noether. This account was meant to illustrate, from the perspective offered by that specific theory, the rise of a central image of twentieth-century mathematical knowledge according to which mathematics is seen as a science of structures. The second part analyzed three reflexive attempts to elucidate that image of knowledge by means of formal mathematical theories. Ore's theory of **structures** had already disappeared from the world of mathematical research in the early fifties. Bourbaki's work at large constituted a mathematical achievement of outstanding value and reached an enormous, worldwide popularity, but the actual contribution of the theory of *structures* to the mathematical elucidation of the idea of structure was rather marginal, and certainly below the expectations of its creators. Category theory, in its turn, emerged after the sixties as an autonomous mathematical discipline. It developed its own particular tools, created its own sub-disciplines, and defined its urgent open problems. The theory afforded a relatively successful alternative approach to several open foundational issues in mathematics. At the same time, some of its limitations became increasingly manifest over the years. But how did the developments related to these three reflexive theories concomitantly affected the images of mathematics? This is discussed in the present, concluding chapter of the book.

9.1   Categories and the Structural Image of Mathematics

Category theory has often been incorporated into the picture furnished in historical accounts that take for granted the identification of mathematical structures with Bourbaki's structures (and *structures*). Since category theory and Bourbaki's *structures* are two attempts to formalize the non-formal idea of a mathematical structure, the former has occasionally been presented as an improvement of the latter, and the two theories together have been described as ensuing stages in the linear progress towards the reflexive elucidation of the essence of structure. André Weil, for instance, described this process as follows:

> About 1936 the collaborators of N. Bourbaki agreed to adopt the notion of structure, and its associated one of isomorphism, as a fundamental principle for the classification of mathematical theories; the notion of category, rather richer in mathematical content, would not be separated until much later.[1]

In the same vein, mathematical structures and their formal counterparts, *structures* and later categories, have been described as akin to structural trends outside mathematics, and in fact as their source of inspiration. All this has been done, more often than not, without any explanation, beyond the metaphorical level, of the actual connection between structural theories in mathematics and those outside mathematics. The following example is taken from an article which attempts to explain the role of category theory in the foundations of mathematics:

> With the rise of abstract algebra ... the attitude gradually emerged that the crucial characteristic of mathematical structures is not their internal constitution as set-theoretical entities but rather the relationship among them as embodied in the network of morphisms. This attitude, strikingly reminiscent of the operational structuralism associated with linguistics and psychology, was exemplified most strongly by the Bourbaki school of France, which had proposed a "structuralist" account of mathematics as far back as the 1930s.
>
> However, although the account of mathematics they gave in their Eléments was manifestly structuralist in intention, in actuality they still defined structures as sets of a certain kind, thereby failing to make them truly independent of their 'internal constitution'. (J.L. Bell 1981, 351)

The conceptual link allegedly established on this account between the essence of category theory and that of structuralist theories in linguistics and in psychology consists in depicting them all as disciplines whose subject matter are the interrelations among objects ("the network of morphisms"). Nevertheless, it is hard to attach to this description some actual meaning at the mathematical level. Moreover, if that is indeed the essence of the notion of

---

1. Weil 1978, 538: "Vers 1936, les collaborateurs de N. Bourbaki s'étaient mis d'accord pour adopter la notion de structure, et cell d'isomorphisme qui lui est liée, come principe fondamental de la classification des théories; la notion de catégorie, d'un contenu mathématique beaucoup plus riche, ne devait se dégager que bien plus tard."

structure in mathematics and elsewhere, it might be asked in what sense did the *structural* account of Bourbaki prefigure, however partially or vaguely, the centrality of the "network of morphisms"? As was seen in the preceding chapters there is no such connection between the stress laid by categories upon links between objects and Bourbaki's *structures*. The former is not a refined, improved version of the latter.[2]

The above quotation implies that with the rise of abstract algebra a clear-cut concept of mathematical structure appeared, and that in the attempt to formalize that concept it was gradually understood that it is necessary to overlook the internal constitution of the objects in behalf of their external links. But as was seen in the preceding chapters the idea behind Bourbaki's *structures* concerns only internal constitution. Moreover, even the very general and rather unuseful concept of morphism defined in *Theory of Sets* is ancillary to that internal constitution. Analyzing morphisms as the central characteristic of a mathematical domain was among the original direct motivations of category theory; yet the insight that the concepts developed by the theory can contribute to the understanding of the structural essence of mathematics in terms of the significance of the network of morphisms was a later one. Again, it is pertinent to point out that the first attempt to elucidate the idea of structure by overlooking internal constitution appeared in Ore's program. This forgotten program, and its indirect influence on the development of categorical ideas, which was certainly greater than Bourbaki's, are never mentioned in quotations like the one above. The description of category theory as the natural continuation of *structure* theory is only a result of hindsight, rather than a faithful description of the actual historical process.

When seen as successive stages in a necessary historical development, Bourbaki's theory of *structures* and category theory have also been credited with the gradual process of establishing a full harmony in the formerly chaotic world of mathematical ideas. That judgement is expressed in the following remarks of Imre Toth:

---

2. A further instance of seeing a direct development from *structures* to categories is implicit in Wussing 1989, 283 (Italics added): "In den 'Eléments de Mathématique', die ab 1939 erschienen, versuchten sie, die gesamte Mathematik durch weitere Axiomatisierung und Formalisierung neu aufzubauen. Ausgehend von den Begriffen der Menge und der Funktion entwickelten sie die Mathematik als Lehre von der Strukturen *und den Abbildungen zwischen ihnen*." See also Choquet 1963, 9-10.

The increasingly chaotic amount of mathematical theorems and theories should be transformed into a *cosmos* in the original sense of the word, into a *beautiful order*. The leading principle of this architecture of mathematical theories was the concept of structure, introduced by Bourbaki, that has since developed a brilliant career inside and outside mathematics, and that only recently has been superseded by Saunders Mac Lane and Samuel Eilenberg's synthetically-ordered concept of a category. (Toth 1987, 143. Italics in the original)[3]

At this stage there should be no need to add further comments on this account; in fact, it is likely that even most Bourbaki members would not accept its idyllic spirit. What is surprising is that close to the publication of the first works on categories, and despite later opinions, Dieudonné himself expressed a view of mathematics similar to the above quoted one, in a lecture delivered in 1961 at the University of Münster, in which he declared:

In the years between 1920 and 1940 there occurred, as you know, a complete reformation in the classification of different branches of mathematics, necessitated by a new conception of the essence of mathematical thinking itself, which originated from the works of Cantor and Hilbert. From the latter there sprang the *systematic axiomatization* of mathematical science in its entirety and the fundamental concept of *mathematical structure*. What you may perhaps be unaware of is that mathematics is about to go through a second revolution at this very moment. This is the one which is in a way completing the work of the first revolution, namely which is releasing mathematics from the far too narrow conditions imposed by 'set'; it is the theory of *categories and functors*, for which estimation of its range or perception of its consequence is still too early.[4]

Some years later (as pointed out above in § 8.5) this initial enthusiasm diminished and Dieudonné was then less eager to concede to categories a much too central role in the overall picture of mathematics.

There is a further accepted image of mathematics involving category theory, and which is related to foundational research in mathematics.

---

3. The reader should notice that the convention introduced in the present book of using italics for Bourbaki's formal concept of *structures* is not intended in this quotation. This is precisely the problem with the ambiguity between the formal and the non-formal senses of the term. One does not know here to which of the senses is Toth referring, or whether he is aware of the distinction at all. In order to avoid misunderstandings from our point of view, the italics have been omitted in the translation.

4. Quoted from Fang 1970, 95-96. Italics appear in the original, but of course, *structure* in italics is not meant to signify the formal concept, as in our convention. As in other cases, however, it is not clear what Dieudonné himself means here by "structure."

Chapter 8 briefly discussed the problems raised by the need to consider very big collections in the central definitions of category theory. Mac Lane proposed several possible solutions to this problem, but Lawvere was the first to propose a radical solution in which the whole picture would be turned upside down, and categories would substitute sets as the basic concept laying at the foundations of mathematics. Following Lawvere's pioneering work, the categorical study of the foundations of mathematics developed into one of the central branches of category theory, under the name of Topos theory. This development certainly opened wide horizons both for category theory and for foundational research. Many problems remained open however, and category theory cannot yet claim to have provided a single, axiomatically defined system which may provide a unified conceptual foundation for the whole of mathematics. Nevertheless, several authors have attempted to indicate new directions in which, by exploiting the insights afforded by categorical research, the philosophy of mathematics may be funneled.

Lawvere himself proposed in an article of 1969 to connect the concept of duality, and other categorical concepts, with epistemological issues related to the philosophy of mathematics. In order to do that, he identified two "dual aspects" of mathematical knowledge—the conceptual and the formal aspects—which appear in many domains of mathematics. Thus for instance, the pair polynomial/curve in geometry. This duality, claimed Lawvere, is manifest also in the foundations of mathematics, but researchers had hitherto limited themselves to one of the two aspects: the formal one, embodied in logic. Now Lawvere proposed to dedicate efforts to develop the second aspect, the conceptual one, embodied in category theory. This proposal, however, remained at the programmatic level and no one seems to have developed it further.

A similar direction was proposed by Engeler and Rohrl who saw the expected contribution of category theory to foundational research in its capability to concentrate on mathematical "structure", as opposed to mathematical "content."[5] This capability would account for the many applications accorded to categories in various mathematical domains. However, here as in other cases, the authors did not actually explain what is that "mathematical structure" upon which categories concentrate, and how they advance our understanding of foundational issues.

---

5. See Engeler and Rohrl 1969. Similar views are expressed in Bell 1981.

Opinions diverge as to the actual possibility of fully working out a categorical foundation of mathematics, as many had hoped for. Solomon Feferman has gone further than anyone else in claiming that the theory *cannot* fulfil those hopes; he has provided *mathematical* arguments to explain the foundationalist limitations of category theory.[6] Feferman totally opposes the opinion that categorical concepts are in any sense "prior" to those of set theory.

In the first place, Feferman claims that the concepts of "operation" and "collection" have *psychological priority* over structural concepts. In other words, in Feferman's opinion one is not able to think of a "structure" before having clear conceptions of "collection" and "operation." Naturally, this opinion is debatable,[7] but Feferman also provided mathematical arguments to show that the above mentioned priority is also a *logical* one. Structural concepts, he concluded, are defined in terms of set-theoretical ones, not the other way round. Feferman concedes that category theory has provided many important insights about the nature of mathematics, but he prefers to see this as a further instance of what he defined elsewhere as "Working Foundations", namely, "a direct continuation by more conscious, systematic means of foundational moves which have been carried on within mathematics."[8]

Feferman's mathematical results are relevant to the arguments put forward throughout this book: the contribution of category theory to the philosophy of mathematics lies in its clarification of several images of mathematical knowledge related to the idea of structure, in its expansion of the body of reflexive knowledge through the development of a new mathematical discipline, and in its providing many new meaningful mathematical theorems. In no sense, however, has category theory provided, to this day, a definite, or even a provisionally satisfactory answer to the question of what is a "mathematical structure."

Neither does category theory provide ultimate foundations for mathematics. On the contrary, developments such as the rise of topos theory seem to have underscored the difficulty inherent in the search of such foundations. Thus, before the rise of category theory and the recognition of the foundationalist possibilities it offered, there existed a single candidate, sets, to provide the foundational ground for mathematics; the question was how to establish

---

6. In Feferman 1977.

7. See, e.g., Mac Lane 1988, 343, for a reply to Feferman's claims.

8. Feferman 1985, 229. See also Feferman 1984. See Corry 1989, 429-430, for further comments on the views exposed in these articles of Feferman.

set-theory axiomatically in order to attain the desired theoretical foundation. Now since the rise of category theory, one must first choose between at least two candidates for that role, sets and categories, before one can proceed to elaborate the desired axiomatic basis according to one's choice. It is still to be decided whether the existence of two such candidates has improved the stakes to solve the question of the foundations of mathematics, or whether, on the contrary, it reinforces the claims of the "quasi-empiricist" trends in the philosophy of mathematics, that such a solution does not exist.[9]

9.2   Categories and the Essence of Mathematics

In discussing the current state of research in 1939, Mac Lane characterized algebra as that branch of mathematics that investigates "the explicit structure of postulationally defined systems closed with respect to one or more rational operations."[10] In order to explain what he meant by the latter expression, Mac Lane gave an explicit inventory of the questions which stood at the center of interest of algebra at the time. Thus, the best way Mac Lane found to explain what "structural research" is, was to describe the doings of the algebraists at the time: What the algebraist are presently doing, implied Mac Lane, that is "structural research in algebra.

Mac Lane's article of 1939 was republished in 1963 as part of a collection edited by Adrian A. Albert. The collection included a new article of Mac Lane under the title "Some additional advances in Algebra." This time he described developments that had taken place in algebra since his earlier piece. These developments, he wrote, had brought about "a second revolution in the character of algebra, fully as important as the dramatic growth of abstract algebra in the decade 1925-1935."[11] In Mac Lane's account, at variance with Dieudonné's account quoted above, this revolution was not restricted to the rise of the theory of categories, or to any other single development. Rather it covered well-established domains of algebra (finite groups, simple groups, local rings, modules), as well as algebraic domains that arose only after 1939 (category theory, homological algebra, Hopf algebras). Remarkably, Mac Lane did mention the unifying capabilities of categories, but in order to illustrate his claim

---

9. The "quasi-empiricist" approach, as opposed to the "foundationalist" approach in the philosophy of mathematics is discussed in Tymoczko 1985 and Kitcher 1988.

10. Mac Lane 1939. This article is discussed in detail above, in the introduction to Part Two.

11. Mac Lane 1963, 35.

he chose to compare the theory with none other than lattice theory! He claimed that "like the notion of lattice the concept of category theory belongs to universal algebra."[12] This comparison testifies that for mathematicians interested in unifying theories of algebra, lattice theory remained a standard of reference until at least 1963.

But the really interesting point in Mac Lane's article of 1963 lies in his new answer to the old question: "What is algebra?" In 1939—wrote Mac Lane in 1963—he had attempted to characterize algebra as "the study of 'structure' of systems defined by suitable postulates on rational operations", but now that attempt appeared to him "naive and somewhat limited." Thus he wrote:

> Portions of abstract algebra can, indeed, be construed as investigations of such structure theorems. But as the present article indicates, the development of the subject is much more varied. Older questions, such as those about finite group theory, return to the center of interest with the development of new ideas and techniques; they cannot always simply be categorized as 'structure theory'. New types of algebraic systems arise from the application of algebra in geometry, topology, and analysis ... The types of algebraic systems to be analyzed are suggested and imposed by various other mathematical areas, and the development of algebras can be understood only within the fabric of all mathematics. (Mac Lane 1963, 54)

Mac Lane was referring here, among other things, to the renewed interest in the classification of finite groups, which took place after 1963, following the success of W. Feit and J.G. Thompson (1932-) in proving a conjecture advanced about hundred years before by William Burnside, namely, that every finite group of odd order is solvable.[13] This result had brought about an

---

12. Mac Lane 1963, 48.

13. Mac Lane felt personally connected with this significant trend, since Thompson had participated in a course on group theory taught by Mac Lane in 1956-57, and had later taken his Ph.D. under Mac Lane before turning to his own work on groups. These details were communicated to me by Professor Mac Lane himself in a conversation in San Sebastián (Spain), in September 1990. It is interesting to compare Mac Lane's opinions as quoted throughout this book, with Frank Nelson Cole's assessment of the state of research on group theory as early as 1893. Thus Cole wrote (E.H. Moore et al. (eds.) 1893, 40-41): "In an abstract and intricate theory like that of groups, too much must not be expected in the way of general development from the accumulation and study of individual examples. No amount of such experimentation could have led to our modern knowledge. Progress is from abstract to abstract. Nevertheless, in the absence of a general method, something may be accomplished by the tentative, step-by-step process, especially within moderate limits where the labor involved is not commensurable with the value of the result." (Quoted from Parshall and Rowe 1994, 321.)

intense research activity, after many years during which the field had been relatively dormant. This research eventually led to what was believed to be a full
classification of finite groups. On the face of it, there is no apparent reason
why such research could not be included under the description of "the study
of the explicit structure of postulationally defined systems closed with respect
to one or more rational operations", as Mac Lane had defined 'structural
research' in 1939. Yet Mac Lane explicitly remarked here that algebraic
research was much less structural in 1963 than in 1939, clearly using term in
its non-formal sense. This kind of research of finite groups was different from
that practiced in other algebraic disciplines at the time, in that it spanned a
wide panoply of *ad-hoc* methods that overrode the standardized techniques,
questions and expected answers evenly applied in other algebraic domains
since the advent of modern, structural algebra, and in particular since the publication of van der Waerden's *Moderne Algebra*.[14]

Mac Lane's description of the structural character of algebra by 1939 is,
of course, debatable and it raises many questions. Does his list really exhaust
all the issues that algebraists in that period saw as the urgent ones in algebra?
Did it represent a real change when compared to what preceded it? Exactly
what are the features in his description that transform algebra into a "structural" discipline? Did a real change take place again between 1939 and 1963
that justified Mac Lane's renewed appraisal of the character of algebra? But
what really matters for the present discussion is that Mac Lane *did* advance a
non-formal description of the structural character of algebra in 1939, and then
in 1963 he claimed that his older characterization was no longer valid to the
same extent. Thus, contrary to other accounts quoted above, in Mac Lane's
view research in algebra had turned "less structural" rather than "more structural" in the years immediately following the rise of category theory.

Mac Lane mentioned in his article of 1963 the unifying power of category
theory, while comparing it with that of lattice theory. But as part of his claim
that the structural characterization of algebra is an insufficient account of this

---

14. It is worth pointing out that no separate chapter on the theory of finite groups is included in
Bourbaki's treatise. According to Dieudonné (1970, 16) the reason for this is obviously not one of
value judgement concerning the importance of the theory in mathematics, but rather that "we cannot
say that we have a general method of attack" in the theory. Obviously, from the formal point of view
a finite group is an immediate example of an algebraic *structure*. Thus finite groups provide an interesting example of a concept that fitted well the formal concepts introduced by Bourbaki, but, at the
same time, it did not fit his actual, working (non-formal) structural images of mathematics.

mathematical domain, he explained the limitations of the categorical language in offering a unified conceptual framework for deriving algebraic knowledge in its manifold aspects. In fact, Mac Lane's position on this issue changed over the years. These changes are manifest in the successive editions of the popular textbook (co-authored by Mac Lane and Garrett Birkhoff) *A Survey of Modern Algebra*.

The *Survey* was the first book written and published in the USA following the new, structural image of mathematics, and it soon became the standard American textbook on the issue. In the introduction to the first edition of the *Survey*, published in 1941, the authors reviewed, as usual, the developments in algebra leading to the actual state of the theories exposed in the book. In spite of the deep transformation undergone by algebra over the past decades, the authors thought it important for the student to learn the direct connections between the new and the classical theories of algebra, on the one hand, and of the new algebra and other fields of mathematics, on the other. However, since the new concepts of algebra were increasingly absorbed into day-to-day mathematical discourse, the stress on those connections gradually diminished and, after several editions of the book, they totally disappeared.

From 1967 on, after two re-editions of the *Survey,* in 1953 and in 1965, the book began to appear under the new name *Algebra*, and the name of the authors now appeared as Mac Lane & Birkhoff, instead of Birkhoff & Mac Lane. This latter change was meant to indicate a modification in the relative weight of each of the authors in the writing of the various chapters of the book.[15] The omission of the word *Modern* in the title indicated that the methods and concepts introduced in the *Survey* had been gradually accepted as mainstream by the mathematical community; a parallel process was mentioned above in § 1.3 concerning van der Waerden's textbook from 1955 on, and Dickson's textbook from 1959 on. Thus, if in the early editions of Mac Lane and Birkhoff's book, concepts such as ideals, fields, groups, etc., appeared as the innovative concepts of algebra, in 1963 they were presented as "standard material." Category theory was now a mature discipline, being increasingly adopted as a unifying language. In the introduction to the 1967 edition, a tension is manifest between the possibility of presenting algebra through the new generalizing concepts, on the one hand, and the didactic con-

---

15. The collaboration of Mac Lane and Birkhoff in writing their textbooks is described in Birkhoff 1977. See also Birkhoff & Mac Lane 1992; Mac Lane 1997.

siderations, on the other hand, which seemed to dictate the adoption of an intermediate approach. Thus in the introduction one reads:

> It is now clear that we study not just a single algebraic system (a group or a ring) by itself, but that we also study the homomorphisms of these systems.... This book proposes to present algebra for undergraduates on the basis of these new insights. In order to combine the standard material with the new, it seemed best to make a wholly new start. At the same time, just as in our *Survey,* we hold that the general and abstract ideas needed should grow naturally from concrete instances. With this in view, it is fortunate that we do not need to begin with the general notion of category. (Mac Lane & Birkhoff 1967, *vi*)

And indeed, the theory of categories is introduced in the book, but only in chapter 15, after all the basic theories of algebra were discussed in detail.[16] However, as is usually the case in the history of mathematics, the general and abstract of yesterday turn into the concrete and the particular of today, and thus, in the 1988 (revised) edition of *Algebra* one finds a discussion on category theory already in the opening chapter.

These successive editions reflect Mac Lane's changing attitudes over the years concerning the question of the nature of algebra and the adequate approach to the exposition of its contents. In fact, in his retiring presidential address to the American Mathematical Society in January 1975, Mac Lane himself acknowledged these changes, and the impossibility of providing definitive answers to these questions. In the closing passage of the address he claimed:

> This essay has recounted a number of instances in which the development of algebraic ideas has been stimulated by problems or queries arising outside algebra: in particular in logic or in geometry ... On some occasions I have been tempted to try to define what algebra is, or should be—most recently in concluding a survey (Mac Lane 1963) on *Recent advances in algebra*. But no such formal definition holds valid for long, since algebra and its various subfields steadily change under the influence of ideas and problems coming not just from logic and geometry, but from analysis, other parts of mathematics, and extra mathematical sources. (Mac Lane 1976, 36)

Obviously Mac Lane's changing opinions reflect the development of the body of algebra from Noether's day to the present. In particular, Mac Lane has

---

16. It may be worth mentioning that contrary to categories, in this, like in all other standard textbooks on algebra, *structures* and **structures** are not treated or otherwise mentioned as possible unifying frameworks for algebra.

seen in categories a useful tool for generalized formulations in algebra, but by no means *the* ultimate unifying concept of algebra. Mac Lane sees in algebra, and more generally in mathematics at large, a much more multi-faceted system of ideas and phenomena that a single theory, flexible and rich though it may be, could ever fully exhaust. Moreover, even the description of algebra as the discipline of algebraic structures, understood in the broadest sense of the term, was sometimes considered too narrow by Mac Lane. But interestingly, in a more recent attempt to formulate a more clearly articulated philosophical position on the nature and contents of mathematics, he returned to one of the leading ideas of category theory, and has built his argumentation around it. The remainder of the present section analyzes in some detail Mac Lane's proposal as it appears in his book *Mathematics: Form and Function* (1987a).[17]

Since the end of the seventies, Mac Lane has been trying to formulate a systematic program "for the revival of the philosophy of mathematics." The basic tenet of his program is that in order to elucidate fully the nature of mathematics it is necessary to consider mathematics in all of its extension instead of concentrating on certain foundational, and sometimes elementary aspects such as the traditional philosophy of mathematics has tended to do. Such a comprehensive perspective on mathematics led Mac Lane to claim that the essential feature of mathematical knowledge is the ramified network of interconnections among the manifold manifestations of the discipline and its various sub-disciplines. Thus, whereas category theory had focused on the connections among objects of a given kind, and among different kinds o mathematical objects, as the key to the *reflexive* inquiry into the nature of mathematical systems in strict mathematical terms (albeit with considerable limitations), Mac Lane now proposed to address philosophical questions regarding the nature and aims of mathematics based on the *non-formal* description of the many connecting links among different ideas and branches of mathematics.

Mac Lane mentioned several philosophical questions central to his inquiry that neither existing mathematical theories (category theory included) nor classical philosophical discourse on mathematics have hitherto helped, in his opinion, to elucidate. Among them are the following: How are the concepts of

---

17. The ideas developed in that book had been variously discussed in earlier articles. See Mac Lane 1976, 37: "The progress of mathematics does indeed depend on many interlocking, unexpected and multiform developments." See also Mac Lane 1980; 1981; 1987.

mathematics (and in particular the various mathematical structures) created? How are we to choose the most important issues for research in mathematics? What sets mathematics apart from other scientific disciplines such as physics? What is in fact the subject matter of mathematics?

In *Mathematics: Form and Function* Mac Lane gives the reader an up-to-date picture of mathematics in all its manifestations, as he himself sees it. In fact, a great part of the book consists of an account of the current state of research in many branches of mathematics, while the philosophical conclusions allegedly arising from this account are left for its closing chapter. Although in his description Mac Lane dedicated only a few paragraphs to the concepts of structure and of category, it is remarkable that without explicitly mentioning Bourbaki, Mac Lane explained that "many Mathematical notions can be described as set-with-structure"[18] and he also tacitly adopted the scheme of the mother-structures as a faithful account of considerable portions of mathematics. Among those notions that can be described as sets-with-structure he included ordered sets, algebraic structures and topological spaces. And he further added:

> There are also structures of a mixed kind. For example there are ... topological groups. As in this case, most composite axiomatic structures (combinations of two kinds of structures on the same set) involve one or more axioms expressing the formal connection between the two structures. (Mac Lane 1987a, 33)

In a different passage Mac Lane claimed that many mathematicians presently use the set-theoretical language to deal with mathematical structures, but this language is not enough to encompass the wealth of phenomena under inquiry today in mathematics. Moreover:

> Now much of Mathematics is dynamic, in that it deals with morphisms of an object L into another object of the same kind. Such morphisms (like functions) form categories, and so the approach via categories fits well with the objective of organizing and understanding mathematics. That, in truth, should be the goal of a proper philosophy of Mathematics. (Mac Lane 1987a, 359)

Naturally, when stressing morphisms, one is stressing categories. Thus it may come as a surprise to find that the organization and the understanding of mathematics that Mac Lane offers the readers is not based on the scheme of categories and functors. Instead, as he had done in the past, Mac Lane characterizes the diverse branches of mathematics, not by resorting to formal math-

---

18. Mac Lane 1987a, 34.

ematical definitions, but rather in a non-formal fashion by providing lists of characteristic features of the leading problems addressed by each domain. Moreover, mathematics as a whole, its origins, the process of creation of its concepts, and the definition of its subject-matter, all these matters are elucidated by him in a similar way, namely, by inventorying its characteristic features: an inventory of the human and scientific activities leading to the creation of mathematical concepts, an inventory of the possible connections that may be established among different mathematical domains, and so on. In these inventories, neither categories nor any other formal concept play any role.

Now it is not an assessment of the philosophical import of Mac Lane's position and argumentation which concerns us here, but rather an examination of his images of mathematics as manifest in the book, and their relationship to the reflexive elucidations of the notion of a mathematical structure. One must, however, point out that the overall contribution of Mac Lane's presentation here does not satisfactorily fulfil the expectations that his claims raise. For instance, Mac Lane had claimed that one of the most important contributions of the detailed analysis of "the form and function of mathematics" will be the possibility of objectively establishing the more effective and the more promising research directions. His conclusion is that future research directions must take into account several factors:

> Mathematics, in our description, rests on ideas and problems arising from human experience and scientific phenomena and consisting in many successive and interconnected steps in formalizing and generalizing these inputs. Thus Mathematical research can be directed in a wide variety of overlapping ways:
>
> (a) Extracting ideas and problems from the (scientific) environment;
> (b) Formulating ideas;
> (c) Solving externally posed problems;
> (d) Establishing new connections between Mathematical concepts;
> (e) Rigorous formulation of concepts;
> (f) Further development of concepts (e.g., new theorems);
> (g) Solving (or partially solving) internal Mathematical problems;
> (h) Formulating new concepts and problems:
> (i) Understanding aspects of all the above. (Mac Lane 1987a, 449-450)

How one is to proceed from the contents of this list to the actual determination of research directions—this is not explained or suggested by Mac Lane, and it probably can not be done. One thus finds it difficult to identify the specific contribution to the understanding of the essence of mathematics obtained from Mac Lane's painstaking description. In fact, the only clear-cut conclusion that arises from Mac Lane's book is that there is indeed a very strong, deep and manifold conceptual interconnection among many ideas and domains of mathematics, and that a further examination of these connections is likely to lead to many new important results. As a consequence, the working mathematician should not specialize his research too narrowly, for this may lead him to ignore many important insights provided by specialties other than his own. But of course this conclusion was largely acknowledged by Mac Lane himself, and by many other mathematicians (in particular by Bourbaki in his collective work and in the works of its individual members), well before Mac Lane began to articulate his philosophical program. In fact, a long-standing attitude against narrow, overly-specialized research was a main motive in the images of mathematical knowledge of Bourbaki, and clearly an important part of Hilbert's legacy. Writing in the late 1980s Mac Lane still seems to have reasons for concern regarding what he sees as an unhealthy trend to over-specialize among contemporary mathematicians. But naturally the question arises, whether the analysis presented in Mac Lane's book, if it is indeed intended to warn against such trends and eventually reverse them, is bound to produce its desired effect. Moreover, Mac Lane's statements throughout the book, as well as in his articles dealing with the same issue, create the impression that Mac Lane intended to attain much more than just such a calling.

One can summarize the present discussion by saying that Mac Lane's book may be justly described as a systematic exposition of an image of knowledge that arises from a deep and long-standing personal acquaintance with several fields of research in mathematics. The intricate and surprising interconnections among seemingly distant mathematical ideas always posed a special interest for Mac Lane, as for many other leading mathematicians besides him. This is true of his research on several specific cases of inquiry into such connections (algebraic topology, homological algebra, logic), and it is also true as a mathematical problem in its own, that can be formally enunciated in terms of categories and functors. This very issue became the basis upon which Mac Lane attempted to present what he saw as a viable alternative for reviving the philosophy of mathematics. From his close acquaintance with the many attempts to generalize and unify under a single conceptual framework as many aspects as possible of mathematical reality, Mac Lane concluded that all those attempts had come short of fulfilling their aims.[19] Category theory, in spite of its meaningful achievements, is counted by Mac Lane himself among such unsuccessful attempts.[20]

9.3   What is Algebra and what has it been in History?

The foregoing chapters examined changes undergone by the body and the images of algebra since the mid-nineteenth century up to the mid-twentieth. It was seen how, over that period, the mathematical discipline called algebra was transformed from a discipline dealing with the theory of polynomial and algebraic forms, and with questions of factorization of (algebraic) numbers, into the discipline of algebraic structures. Later on, some of the formal reflexive attempts to elucidate the concept of structure—the theories of **structures**, categories, and Universal algebra—have been also considered to belong to algebra. In fact, algebraic methods became increasingly pervasive in mathematics throughout the century, and disciplines like algebraic topology, algebraic geometry and the theory of algebraic functions received an unprecedented impulse. On the other hand, as was seen in the preceding section, the views of a leading algebraist, Saunders Mac Lane, concerning the essence of the discipline changed over the last decades. In view of this situation several questions arise: What is that discipline called algebra with which so many diverse mathematical activities have been associated? Is the study of algebraic structures *the* essence of algebra? What is the role of algebra within the whole fabric of mathematics? In the present, closing section of the book it is suggested how the body/images of knowledge scheme, and the account presented here may help addressing these questions.

The question about the nature of algebra has been addressed in various contexts. It has been raised in the context of historical debates, such as the debate on the so-called "geometrical algebra" of the Greeks. This debate centered around the closely related question, whether, and to what degree, was there "algebraic thinking" in Greek mathematics.[21] As we saw in the foregoing chapters, the question about the nature of algebra also arose as part of var-

---

19. Mac Lane 1992 also elaborates the view that "the protean character of mathematics" cannot be subsumed under any comprehensive philosophical or historical scheme.

20. Cf. Mac Lane 1987a, 406-407: "[S]et theory and category theory may be viewed as proposals for the organization of Mathematics.... Neither organization is fully successful... We conclude that there is no simple and adequate way of conceptually organizing all of Mathematics."

21. A short account of the essence of this debate and the recent relevant bibliography appear in Berggren 1984, 397-398.

ious "reflexive" attempts to elucidate the concept of structure in mathematical terms, in the framework of which, the characterization of algebraic structures, as a well-defined class from among all mathematical structures, was sought. The achievements and shortcomings of those theories in clarifying the possible answers to this question were discussed above. Moreover, the difficulties inherent in describing the historical development of algebra from the perspective offered by those theories, especially in the case of Bourbaki, were also mentioned. Thus, explaining in terms of "mathematical structures" (whatever sense one ascribes to the term) the common mathematical grounds, and, more particularly, looking for the common algebraic structures investigated in Babylonian texts, in Greek geometry, in Renaissance techniques of equation-solving, in the research of polynomial and algebraic forms from the seventeenth- to the nineteenth-century, and finally, in the research of twentieth-century algebraists, turns out to be, to a very great extent, historically misleading. Without entering here deep into the specific debates concerning the nature of algebra, it seems that a common difficulty that has been manifest in all of them is the attempt to define, by either of the sides involved, the "essence" of algebraic thinking throughout history. Such an attempt appears, from the perspective offered by the views advanced throughout the present book, as misconceived. In trying to explain why it is so, and how can indeed any viable definition of algebra be advanced, we will rely on the body/images of knowledge scheme.

Consider the following list of mathematical statements:

1. $3^2 + 4^2 = 5^2$

2. The product of a quantity by a sum of quantities equals the sum of the products of the first quantity by each of the factors of the sum
$$x \cdot (a + b + c + \dots) = x \cdot a + x \cdot b + x \cdot c + \dots$$

3. The equation $x^2 + (x + 1)^2 = (x + 2)^2$ has solutions which may be obtained through the application of a general, simple procedure valid for all quadratic polynomial equations.

4. There is no general formula in radicals for solving the general polynomial equation of degree greater than four.

5. The alternating group A_n is simple when $n > 4$.

6. The "forgetful" functor from *Set* to *Grp* and the functor from *Set* to *Grp*, sending any set X to the free group $\mathfrak{I}_X$, with basis X, are adjoint functors.

These six statements belong to the accepted body of mathematical knowledge, and obviously they are mutually compatible. From our point of view, they all may be classified today as "algebraic knowledge." (Admittedly, statement 1. is in fact an arithmetical one, and there certainly may be some opposition to classify it here as "algebra." But it has been included in the list just in order to make the argument more persuasive, and, since arithmetic is after all a prerequisite—historically and didactically speaking—to algebra, it should not cause great problems to the reader to accept it as "algebraic.") But then, what are the grounds for classifying all the above statements as algebra?

The above account of the rise of the structural approach to algebra has shown that the grounds for internal classification of mathematical domains, being the result of a historically determined process, are contingent. A different historical process from the one that actually took place could conceivably have produced different stages in the organization of mathematical knowledge, and the above statements, still without producing contradictions in the body of knowledge, could have been classified under different headings at different times and also perhaps now. But the verdict that under the presently accepted classification, the above statements are considered as algebraic, while non-algebraic ones include approximation algorithms for polynomial equations, or theorems concerning the roots of polynomials, in whose proofs ε–δ arguments are used, this verdict is an image of knowledge created by mathematicians and presently shared by them, and it is thus subject to historical change.

It therefore seems that the question of the essence of algebra is an ill-posed one. One might more properly ask "What is the algebra of Fermat, Descartes and Viète?", or "What is van der Waerden's algebra?", or even "What was the algebra of the Greeks?." One could then claim, according to one's answers to those questions, that the Greeks were, or were not, doing algebra like it was later done in the seventeenth century, or like it is done in the twentieth-century. But there is no meaning to the claim that the Greeks were not doing "algebra", period. In view of this, the only suitable definition for algebra throughout history seems to be the sociological one. This idea is very clearly formulated in a rather unknown article, as follows:

> [We] have no definition of Algebra, no central idea or unifying principle, not even a well centered picture of the whole, but rather an enumeration of algebraic topics in the literature of the epoch, even though this enumeration is in some

sense natural but not uniquely determined order.... In final analysis one is reduce
to the seemingly meaningless circular statement that "Algebra is what algebraists
do or declare to be Algebra, and algebraists are people doing Algebra or declaring themselves to be algebraists." (Tamari 1978, 198-199)

It might be claimed that such an answer evades the question and implies a
relativistic view of mathematical knowledge. However, a statement like the
one appearing in the last quotation does not concern the essence of mathematics, or the epistemological status of mathematical knowledge or its subject
matter, but only the organization of mathematics into subdisciplines. This
organization is a typical image of knowledge and, as the preceding chapters
clearly show, it is sometimes, but certainly not always possible to provide a
precise, explicit mathematical elucidation of the images of mathematics.

The rise of the structural approach was an impressive (one is almost
tempted to say "revolutionary")[22] change in the definition of disciplinary
boundaries within mathematics, which brought to the center of interest the
idea that algebra, and later mathematics in general, deal with different kinds
of structures. It was seen throughout the chapters of the book that this change
was preceded by a significant growth in the body of knowledge; by a steady
flow of new results concerning, among others, factorization in fields of algebraic numbers and factorization of polynomials forms, and on the other hand,
by the introduction of new techniques for the analysis of postulational systems. But it was likewise seen that this growth of the body of mathematical
knowledge did not in itself determine the change of images of knowledge that
implied the adoption of a structural approach to algebra. Rather, the latter corresponded to specific choices made by certain mathematicians, concerning the
kinds of ideas, interesting questions, appropriate conceptual tools, and legitimate answers, that should be pursued in algebraic research. Moreover, starting
in the 1930s this change was followed by attempts to elucidate reflexively that
idea of structure. Such attempts have failed so far to provide a thorough explanation of the organization of mathematics in terms of structures. It is arguable
that in the future we shall have a reflexive, and therefore non-contingent organization of mathematical knowledge, but as far as we know today, and authoritative opinions to the contrary notwithstanding, it is impossible to give a
definite mathematical answer to the question "What is the *essence* of algebra?".

---

22. See Corry 1993, 114-117.

Bibliography

The following abbreviations are frequently used:

AFLN	Arbeitsgemeinschaft für Forschung des Landes Nordheim-Westfalen
AHES	Archive for History of Exact Sciences
AJM	American Journal of Mathematics
AM	Annals of Mathematics
AMM	American Mathematical Monthly
AMS	American Mathematical Society
AMSHU	Abhandlungen aus dem Math. Seminar der Hamburgischen Universität
BA	Berliner Akademie der Wissenschaften
BAMS	Bulletin of the American Mathematical Society
BJPS	British Journal for the Philosophy of Science
BSMA	Bulletin des sciences mathématiques et astronomiques
DMJ	Duke Mathematical Journal
FM	Fundamenta Mathematicae
GN	Nachrichten von der König. Ges. der Wissenschaften zu Göttingen
HM	Historia Mathematica
HPL	History and Philosophy of Logic
JDMV	Jahresbericht der Deutschen Mathematiker-Vereiningung
JMPA	Journal de mathématiques pures et appliqueés
JPAA	Journal of Pure and Applied Algebra
JRAM	Journal für die reine und angewandte Mathematik
JSL	Journal of Symbolic Logic
LMS	London Mathematical Society
MA	Mathematische Annalen
MI	The Mathematical Intelligencer
ML	Modern Logic
MR	Mathematical Reviews
MZ	Mathematische Zeitschrift
NTM	Schriftenreihe für Geschichte der Naturwiss. Technik und Medizin
PCPS	Proceedings of the Cambridge Philosophical Society
PICM	Proccedings of the International Congress of Mathematicians
PNAS	Proceedings of the National Academy of Sciences USA
PS	Philosophy of Science
SA	Scientific American
SHPS	Studies in History and Philosophy of Science
SIC	Science in Context
SM	Scripta Mathematica
TAMS	Transactions of the American Mathematical Society
ZMP	Zeitschrift für Mathematik und Physik

ALBERS, D.J. AND G.L. ALEXANDERSON (EDS.)
1985   *Mathematical People*, Boston, Birkhäuser.
ALBERT, A.A.
1943   "Quasigroups, I", *TAMS* 54, 507-519.
1944   "Quasigroups, II", *TAMS* 55, 401-419.
1955   "Leonard Eugene Dickson 1874-1954", *BAMS* 61, 331-345.
ALBERT, A.A. (ED.)
1963   *Studies in Modern Algebra*, New Jersey, Englewood Cliffs.
ALEXANDROV, P.S.
1935   "In Memory of Emmy Noether" (English by N. & A. Koblitz), in Noether *GA*, 1-11.
ALEXANDROV, P.S. AND H. HOPF
1935   *Topologie*, Berlin, Springer.
ANON.
1970   "Oystein Ore, 1899-1968", *Journal of Combinatory Theory* 8, i-iii.
ARTIN, E.
1953   Review of Bourbaki (1939-), Book II, Chs. 1-7 (1942-52), *BAMS* 59, 474-479.
ARTIN E. AND O. SCHREIER
1926   "Algebraische Konstruktion reeller Körper", *AMSHU* 5, 85-99.
1927   "Eine Kennzeichnung der reell abgeschlossenen Körper", *AMSHU* 5, 225-231.
ARTIN, M.
1991   "Zariski's Papers on Holomorphic Functions", in Parikh (1991), 215-222.
ASCHBACHER, M. ET AL.
1999   "Michio Suzuki (1926-1998)", *Notices AMS* 46 (6), 543-550.
ASKEY, R.
1988   "How Can Mathematicians and Mathematical Historians Help Each Other", in Aspray
          & Kitcher (eds.) (1988), 201-217.
ASPRAY, W.
1991   "Oswald Veblen and the Origins of Mathematical Logic at Princeton", in T. Drucker
          (ed.) *Perspectives on the History of Mathematical Logic*, Boston, Birkhäuser, 54-70.
ASPRAY, W. AND P. KITCHER (EDS.)
1988   *History and Philosophy of Modern Mathematics*, Minnesota Studies in the Philosophy
          of Science, Vol. XI, Minneapolis, University of Minnesota Press.
AUBIN, D.
1997   "The Withering Immortality of Nicolas Bourbaki: A Cultural Connection at the Conflu-
          ence of Mathematics, Structuralism, and the Oulipo in France", *Science in Context* 10,
          297-343.
AYOUB, R.G.
1980   Paolo Ruffini's Contributions to the Quintic, *AHES* 23, 253-277.
BAGEMIHL, R.
1958   Review of Bourbaki (1939-), Book I, Ch. 3, *BAMS* 64, 390.
BAKER H.F.
1938   "Francis Sowerby Macaulay", *JLMS* 13, 157-160.
BARTLE, R.G.
1955   "Nets and Filters in Topology", *AMM* 62, 551-557.

BASHMAKOVA, I. ET AL.

1992 "Algebra and Algebraic Number Theory", in A.N. Kolmogorov & A.P. Yushkewich (eds.) *Mathematics of the 19th Century*, Basel, Birkhäuser (English trans. from the Russian Original, 1978), 35-136.

BEAULIEU, L.

1989 *Bourbaki. Une histoire du groupe de mathématiciens français et de ses travaux, 1934-1944*, 2 vols., Unpublished Diss., Université de Montréal.

1990 "Proof in Expository Writings - Some Examples from Bourbaki's Early Drafts", *Interchange* 21, 35- 45.

1993 "A Parisian Café and Ten Proto-Bourbaki Meetings (1934-35)", *MI* 15, 27-35.

1994 "Dispelling the Myth: Questions and Answers About Bourbaki's Early Work, 1934-1944", in Ch. Sasaki et al. (eds.) (1994), 241-252.

BECK, H.

1926 *Einführung in die Axiomatik der Algebra*, Berlin/Leipzig, De Gruyter.

BEHNKE, H.

1956 "Der Strukturwandel der Mathematik in der ersten Hälfte des 20. Jahrhunderts", *AFLN* 27, 7-40.

BEHNKE, H. AND G. KÖHTE

1963 "Otto Toeplitz zum Gedächtnis", *JDMV* 66, 1-16.

BELL, E.T.

1927 "On the Arithmetic of Logic", *TAMS* 29, 597-611.

1938 "Fifty Years of Algebra in America. 1888-1938", in R.C. Archibald (ed.) *Semicentennial Addresses of the AMS*, Vol. 1, New York, AMS, 1-34.

1945 *The Development of Mathematics*, 2d ed., New York, McGraw-Hill.

BELL, J.L.

1981 "Category Theory and the Foundations of Mathematics", *BJPS* 32, 349-358.

BENIS-SINACEUR, H.

1984 "De D. Hilbert à E. Artin: Les différents aspects du dix-septième problème et les filiations conceptuelles de la théorie des corps réels clos", *AHES* 29, 267-286.

1987 "Structure et concept dans l'epistémologie mathématique de Jean Cavaillès", *Revue d'histoire des Sciences* 6, 5-30.

1991 *Corps et Modèles*, Paris, Vrin.

BERGGREN, J.L.

1984 "History of Greek Mathematics: A Survey of Recent Research", *HM* 11, 394-410.

BERNAYS, P.

1935 "Hilberts Untersuchungen über die Grundlagen der Arithmetik", in Hilbert *GA* Vol. 3., 178-191.

BERNKOPF, M.

1966 "The Development of Function Spaces with Particular Reference to their Origins in Integral Equation Theory", *AHES* 4, 1-19.

BIERMANN, K.R.

1971 "Richard Dedekind", *DSB* 4, 1-4.

1988 *Die Mathematik und ihre Dozenten an der Berliner Universität 1810-1933*, Berlin, Akademie Verlag.

BIRKHOFF, G.H.
1933 "On the combination of Sub-Algebras", *PCPS* 29, 441-464.
1934 "On the Lattice Theory of Ideals", *BAMS* 40, 613-619.
1935 "On the Structure of Abstract Algebras", *PCPS* 31, 433-454.
1938 "Lattices and their Applications", *BAMS* 44, 793-800.
1948 *Lattice Theory*, AMS Coll. Publ., 2d edn., Providence, AMS.
1970 "What Can Lattices Do for You?", in Abbot (ed.) (1970), 1-40.
1973 "Current Trends in Algebra", *AMM* 80, 760-782.
1974 Review of Novy (1973).
1977 "Some Leaders in American Mathematics", in D. Tarwater (ed.) (1977), 25-78.

BIRKHOFF, G. AND KREYSZIG, E.
1984 "The Establishment of Functional Analysis", *HM* 11, 258-321.

BIRKHOFF, G. AND S. MAC LANE
1941 *A Survey of Modern Algebra*, New York, MacMillan.
1992 "A Survey of Modern Algebra - The Fiftieth Anniversary of its Publication", *MI* 14, 26-31.

BISHOP, E.
1975 "The Crisis in Contemporary Mathematics", *HM* 2, 507-517.

BLISS, G.A.
1933 "Eliakim Hastings Moore", *BAMS* 39, 831-838.

BLUMBERG, R.
1913 "Research Announcement", *BAMS* 19, 58-59.

BLUMENTHAL, O.
1935 "Lebensgeschichte", in Hilbert *GA* Vol. 3, 387-429.

BOAS, R.
1970 "Nicolas Bourbaki", *DSB* 2, 351-353.
1974 "Edward V. Huntington", *DSB* 6, 570.
1986 "Bourbaki and Me", *MI* 8, 84.

BOOS, W.
1985 "'The True' in Frege's «Über die Grundlagen der Geometrie»", *AHES* 34, 141-192.

BOREL, A.
1998 "Twenty-Five Years with Nicolas Bourbaki", *Notices AMS* 45(3), 373-380.

BORSUK, K.
1936 "Über den Lusternik-Schnirelmannschen Begriff der Kategorie", *FM* 26, 123-136.

BOS, H.J.M.
1981 "On the Representation of Curves in Descartes' *Géometrie*", *AHES* 24, 295-338.
1993 *Lectures in the History of Mathematics*, Providence, AMS/LMS.

BOURBAKI, N.
1939- *Eléments de mathématique*, 10 vols., Paris, Hermann.
1949 "The Foundations of Mathematics", *JSL* 14, 1-8.
1950 "The Architecture of Mathematics", *AMM* 67, 221-232.
1966 *General Topology*, 2 vols., Paris, Hermann.
1968 *Theory of Sets*, Paris, Hermann.

1969   *Eléments d'histoire des mathématiques* (Deuxième édition revuée, corrigée, augmentée), Paris, Hermann.

1972   *Commutative Algebra*, Paris, Hermann.

1972a   "Univers", in E. Artin, A. Grothendieck and J.L. Verdier (eds.) *Théorie des topos et cohomologie étale des schémas* (SGA 4), Heidelberg, Springer.

1973   *Algebra*, Paris, Hermann.

1974   *Eléments d'histoire des mathématiques* (Nouvelle èdition augmentée), Paris, Hermann.

1980   *Homological Algebra*, Paris, Hermann.

BRAUER, R.

1967   "Emil Artin", *BAMS* 73, 27-43.

BREWER J.W. AND M.K. SMITH (EDS.)

1981   *Emmy Noether, A Tribute to her Life and Work*, New York/Basel, Marcel Dekker, Inc.

BRIGAGLIA, A.

1984   "La Teoria Generale delle Algebre in Italia, 1919-1937", *Riv. Sto. Sci.* 1 (2), 199-237.

BROWDER, F.E.

1975   "The Relation of Functional Analysis to Concrete Analysis in 20th Century Mathematics", *HM* 2, 577-590.

BROWDER, F.E. (ED.)

1970   *Functional Analysis and Related Fields*, Berlin, Springer,

1976   *Mathematical Problems Arising from Hilbert Problems*, Proceeding of Symposia in Pure Mathematics, Vol. 28, Providence, AMS.

BRUCK, R.H.

1944   "Simple Quasigroups", *BAMS* 50, 769-781.

1949   "Contributions to the Theory of Loops", *TAMS* 60, 245-354,

BUCHSBAUM, D.

1955   "Exact Categories and Duality", *TAMS* 80, 1-34.

BUNGE, M.

1961   "Laws of Physical Laws", *American Journal of Physics* 29, 518-529.

BURNSIDE, W.

1897   *Theory of Groups of Finite Order*, Cambridge, Cambridge University Press.

CARTAN, H.

1937   "Théorie des filtres", *C.R. Acad. sci.* 205, 595-598.

1937a   "Filtres et ultrafiltres", *C.R. Acad. sci.* 205, 777-779.

1946   "Méthodes modernes en topologie algébrique", *Commentarii Math. Helvetici* 18, 1-15.

1958   "Structures algébriques", in Cartan et al. (1958), 5-15.

1980   "Nicolas Bourbaki and Contemporary Mathematics", *MI* 2, 175-180. (English transl. by Kevin Lenzen of "Nicolas Bourbaki und die heutige Mathematik", *AFLN* 76, 5-18.)

CARTAN, H. AND S. EILENBERG

1956   *Homological Algebra*, Princeton, Princeton University Press.

CARTAN, H. ET AL.

1958   *Structures algébriques et structures topologiques*, Paris, Bulletin de l'association des professeurs de mathématiques de l'enseignement public.

CARTIER, P.

1998   "The Continuing Silence of Bourbaki", *MI* 20 (1), 22-28.

CAYLEY, A.
  1854   "On the Theory of Groups, as Depending on the Symbolic Equation $\theta^n=1$", *Philosoph-ical Magazine* 7, 123-130.
  1878   "The Theory of Groups; Graphical Representations", *AJM* 1, 403-405.

CHANDLER, B. AND W. MAGNUS
  1982   *The History of Combinatorial Group Theory: A Case Study in the History of Ideas*, New York, Springer.

CHANG, C.C.
  1974   "Model theory 1945-1971", in L. Henkin et al. (eds.) *Proceedings of the Tarski Sympo-sium*, Providence, AMS, 173-186.

CHANG, C.C. AND J. KEISLER
  1990   *Model Theory*, 3d. ed., Amsterdam, North-Holland.

CHEVALLEY, C.
  1956   *Fundamental Concepts of Algebra*, New York, Academic Press.

CHOQUET, G.
  1963   "Die Analysis und Bourbaki", *Math.-Phys. Semesterberichte* 9, 1-21.

CHOUCHAN, M.
  1995   *Nicolas Bourbaki - Faits et légendes*, Paris, Éditions du Choix.

COHN, P.M.
  1965   *Universal Algebra*, New York, Harper & Row. (2nd. rev. ed. 1981, Dordrecht, Reidel.)

COLLISON, M.J.
  1977   "The Origins of the Cubic and Biquadratic Reciprocity Laws", *AHES* 16, 63-69.

CONTRO, W.
  1976   "Von Pasch bis Hilbert", *AHES* 15, 283-295.

CORRY, L.
  1989   "Linearity and Reflexivity in the Growth of Mathematical Knowledge", *SIC* 3, 409-440.
  1993   "Kuhnian Issues, Scientific Revolutions and the History of Mathematics", *SHPS* 24, 95-117.
  1994   "La Teoría de las Proporciones de Eudoxio vista por Dedekind", *Mathesis* 10, 1-24.
  1997   "David Hilbert and the Axiomatization of Physics (1894-1905)", *AHES* 51: 89-197
  2000   "The Empirical Roots of Hilbert's Axiomatic Method", in V. F. Hendricks et al. (eds.) *Proof Theory: History and Philosophical Significance*, Dordrecht, Kluwer, 35-54.

CRILLY, A.J.
  1986   "The Rise of Cayley's Invariant Theory (1841-1862)", *HM* 13, 241-254.

CROWE, M.
  1967   *A History of Vector Analysis*, Notre Dame, Notre Dame University Press.
  1975   "Ten 'Laws' concerning Patterns of Change in the History of Mathematics", *HM* 2, 161-166. (Repr. in Gillies (ed.) (1992), 15-20.)

CURTIS, C.W.
  1999   *Pioneers of Representation Theory: Frobenius, Burnisde, Schur, and Brauer* (History of Mathematics, Vol. 15), Providence, AMS/LMS.

DA COSTA, N.C.A.

1986 "Tarski, Sebastião e Silva e o conceito de estrutura", *Boletim da Sociedade Paranaense de Matemática*, 7, 137-145.

DA COSTA, N.C.A. AND R. CHUAQUI

1988 "On Suppes' Set Theoretical Predicates", *Erkenntnis* 29, 95-112.

DAHAN, A.

1980 "Les Travaux de Cauchy sur les Substitutions. Étude de son approche du concept de groupe", *AHES* 23, 279-319.

DARBOUX, G.

1875 "Sur la composition des forces en statique", *BSMA* 8, 281-288.

DAUBEN, J.

1994 "Mathematics: An Historian Perspective", in Sasaki et al. (eds.) (1994), 1-14.

DAVIS, P.J. AND R. HERSH

1981 *The Mathematical Experience*, Boston, Birkhäuser.

1986 *Descartes' Dream. The World According to Mathematics*, San Diego, Harcourt Brace Jovanvich.

DEDEKIND, R.

Werke *Gesammelte mathematische Werke*, 3 vols., ed. by R. Fricke, E. Noether und O. Ore, Braunschweig (1930-1932). (Chelsea reprint, New York, 1969.)

1854 "Über die Einführung neuer Funktionen in der Mathematik", *Werke* Vol. 3, 428-438.

1871 "Über die Komposition der binären quadratischen Formen", *Werke* Vol.3, 223-261.

1872 *Stetigkeit und irrationale Zahlen*, Braunschweig. (*Werke* Vol. 3, 315-334.)

1876-7 "Sur la théorie des nombres entiers algébriques", *BSMA* 11 (1876), 278-288; 12 (1877), 17-41, 69-92, 144-164, 207-248. (Partially in *Werke* Vol.3, 262-296.)

1877 "Über die Anzahl der Ideal-Classen in den verschiedenen Ordnungen eines endlichen Körpers", *Festschrift der Technischen Hochschule Braunschweig zur Säkularfreier des Geburtstages von C.F. Gauß*, Braunschweig. (*Werke* Vol. 1, 105-157.)

1879 "Über die Theorie der ganzen algebraischen Zahlen", *Werke* Vol. 3, 297-313.

1882 "Über die Diskriminanten endlicher Körper", *Abhandlungen der Königlichen Gesellschaft der Wissenschaften zu Göttingen* 19, 1-56. (*Werke* Vol. 1, 351-396.)

1888 *Was sind und was sollen die Zahlen?*, Braunschweig. (*Werke* Vol. 3, 335-391.)

1894 "Über die Theorie der ganzen algebraischen Zahlen", in Dirichlet 1894, 434-657.

1895 "Über die Begründung der Idealtheorie", *GN* (1895), 106-113. (*Werke* Vol. 2, 50-58.)

1897 "Über Zerlegungen von Zahlen durch ihre grössten gemeinsamen Teiler", *Festschrift der Technischen Hochschule Braunschweig*. (*Werke* Vol. 2, 103-147.)

1897a "Über Gruppen, deren sämtliche Teiler Normalteiler sind", *MA* 48, 548-561. (*Werke* Vol. 2, 87-101.)

1900 "Über die von drei Moduln erzeugte Dualgruppe", *MA* 53, 371-403. (*Werke* Vol. 2, 236-271.)

1901 "Über die Permutationen des Körpers aller algebraischen Zahlen", *Abh. Ges. Wiss. Göttigen 1901 (Festschrift)*.

1964 *Über die Theorie der ganzen algebraischen Zahlen* (Foreword by B.L. van der Waerden), Braunschweig.

1981 "Eine Vorlesung über Algebra", in Scharlau (ed.) (1981), 59-100.

DEDEKIND, R. AND H. WEBER
1882 "Theorie der algebraischen Funktionen einer Veränderlichen", *JRAM* 92, 181-290. (Dedekind *Werke* Vol. 1, 238-350.)

DICK, A.
1981 *Emmy Noether, 1882-1935.* Boston, Birkhäuser. (English trans. by H.I. Blocher of the German original (1970).)

DICKSON, L.E.
1903 "Definition of a Field by Independent Postulates", *TAMS* 4, 13-20.
1903a "Deafinition of a Linear Associative Algebra by Independent Postulates", *TAMS* 4, 21-27.
1905 "Definition of a Group and a Field by Independent Postulates", *TAMS* 6, 198-204.
1920 *History of the Theory of Numbers. Vol II: Diophantine Analysis,* Washington, The Carnegie Institution.
1926 *Modern Algebraic Theories,* Chicago, Benjamin H. Sanborn. (Repr. 1959: *Algebraic Theories.*)

DICKSON, L.E. ET AL.
1923 *Algebraic Numbers,* New York, Dover.

DIEDERICH, W.
1989 "The Development of Structuralism", *Erkenntnis* 30, 363-386.

DIEUDONNÉ, J.
1969 "Richard Dedekind", in *Enciclopaedia Universalis* 5, 373-375.
1970 "The Work of Nicolas Bourbaki", *AMM* 77, 134-145.
1979 "The Difficult Birth of Mathematical Structures. (1840-1940)", in U. Mathieu and P. Rossi (eds.) *Scientific Culture in Contemporary World*, Milan, Scientia, 7-23.
1982 "The Work of Bourbaki in the Last Thirty Years", *Notices AMS* 29, 618-623.
1982a *A Panorama of Pure Mathematics. As seen by Nicolas Bourbaki,* New York, Academic Press. (English trans. by I.G. Mac Donald of *Panorama des mathématiques pures. Le choix bourbachique* (1977).)
1984 "Emmy Noether and Algebraic Topology", *JPAA* 31, 5-6.
1985 *History of Algebraic Geometry. An Outline of the Historical Development of Algebraic Geometry,* Monterrey, Wadsworth. (English transl. by J.D. Sally of *Cours de géometrie algébrique I* (1974), Paris, PUF.)
1987 *Pour l'honneur de l'esprit humain. Les mathématiques ajourd'hui,* Paris, Hachette.

DIEUDONNÉ, J. (ED.)
1978 *Abrégé d'histoire des mathématiques, 1700-1900,* 2 vols., Paris, Hermann.

DIEUDONNÉ, J. AND J.B. CARRELL
1971 *Invariant Theory, Old and New,* New York, Academic Press.

DIEUDONNÉ, J. AND J. TITS
1987 "Claude Chevalley (1909-1984)", *BAMS* 17, 1-7.

DIRICHLET, P.G.L.
1825 "Mémoire sur l'impossibilité des quelques équations indeterminées des cinquième degré", *JRAM* 3, 354-375.
1864 "Zur Theorie der complexen Einheiten", in *Werke* (ed. L. Kronecker) Vol. 1, 640-644.

1894   *Vorlesungen über Zahlentheorie* (4th. ed.), edited and with supplements by R. Dedekind, Braunschweig (1st. ed. 1863; 2d. ed. 1871; 3d. ed. 1879). (Chelsea reprint, New York, 1968.)

DIPERT, R.R.
1991   "The Life and Work of Ernst Schröder", *ML* 1, 119-139.

DOLD-SAMPLONIUS, Y.
1997   "In Memoriam: Bartel Leendert van der Waerden (1903-1996)," *HM* 24, 125-130.

DORIER, J.L.
1995   "A General Outline of the Genesis of Vector Space Theory", *HM* 22, 227-261.

DORWART, H.
1989   "Mathematics at Yale in the Nineteenth Twenties", in P. Duren (ed.) (1988-9) Vol. 2, 87-98.

DRESHER, M. AND O. ORE
1938   "Theory of Multigroups", *AJM* 60, 705-733.

DUBREIL, P. AND M.L. DUBREIL-JACOTIN
1939   "Théorie des relations d'équivalence", *JMPA* 18, 63-95.
1964   *Leçons d'algèbre moderne*, Paris, Dunod.

DUFFIN, R.J.
1938   "Structure Elements of Quasi-Groups" (Abstract), *BAMS* 44, 333.

DUFFIN, R.J. AND R.S. PATE
1943   "An Abstract Theory of the Jordan-Hölder Composition Series", *DMJ* 10, 743-750.

DUGAC, P.
1973   "Eléments d'analyse de Karl Weierstrass", *AHES* 10, 41-176.
1976   *Richard Dedekind et les Fondements des Mathématiques*, Paris, Vrin.
1978   "Fondements de l'Analyse", in Dieudonné (ed.) (1978), 335-392.

DUREN, P. (ED.)
1988-9   *A Century of Mathematics in America*, 3 Vols., Providence, AMS.

DYCK, W.V.
1882   "Gruppentheoretische Studien", *MA* 20, 1-44.
1883   "Gruppentheoretische Studien II. Über die Zusammensetzung einer Gruppe discreter Operationen, über ihre Primitivität und Transitivität", *MA* 22, 70-108.

EDWARDS, H.M.
1975   "The Background of Kummer's Proof of Fermat's Last Theorem for Regular Primes", *AHES* 14, 219- 236.
1977   *Fermat's Last Theorem - A Genetic Introduction to Algebraic Number Theory*, New York, Springer.
1977a   "Postscript to 'The Background of Kummer's Proof of Fermat's Last Theorem for Regular Primes'", *AHES* 14, 219-236.
1980   "The Genesis of Ideal Theory", *AHES* 23, 321-378.
1983   "Dedekind's Invention of Ideals", *Bull. LMS* 15, 8-17.
1987   "An Appreciation of Kronecker", *MI* 9, 28-35.
1988   "Kronecker's Place in History", in Aspray & Kitcher (eds.) (1988), 139-144.
1990   *Divisor Theory*, Boston, Birkhäuser.
1992   "Mathematical Ideals, Ideals and Ideology", *MI* 14(2), 6-18.

EDWARD, H., O. NEUMANN, AND W. PURKERT
1982   "Dedekinds 'Bunte Bemerkungen' zu Kroneckers 'Grudzüge'", *AHES* 27, 49-85.
EHRESMANN, CH.
1981   *Ouevres complètes et comentées*, ed. by A. C. Ehresmann, Amiens.
1957   "Gattungen von lokalen Strukturen," *JDMV* 60, 49-77. (Repr. in Ehresmann (1981) II-1, 125-153.)
1965   *Categories et Structures*, Paris, Dunod.
EILENBERG, S.
1936   "Transformations continues en circonférence et la topologie du plan", *FM* 26, 61-112.
1936a   "Sur les espaces multicohérents", *FM* 27, 153-190.
1942   Review of Bourbaki (1939-), Book I, Fascicule de resultats (1939), *MR* 3, # 55.
1945   Review of Bourbaki (1939-), Book II, Ch. 1, *MR* 6, 113.
EILENBERG, S. AND S. MAC LANE
1942   "Natural Isomorphisms in Group Theory", *PNAS* 28, 537-543.
1942a   "Group Extensions and Homology", *AM* 43, 757-831
1945   "General Theory of Natural Equivalences", *TAMS* 28, 231-294.
1986   *Eilenberg-Mac Lane: Collected Works*, Orlando, Fla., Academic Press.
EILENBERG, S. AND N. STEENROD
1945   "Axiomatic Approach to Homology Theory", *PNAS* 31, 117-120,
1952   *Foundations of Algebraic Topology*, Princeton, Princeton University Press.
EISENSTEIN, F.G.
Werke   *Mathematische Werke*, New York, Chelsea Reprints (1975).
1844   "Lois de Reciprocité", *JRAM* 28, 53-67. (*Werke*, 126-140.)
1845   "Applications de l'algèbre a l'artihmétique transcendante", *JRAM* 29, 177-184. (*Werke*, 291-298.)
1847   "Genaue Untersuchung der unendlichen Doppelproducte...", *JRAM* 35, 153-274. (*Werke*, 357-479.)
1850   "Über einige allgemeine Eigenschaften der Gleichung...", *JRAM* 39, 224-287. (*Werke*, 556-619.)
1850a   "Beweis der allgemeinsten Reciprocitätsgesetze zwischen reellen und complexen Zahlen", *Monatsberichte BA*, 189-198. (*Werke*, 712-721.)
ELKANA, Y.
1981   "A Programmatic Attempt at an Anthropology of Knowledge", in E. Mendelsohn and Y. Elkana (eds.) *Sciences and Cultures*, Sociology of the Sciences, Vol. 5, Reidel, Dordrecht, 1-70.
1986   "The Emergence of Second-order Thinking in Classical Greece", in S.N. Eisenstadt (ed.) *Axial Age Civilizations*, Albany, State University of New York Press, 40-64.
ELLISON, W. AND F. ELLISON
1978   "Théorie des nombres", in Dieudonné (ed.) (1978), 151-236.
ENGELER, E. AND H. ROHRL
1969   "On the Problem of Foundations of Category Theory", *Dialectica* 23, 58-66.
FABER, G.
1935   "Walther v. Dyck", *JDMV* 45, 89-98.
FANG, J.
1970   *Bourbaki: Towards a Philosophy of Modern Mathematics I*, New York, Paideia Press.

FEFERMAN, S.

1977 "Categorical Foundations and Foundations of Category Theory", in R. Butts and J. Hintikka (eds.) *Logic, Foundations of Mathematics and Computability Theory*, Dordrecht, Reidel, 149-169.

1984 "Foundational Ways", in *Perspectives in Mathematics*, Anniversary of Overwolfach, Basel, Birkhäuser, 147-158.

1985 "Working Foundations", *Synthese* 62, 229-254.

FENSTER, D.D.

1997 "Role Modelling in Mathematics: the Case of Leonard Eugene Dickson", *HM* 24, 7-24.

1998 "Leonard Eugene Dickson and his Work in the Arithmetics of Algebra", *AHES* 52, 119-159.

1999 "The Development of the Concept of an Algebra: Leonard Eugene Dickson's Role", *Rendiconti Circ. Mat. di Palermo, Ser. II*, Suppl. 61, 59-122.

FERREIRÓS, J.

1993 "On the Relations between Georg Cantor and Richard Dedekind", *HM* 20, 343-363.

1999 *Labyrinth of Thought. A History of Set Theory and its Role in Modern Mathematics* (Science Networks, Historical Studies: Vol. 23), Basel, Boston, Berlin, Birkhäuser. (A earlier, reduced version appeared as: *El nacimiento de la teoría de conjuntos*, Ediciones de la Universidad Autónoma de Madrid, 1993).

FINGERMAN, J.J.

1981 *The Historical and Philosophical Significance of the Emergenece of Point-Set Topology*, Unpublished Diss., The University of Chicago.

FISHER, C.S.

1966 "The Death of a Mathematical Theory: A Study in the Sociology of Knowledge", *AHES* 3, 137-159.

FORMAN, P.

1974 "Carl Runge", *DSB* 11, 610-615.

FOWLER, D.

1992 "Dedekind's Theorem $\sqrt{2} \times \sqrt{3} = \sqrt{6}$", *AMM* 99, 725-733.

FOX, R.H.

1941 "On the Lusternik-Schnirelmann Category", *AM* 42, 333-370.

FRAENKEL, A.

1912 "Axiomatische Begründung von Hensels p-adischen Zahlen", *JRAM* 141, 43-76.

1914 "Über die Teiler der Null und die Zerlegung von Ringen", *JRAM* 145, 139-176.

1916 *Über gewisse Teilbereiche und Erweiterungen von Ringen*, Leipzig, Teubner.

1921 "Über einfache Erweiterungen zerlegbarer Ringe", *JRAM* 151, 121-166.

1938 "Alfred Loewy (1873-1935)", *SM* 5, 17-22.

1967 *Lebenskreise. Aus den Erinnerungen eines jüdischen Mathematikers*, Stuttgart, Deutsche Verlags-Anstalt.

FREI, G.

1993 "Bartel Leendert van der Waerden zum 90. Geburtstag", *HM* 20, 5-11.

1994 "The Reciprocity Law from Euler to Eisenstein", in Sasaki et al. (eds.) (1994), 67-90.

FREI, G. (ED.)

1985 *Der Briefwechsel David Hilbert-Felix Klein (1886-1918)*, Göttingen, Vandenhoeck & Ruprecht.

FREI, G. AND U. STAMMBACH

1992   *Hermman Weyl und die Mathematik and der ETH Zürich, 1913-1930*, Basel, Birkhäuser.

FREUDENTHAL, H.

1956   "Zur Geschichte der Grundlagen der Geometrie. Zugleich eine Besprechung der 8. Auflage von Hilberts Grundlagen der Geometrie", *Nieuw Archief v. Wiskunde* 5, 105-142.

1973   *Mathematics as an Educational Task*, Dordrecht, Reidel.

1974   "Adolf Hurwitz", *DSB* 6, 570-572.

1974a   "The Impact of von Staudt's Foundations of Geometry", in R. Cohen et al. (eds.) *For Dirk Struik*, Dordrecht, Reidel, 189-200.

1981   Review of Dieudonné (ed.) (1978), *ISIS* 72, 660-661.

FREYD, P.

1964   *Abelian Categories: An Introduction to the Theory of Functors*, New York, Harper and Row.

FRICKE, R.

1924   *Lehrbuch der Algebra - verfasst mit Benutzung vom Heinrich Webers gleichnamigem Buche*, Braunschweig, Vieweg.

FRIED, M. AND S. UNGURU

2001   *Apollonius of Perga's 'Conica'. Text, Subtetx, Context*, Leiden, Brill.

FRIEDMANN, J.

1975   *L'Origine et le Développement de Bourbaki*, Unpublished Diss., Centre Alexandre Koyré, EPHESS, Paris.

FROBENIUS G. AND L. STICKELBERGER

1879   "Über Gruppen von vertauschbaren Elementen", *JRAM* 86, 217-262.

FUCHS, W.

1982   "Zum 150-jährigen Geburtstag von Richard Dedekind", *Gauss-Gesellschaft Mitteilungen* 19, 79-97.

FUETER, R.

1905   "Die Theorie der Zahlstrahlen", *JRAM* 130, 197-237.

1907   "Die Theorie der Zahlstrahlen. II", *JRAM* 132, 255-269.

1917   *Synthetische Zahlentheorie*, Berlin/Leipzig, Göschensche Verlag.

GABRIEL, G. ET AL. (EDS.)

1976   *Gottlob Frege - Wissenschaftlischer Briefwechsel*, Hamburg, Felix Meiner.

GANDY, R.O.

1959   Review of Bourbaki (1949), *JSL* 24, 71-73.

GAUSS, C.F.

Werke   *Werke*, 12 Vols., Göttingen, Königlichen Gesellschaft der Wissenschaften (1863-1927).

1801   *Disquisitiones Arithmeticae*, Leipzig (*Werke* Vol. 1).

1831   "Theoria residuorum biquadraticorum. Commentatio secunda", *Göttingische Gelehrte Anzeigen* (23. IV.1831). (*Werke* Vol. 2, 169-178.)

1832   "Theoria residuorum biquadraticorum. Commentatio secunda", *Commentationes Societatis Regiae Scientarum Gött.* 7 (*Werke* Vol. 2, 95-148.)

GAUTHIER, Y.
1969 "La notion theorétique de structure", *Dialectica* 23, 217-227.
1976 *Fondements des Mathématiques*, Montreal, Les Presses de l'Universite de Montreal.
1972 Review of Bourbaki (1939-), Book I, *Canadian Mathematical Bulletin* 15, 623-626.

GEORGE, E.
1939 "Über den Satz von Jordan-Hölder", *JRAM* 180, 110-120.

GEYER, W.D.
1981 "Die Theorie der algebraischen Funktionen einer Veränderlichen nach Dedekind und Weber", in Scharlau (ed.) (1981), 109-131.

GILLIES, D. (ED.)
1992 *Revolutions in Mathematics*, Oxford, Clarendon Press.

GILLOIS, M.
1965 "Rélation d'identité en génétique", *Annales de l'institut H. Poincaré* 2, 1-94.

GILMER, S.
1981 "Commutative Ring Theory", in Brewer & Smith (eds.) (1981), 131-143.

GODEMENT, R.
1958 *Théorie de faisceaux*, Paris, Hermann.

GOLDSTEIN, CH.
1992 "On a Seventeenth Century Version of the 'Fundamental Theorem of Arithmetic'", *HM* 19, 177-187.

GORDAN, P.
1868 "Beweis, dass jede Covariante und Invariante einer binären Form eine ganze Function mit numerischen Coefficienten einer endlichen Anzahl solchen Formen ist", *JRAM* 69, 323-354.
1893 "Über einen Satz von Hilbert", *MA* 42, 132-142.

GRABINER, J.
1975 "The Mathematician, the Historian, and the History of Mathematics", *HM* 2, 439-447.

GRATTAN-GUINNESS, I.
1979 Review of Dieudonné (ed.) (1978), *Annals of Science* 36, 653-655.
1990 "Does the History of Science Treat of the History of Science? The Case of Mathematics," *History of Science* 28, 149-173.
1992 "Structure-Similarity as a Cornerstone of the Philosophy of Mathematics", in J. Echeverría et al. (eds.) *The Space of Mathematics*, New York/Berlin, Walter de Gruyter, 93-111.

GRÄTZER, G.
1968 *Universal Algebra*, Princeton, Princeton University Press.

GRAY, J.J.
1984 "A Commentary on Gauss's Mathematical Diary, 1796-1814, with an English Translation", *Expositiones Mathematicae* 2, 97-130.
1990 "Herausbildung von strukturellen Grundkonzepten der Algebra im 19. Jahrhundert", in Scholz (ed.) (1990), 293-323.
1997 "Algebraic Geometry Between Noether and Noether: a Forgotten Chapter in the History of Algebraic Geometry", *Revue d'histoire des math.* 3, 1-48.
2000 *The Hilbert Challenge*, Oxford, Oxford University Press.

GRAY, J.W.
1979   "Fragments of the History of Sheaf Theory", in M. Fourman, C. Mulvey and D. Scott
        (eds.) *Applications of Sheaves*, LNM 753, Berlin, Springer, 1-79.

GRIFFIN, H.
1940   "The Abelian Quasi-groups", *AJM* 62, 725-737.

GROSHOLZ, E.R.
1985   "Two Episodes in the Unification of Logic and Topology", *BJPS* 36, 147-157.

GROTHENDIECK, A.
1951   "Critères de compacité dans les espaces fonctionneles généraux", *AJM* 74, 168-186.
1957   "Sur quelques points d'algèbre homologique", *Tôhoku Math. Journal* 9, 119-221.
1958   "The Cohomology Theory of Abstract Algebraic Varieties" in *PICM* Edinburgh, 103-
        118.

GROTHENDIECK, A. AND J. DIEUDONNÉ
1968   "Les travaux de Alexander Grothendieck", *PICM* Moscow (1966), 21-24.
1971   *Eléments de géométrie algébrique*, Berlin, Springer.

GUEDJ, D.
1985   "Nicolas Bourbaki. Collective Mathematician. An interview with Claude Chevalley",
        MI 7, 18-22. (Originally appeared in *Dèdales*, Nov. 1981.)

GUÉRINDON, J. AND J. DIEUDONNÉ
1978   "L'Algèbre depuis 1840", in Dieudonné (ed.) (1978), 91-127.

GUILLAUME, M.
1978   "Axiomatique et logique", in Dieudonné (ed.) (1978), 315-364.

HALL, F.M.
1966   *An Introduction to Abstract Algebra*, 2 Vols., Cambridge, Cambridge University Press.

HALMOS, P.
1953   Review of Bourbaki (1939-), Book VI, Ch. 1-4 (1952), *BAMS* 59, 249-55.
1955   Review of Bourbaki (1939-), Book I, Ch. 1-2 (1954), *MR* 16, #454.
1956   Review of Bourbaki (1939-), Book I, Ch. 3 (1956), *MR* 17, #1062.
1957   "Nicolas Bourbaki", *SA* 196 (May), 88-99.
1970   Review of Fang (1970), *MR* 40, #7066.

HALSTED, G.B.
1902   "The Betweeness Assumption", *AMM* 9, 98-101.

HAMEL, G.
1905   "Über die Zusammensetzung von Vektoren", *ZMP* 49, 363-371.

HANNAK, J.
1991   *Emanuel Lasker, the Life of a Chess Master*, New York, Dover (Translation from the
        German original, Berlin, 1953).

HASSE, H.
1926   *Höhere Algebra*, 2 vols., Berlin, Sammlung Göschen.
1930   "Die moderne algebraische Methode", *JDMV* 39, 22-34. (Repr. in *MI* 8, 18- 23 (1986).
        English trans. by Abe Shenitzer.)
1932   "Zu Hilberts algebraisch-zahlentheoretischen Arbeiten", in Hilbert *GA* Vol. 1, 528-535.
1949   "Kurt Hensel zum Gedächtnis", *JRAM* 187, 1-13.

1954   *Higher Algebra*, New York, Frederic Ungar. (English trans. of the 3d ed. of Hasse (1926) by Theodor J. Benac.)

1967   "History of Class Field Theory", in J.W.S. Cassels & A. Fröhlich (eds.) *Algebraic Number Theory*, London, Academic Press.

HAUBRICH, R.

1988   "Zur Entstehung der Idealtheorie Dedekinds", Unpublished Diplomarbeit, Georg-August- Universität zu Göttingen.

1993   "Die Herausbildung der algebraischen Zahlentheorie als wissenschaftliche Disziplin", Unpublished manuscript read at the "Fourth Annual Göttingen Workshop on the History of Modern Mathematics".

1996   *Zur Entstehung der algebraischen Zahlentheorie Richard Dedekinds*, Basel, Birkhäuser (Forthcoming).

HAUPT, O.

1929   *Einführung in die Algebra*, Leipzig.

HAUSMANN, B.A. AND O. ORE

1937   "Theory of Quasi-Groups", *AJM* 59, 983-1004.

HAUSDORFF, F.

1914   *Grundzüge der Mengenlehre*, Leipzig, Veit.

HAWKINS, T.

1971   "The Origins of the Theory of Group Characters", *AHES* 7, 142-170.

1974   "New Light on Frobenius' Creation of the Theory of Group Characters", *AHES* 12, 217-243.

2000   *Emergence of the Theory of Lie Groups. An Essay in the History of Mathematics, 1869-1926*, (Sources and Studies in the History of Mathematics and Physcal Sciences), New York, Springer Verlag.

HENSEL, K.

1899   "Über eine neue Begründung der Theorie der algebraischen Zahlen", *JDMV* 6, 83-88.

1908   *Theorie der algebraischen Zahlen*, Leipzig, Teubner.

1910   *Ernst Eduard Kummer und der große Fermatsche Satz*, Marburg.

1913   *Zahlentheorie*, Berlin/Leipzig, Göschensche Verlag.

HENSEL, K. AND G. LANDSBERG

1902   *Theorie der algebraischen Funktionen einer Variablen und ihre Anwendung auf algebraische Kurven und Abelsche Integrale*, Teubner, Leipzig.

HERMANN, R.

1986   "Mathematics and Bourbaki", *MI* 8, 32-33.

HERMITE, C.

Oeuv   *Oeuvres*, 4 Vols., ed. by E. Picard (1905-1917), Paris, Gauthiers-Villars.

HERSH, R.

1971   "Some Proposals for Reviving the Philososphy of Mathematics", *Advances in Mathematics* 31, 31-50. (Repr. in Tymozcko (ed.) (1985), 9-28.)

HEWITT, E.

1956   Review of Bourbaki (1939-), Book IV (1953-55), *BAMS* 62, 507-508.

1966   Review of Bourbaki (1939-), Book VI, Ch. 7-8, *MR* 31, 3539.

HILBERT, D.

GA   *Gesammelte Abhandlungen*, 3 vols., Berlin, Springer, (1932-1935; 2d ed. 1970).

1888-9   "Zur Theorie der algebraischen Gebilde", *GN* I (1888), 450-457; II (1889), 25-34; III (1889) 423-430. (*GA* Vol. 2, 176-198.)

1889   "Über die Endlichkeit des Invariantensystems für binäre Grundformen", *MA* 33, 223-226. (*GA* Vol. 2, 162-164.)

1890   "Über die Theorie der algebraischen Formen", *MA* 36, 473-534. (*GA* Vol. 2, 199-257.)

1893   "Über die vollen Invariantensysteme", *MA* 42, 313-373. (*GA* Vol. 2, 287-344.)

1896   "Über die Theorie der algebraischen Invarianten," in *Mathematical Papers read at the International Mathematical Congress,* Chicago 1893, 116-124. (*GA* Vol. 2, 376-383.)

1897   "Die Theorie der algebraischen Zahlkörper (Zahlbericht)", *DMV* 4, 175-546. (*GA* Vol. 1, 63-363.)

1898   "Über die Theorie der relativ-Abelschen Zahlkörper", *GN* (1898) 370-399. (Also in *Acta Mathematica* 26, 99-132, & *GA* Vol. 1, 483-500.)

1898-9   *Mechanik* (Vorlesung WS 1898-99, Göttingen), Niedersächssiche Staats - und Universitätsbibliothek Göttingen, Cod. Ms. David Hilbert 558.

1899   *Grundlagen der Geometrie*, Leipzig, Teubner.

1899a   "Über die Theorie der relativquadratischen Zahlkörper", *MA* 51, 1-127. (*GA* Vol. 1, 370-482).

1900   "Über den Zahlbegriff", *JDMV* 8, 180-184.

1900a   "Theorie der algebraischen Zahlkörper", in W.F. Meyer (ed.) *Encyclopädie der Mathematischen Wissenschaften,* Vol. 1, Leipzig, Teubner, 675-698.

1901   "Mathematische Probleme", *Archiv f. Math. u. Phys.* 1, 213-237. (*GA* Vol. 3, 290-329.)

1902   "Mathematical Problems" *BAMS* 8, 437-479. (English transl. by M.W. Newson of Hilbert 1901.)

1905   *Logische Principien des mathematischen Denkens*, Ms. Vorlesung SS 1905, annotated by E. Helliger, Bibliothek des Mathematischen Seminars der Universität Göttingen.

1918   "Axiomatisches Denken", *MA* 78, 405-415. (*GA* Vol. 3, 146-156.)

1923   "Die logischen Grundlagen der Mathematik", *MA* 88, 151-165. (*GA* Vol. 3,178-191.)

1930   "Naturerkennen und Logik", *Die Naturwissenschaften* 959-963. (*GA* Vol. 3, 378-387.)

1971   *Foundations of Geometry.* (English transl. by Leo Unger of *Grundlagen der Geometrie* (10th ed. - 1968), revised and enlarged by Paul Bernays), La Salle, Ill., Open Court.

1992   *Natur und Mathematisches Erkennen: Vorlesungen, gehalten 1919-1920 in Göttingen. Nach der Ausarbeitung von Paul Bernays* (Edited and with an English introduction by David E. Rowe), Basel, Birkhäuser.

HILTON, P.

1982   "The Language of Categories and Category Theory", *MI* 4, 79-82.

HIRZEBRUCH, F.

1999   "Emmy Noether and Topology", in Teicher (ed.) 1999, 57-65.

HÖLDER, O.

1889   "Zurückführung einer beliebigen algebraischen Gleichung auf eine Kette von Gleichungen", *MA* 34, 26-56.

1893   "Die Gruppen der Ordnungen p^3, pq^2, pqr, p^4", *MA* 43, 301-412.

1898   "Galois'sche Theorie mit Anwendungen", *Encyclopädie der Mathematischen Wissenschaften,* Vol. 1, Leipzig, Teubner, 481-520.

HOPF, H.

1928 "Eine Verallgemeinerung der Euler-Poincaréschen Formel", *Nach. Ges. der Wiss. zu Gött.*, 127-136.

1966 "Eininge persönliche Erinnerungen aus der Vorgeschichte der heutigen Topologie", in *Colloque de Topologie* (Bruxelles 1964), Centre Belge de Recherches Mathématiques, Paris, Gauthier-Villars, 9-20.

HOWSON, A.G. (ED.)

1973 *Developments in Mathematical Education - Proceedings of the Second International Congress on Mathematical Education*, Cambridge, Cambridge University Press.

HUNTINGTON, E.V.

1902 "Simplified Definition of a Group", *BAMS* 8, 296-300.

1902a "A Second Definition of a Group", *BAMS* 8, 388-391.

1902b "A Complete Set of Postulates for the Theory of the Absolute Continuous Magnitudes", *TAMS* 3, 264-279.

1903 "Two Definitions of Abstract Groups by Sets of Independent Postulates", *TAMS* 4, 27-30.

1903a "Definitions of Fields by Sets of Independent Postulates", *TAMS* 4, 31-37.

1903b "Complete Sets of Postulates for the Theory of Real Quantities", *TAMS* 4, 358-370.

1904 "Set of Independient Postulates for the Algebra of Logic", *TAMS* 5, 288-309.

1905 "A Set of Postulates for Real Algebra, Comprising Postulates for a One- dimensional Continuum and for the Theory of Groups," *TAMS* 6, 17-41.

1905a "Note on the Definitions of Abstract Groups and Fields by Sets of Independent Postulates", *TAMS* 6, 181-97.

1905b "A Set of Postulates for Ordinary Complex Algebra", *TAMS* 6, 209-229.

1913 "A Set of Postulates for Abstract Geometry, Expressed in Terms of the Simple Relation of Inclusion", *MA* 73, 522-559.

1937 "The Method of Postulates", *Philosophy of Science* 4, 299-336.

1938 "The Duplicity of Logic", *SM* 5, 149-157.

HUREWICZ, W.

1936 "Beiträge zur Topologie der Deformationen IV. Asphärische Räume", *Proceedings of the Koninklijke Akademie van Wetenschappen te Amsterdam. Section of Sciences* 39, 215-224.

1941 "On Duality Theorems", *BAMS* 47, 562-562.

INGRAO, B. AND G. ISRAEL

1990 *The Invisible Hand*, Cambridge, Ma., MIT Press (English trans. by Ian McGilvray of *La Mano Invisibile* (1987), Roma-Bari, Gius. Laterra & Figli Spa.)

ISBELL, J.

1966 Review of Lawvere (1966), *MR* 34, #7332.

ISRAEL, G.

1977 "Un aspetto ideologico della matematica contemporanea: il 'bourbakismo'", in E. Donini, A. Rossi and Tito Tonietti (eds.) *Matematica e Fisica: Cultura e Ideologia*, Bari, De Donato Editore, 35-70.

1978 "La matematica assiomatica ed il 'bourbakismo'", in E. Casari, F. Marchetti and G. Israel (eds.) *I Fondamenti della Matematica dall'800 ad Oggi*, Firenze, Guaraldi, 46-67.

1981 "'Rigor' and 'Axiomatics' in Modern Mathematics", *Fund. Scientiae* 2, 205-219.

ISRAEL, G. AND L.L. RADICE
  1976   "Alcune recenti linee di tendenza della matematica contemporanea", in M. Daumas
         (ed.) *Storia della Scienza* (Vol. 2 - "Le scienze mathematiche e l'astronomia"), Roma-
         Bari, Editori Laterza, 162-201.
JACOBI, C.G.
  1839   "Über die complexen Primzahlen, ...", *JRAM* 19, 314-318.
JACOBSON, N.
  1946   Review of Weyl (1939), *BAMS* 52, 592-595.
  1980   *Basic Algebra II*, San Francisco, Freeman and Co.
  1983   "Introduction to the selected works of Emmy Noether", in Noether *GA*, 12- 26.
JOHNSTONE, P.T.
  1982   *Stone Spaces*, Cambridge, Cambridge University Press.
  1983   "The Point of Pointless Topology", *BAMS* 8 (NS), 41-53.
JONSSON, B.
  1959   Review of Bourbaki (1939-), Book I (1957), *MR* 29, #3804.
JONSSON, B. AND A. TARSKI
  1947   *Direct Decomposition of Finite Algebraic Systems*, Notre Dame, Notre Dame Mathe-
         matical Lectures.
JORDAN, C.
  1869   "Mémoire sur les groupes de mouvements", *Annali di Matematica Pura ed Applicata
         (2d. series)* 2, 167-215, 322-345.
  1870   *Traité des substitutions et de équations algébriques*, Paris.
KAKUTANI, N.
  1944   "Free Topological Groups and Infinite Direct Product of Topological Groups", *Proc.
         Imp. Acad. Tokyo* 20, 595-598.
KAN, D.
  1958   "Adjoint Functors", *TAMS* 87, 294-329.
KAPLANSKI, I.
  1953   Review of Bourbaki (1939-), Book II, Ch. 6-7, *MR* 14, 237.
  1960   Review of Bourbaki (1939-), *BAMS* 66, 266.
KELLEY, J.L.
  1956   Review of Bourbaki (1939-), Book IV, *MR* 17, #1109.
KELLEY, J.L. AND E. PITCHER
  1947   "Exact Homomorphism Sequences in Homology Theory", *AM* 48, 682-709.
KENNEDY, H.
  1980   *Peano - Life and Work of Giuseppe Peano*, Dordrecht, Reidel.
  1981   "Giuseppe Peano", *DSB* 10, 441-444.
  1981a "Mario Pieri", *DSB* 10, 605-606.
KIERNAN, B.M.
  1971   "The Development of Galois Theory from Lagrange to Artin", *AHES* 8, 40-154.
KIMBERLING, C.
  1981   "Emmy Noether and Her Influence", in Brewer & Smith (eds.) (1981), 3-65.
KINOSHITA, Y.
  1998   "Nobuo Yoneda (1930-1996)", *Math. Japonica* 47 (1), 155.

KITCHER, P.

1988 "Mathematical Naturalism", in Aspray & Kitcher (eds.) (1988), 293-323.

KLEIN, F.

1873 "Über die sogenannte Nicht-Euklidische Geometrie", *MA* 6, 112-145.

1926-7 *Vorlesungen über die Entwicklung der Mathematik im 19. Jahrhundert*, 2 Vols., ed. by R. Courant and O. Neugebauer, Berlin, Springer. (Chelsea reprint, New York, 1948.)

KLEIN-BARMEN, F.

1929 "Einige distributive Systeme in Mathematik und Logik", *JDMV* 38, 35-42.

1931 "Zur Theorie der abstrakten Verknüpfungen", *MA* 105, 308-323.

KLEINER, I.

1992 "Emmy Noether: Highlights of hre Life and Work", *L'Ens. Math.* 38, 103-124.

KLINE, M.

1972 *Mathematical Thought from Ancient to Modern Times*, New York, Oxford University Press.

1980 *Mathematics. The Loss of Certainty*, New York, Oxford University Press.

KNORR, W.R.

1976 "Problems in the Interpretation of Greek Number Theory", *SHPS* 7, 353-368.

KOBLITZ, N.

1979 *p-adic Numbers, p-adic Analysis and Zeta-Functions*, New York, Springer.

KOLCHIN, E.R.

1951 Review of Bourbaki (1939-), Book II, Chs. 4-5. *MR* 12, #6.

KRAMER, E.

1974 "Noether, Amalie Emmy", *DSB* 10, 137-139.

KRASNER, M.

1938 "Une généralisation de la notion de corps", *JMPA* 17, 367-385.

1939 "Remarque au sujet d'«Une généralisation de la notion de corps»", *JMPA* 18, 417-418.

1950 "Généralisation abstraite de la théorie de Galois", *Coll. int. CNRS* 24, 163- 168.

KREISEL, G.

1976 "What have we learned from Hilbert's Second Problem?", in Browder (ed.) (1976), 93-130.

KRONECKER, L.

1853 "Über die algebraisch auflösbaren Gleichungen", *Monatsberichte der BA*, 365- 374.

KRULL, W.

1922 "Algebraische Theorie der Ringe I", *MA* 88, 80-122.

1923 "Algebraische Theorie der Ringe II", *MA* 91, 1-46.

1924 "Algebraische Theorie der Ringe III", *MA* 92, 183-213.

1925 "Über verallgemeinerte endliche Abelsche Gruppen", *MZ* 23, 161-186.

1928 "Galoissche Theorie der unendlichen algebraischen Erweiterungen", *MA* 100, 687-698.

1935 *Idealtheorie*, Ergebnisse der Mathematik, Bd. 4, Berlin, Springer.

KUMMER, E.E.

Coll *Collected Papers*, 2 Vols., ed. by André Weil, New York, Springer (1975).

1844 "De numeris complexis, qui radicibus unitatis et numeris integris realibus constant, Gratulationschrift der Univ. Breslau zur Jubelfeier der Univ. Königsberg." Reprinted in *JMPA* 12 (1847), 185-212. (*Coll* Vol. 1, 165-192.)

1847   "Zur Theorie der complexen Zahlen", *JRAM* 35, 319-326. (*Coll* Vol. 1, 203- 210.)

1847a   "Über die Zerlegung der aus Wurzeln der Einheit gebildeten complexen Zahlen in ihre Primfactoren", *JRAM* 35, 327-367. (*Coll* Vol. 1, 211-251.)

1847b   "Beweis des Fermat'schen Satzes ...", *Monatsberichte BA*, 132-141 & 305-319. (*Coll* Vol. 1, 274-297.)

1850   "Allgemeine Reciprocitätsgesetze für beliebige höhere Potenzreste", *Monatsberichte BA*, 154-165. (*Coll* Vol. 1, 347-357.)

1850a   "Allgemeiner Beweis des Fermat'schen Satzes, ...", *JRAM* 40, 130-138. (*Coll.* Vol. 1, 336-344.)

1851   "Mémoire sur la théorie des nombres complexes composés de racines de l'unite et de nombres entiers", *JMPA* 16, 377-498. (*Coll* Vol. 1, 363-484.)

1856   "Theorie der idealen Primfactoren der complexen Zahlen, welche aus den Wurzeln der Gleichung $\theta^n=1$ gebildet sind, wenn n eine Zusammengesetzte Zahl ist", *Abh. BA*, 1-47. (*Coll* Vol. 1, 583-629.)

1857   "Über die den Gaussischen Perioden der Kreistheilung entsprechenden Congruenzwurzeln", *JRAM* 53, 142-148. (*Coll* Vol. 1, 574-580.)

1857a   "Einige Sätze über die aus den Wurzeln der Gleichung ...", *Abh. BA*, 1857, 41-74. (*Coll* Vol. 1, 639-672.)

1859   "Über die allgemeinen Reciprocitätsgesetze unter den Resten und Nichtresten der Potenzen, deren Grad eine Primzahl ist", *Abh. BA*, 19-159. (*Coll* Vol.1, 699-839.)

KUROSH, A.
1935   "Durschnittsdarstellungen mit irreduziblen Komponenten in Ringen und sogenannten Dualgruppen", *Mat. Sbornik* 42, 613-16.

1935a   "Eine Verallgemeinerungen des Jordan-Hölderschen Satzes", *MA* 111, 13-18.

1963   *Lectures on General Algebra* (English trans. by K.A. Hirsh), New York, Dover.

KÜRSCHÁK, J.
1913   "Über Limesbildung und allgemeine Körpertheorie", *JRAM* 142, 211-253.

LAGRANGE, J.L.
Oeuv   *Oeuvres de Lagrange*, 14 Vols., ed. by J.A. Serret & G. Darboux, Paris, Gauthiers-Villars (1867-1892).

1769   "Sur la solution des problèmes indéterminés du second degrée", *Mémoires BA* 23, 165-311. (*Oeuvres* Vol. 2, 377-535.)

1770-1   "Réflexions sur la théorie algébrique des équations", in *Oeuvres* Vol. 3, 203-421.

LAITA, L.
1975   *A Study on the Genesis of Boolean Algebra*, Unpubl. Diss., University of Notre Dame.

LAKATOS, I.
1976   *Proofs and Refutations*, Cambridge, Cambridge University Press.

1978   *Philosophical Papers*, 2 vols., Cambridge University Press.

LAM, T.Y.
1981   "Representation Theory", in Brewer and Smith (eds.) (1981), 145-155.

LANE, M. (ED.)
1970   *Introduction to Structuralism*, New York, Basic Books.

LANGLANDS, R.P.
1976   "Some Contemporary Problems with Origins in the *Jugendtraum*", in F.E. Browder (ed.) (1976), 401-418.

LASKER, E.
1905 "Zur Theorie der Moduln und Ideale", *MA* 60, 20-116.

LAWVERE, F.W.
1966 "The Category of Categories as a Foundation for Mathematics", in *Proceedings of the Conference on Categorical Algebra* (La Jolla 1965), New York, AMS, 1-20.
1969 "Adjointness in Foundations", *Dialectica* 23, 281-295.

LAX, P.D.
1989 "The Flowering of Applied Mathematics in America", in P. Duren (ed.) (1988-9) Vol. 2, 455-466.

LEFSCHETZ, S.
1942 *Algebraic Topology*, Providence, AMS.

LOEWY, A.
1915 *Grundlagen der Arithmetik*, Berlin, Veit.

LORENZO, J. DE
1977 *La Matemática y el Problema de su Historia*, Madrid, Tecnos.

LURCAT, F.
1976 "L'espace en mathématiques et en psychologie: remarques d'un physicien", *Cahiers de Psychologie* 19, 257-283.

LUSTERNIK, L.A. AND L. SCHNIRELMANN
1934 *Méthodes topologiques dans les problémes variationelles*, Paris, Hermann.

LÜTZEN, J.
1990 *Joseph Liouville, 1809-1882. Master of Pure and Applied Mathematics* (Sources and Studies in the History of Mathematics and Physcal Sciences: Vol. 15), New York, Springer Verlag.

MACAULAY, F.S.
1913 "On the Resolution of a given Modular System into Primary Systems including some Properties of Hilbert Numbers", *MA* 74, 66-121.
1916 *The Algebraic Theory of Modular Systems* (Cambridge Tracts in Mathematics and Mathematical Physics, 19), Cambridge, Cambridge University Press (New ed. with an introduction by P. Roberts - 1996).
1933 "Modern Algebra and Polynomial Ideals", *PCPS* 30, 27-46.

MAC LANE, S.
1934 "General Properties of Algebraic Systems. (Abstract)", *DMJ* 4, 455-468.
1939 "Some Recent Advances in Algebra", *AMM* 46, 3-19.
1943 "A Conjecture of Ore on Chains in Partially Ordered Sets", *BAMS* 49, 567-568.
1948 Review of Bourbaki (1939-), Book II, Ch. 2, *MR* 9, 406.
1948a "Groups, Categories and Duality", *PNAS* 34, 263-67.
1948b Review of Samuel (1948), *MR* 9, 605.
1950 "Duality for Groups", *BAMS* 56, 485-516.
1956 Review of Cartan and Eilenberg (1956), *BAMS* 62, 615-624.
1959 "Locally Small Categories and the Foundations of Set-Theory", in *Infinitistic Methods*, Warszaw, Pergamon, 25-34.
1963 "Some Additional Advances in Algebra", in Albert (ed.) (1963), 35-58.
1970 "The influence of M.H. Stone on the origins of Category Theory", in Browder (ed.) (1970), 228-235.

1971   *Categories for the Working Mathematician*, New York, Springer.

1976   "Topology and Logic as a Source of Algebra", *BAMS* 82, 1-40.

1976a   "The Work of Samuel Eilenberg in Topology", in A. Heller and M. Tierney (eds.) *Algebra, Topology, and Category Theory - A Collection of Papers in Honor of Samuel Eilenberg*, New York, Academic Press, 133-144.

1980   "The Genesis of Mathematical Structures", *Cahiers Topol. Geom. Diff.* 21, 353-365.

1981   "Mathematical Models: A Sketch for the Philosophy of Mathematics", *AMM* 88, 462-472.

1981a   "Mathematics at the University of Göttingen (1931-1933)", in Brewer and Smith (eds.) (1981), 65-78.

1986   Review of Bourbaki (1969) (1984 English edition), *MR* 86h, 01:005.

1986a   "Topology becomes algebraic with Vietoris and Noether", *JPAA* 39, 305-307.

1987   "Address with occasion of the 1986 Steele prizes award. 93th. annual meeting of the AMS at San Antonio, Texas", *Notices of the AMS* 34, 229-230.

1987a   *Mathematics: Form and Function*, New York/Berlin, Springer.

1988   "Concepts and Categories in Perspective", in P. Duren (ed.) (1988-9) Vol. 1, 323-365.

1992   "The Protean Character of Mathematics", in J. Echeverría et al. (eds.) *The Space of Mathematics*, Berlin/New York, de Gruyter, 3-13.

1997   "Garrett Birkhoff and the *Survey of Modern Algebra*", *Notices AMS* 44 (11), 1438-1439.

1998   "The Yoneda Lemma", *Math. Japonica* 47 (1), 156.

MAC LANE, S. AND G. BIRKHOFF

1967   *Algebra*, New York, MacMillan.

MADDY, P.

1980   "Perception and Mathematical Intuition", *The Philosophical Review* 89, 163- 196.

MAL'CEV, A.

1971   "On the History of Algebra in the USSR during the first 25 Years", *Algebra and Logic* 10, 68-75.

MANDELBROT, B.

1989   "Chaos, Bourbaki, and Poincaré", *MI* 11, 10-12.

MARKOFF, A.

1942   "On Free Topological Groups", *Bull. Acad. Sci. USSR (Ser. Math.)* 9, 3-64.

MASHAAL, M.

2000   *Bourbaki. Une société secrète de mathématiciens*, Pour la science, Fev-Mai 2000.

MATHIAS, A.R.D.

1992   "The Ignorance of Bourbaki", *MI* 14, 4-13.

MAY, K.O.

1975   "What is good history and who should do it?", *HM* 2, 449-455.

MAYER, W.

1929   "Über abstrakte Topologie", *Monats. für Math. und Phys.* 36, 1-42, 219-258.

MCLARTY, C.

1990   "The Uses and Abuses of the History of Topos Theory", *BJPS* 41, 351-375.

MEHRA, J.

1974   *Einstein, Hilbert, and the Theory of Gravitation*, Dordrecht, Reidel.

MEHRTENS, H.

1976   "T.S. Kuhn's Theories and Mathematics: a Discussion Paper on the 'New Historiography' of Mathematics" *HM* 3, 297-320. (Repr. in Gillies (ed.) (1992), 21-41.)

1979   *Die Entstehung der Verbandstheorie*, Hildesheim, Gerstenberg.

1979a   "Das Skelett der modernen Algebra. Zur Bildung mathematischer Begriffe bei Richard Dedekind", in C.S. Scriba (ed.) *Zur Entstehung neuer Denk- und Arbeitsrichtungen in der Naturwissenschaft*, Göttingen, Vandenhoeck & Ruprecht, 25-43.

1982   "Richard Dedekind, der Mensch und die Zahlen", *Ab. Br. Wiss. Ges.* 33, 19- 33.

1990   *Moderne - Sprache - Mathematik*, Frankfurt, Suhrkamp.

MENGER, K.

1928   "Bemerkungen zu Grundlagenfragen IV", *JDMV* 37, 309-325.

1940   "Topology without Points", *The Rice Institute Pamphlet*, 27, 80-107.

MERTENS, F.

1886   "Beweis, dass alle Invarianten und Covarianten eines Systems binärer Formen ganze Functionen einer endlichen Anzahl von Gebilden dieser Art sind", *JRAM* 100, 223-230.

MERTZBACH, U.

1992   "Robert Remak and the Estimation of Units and Regulators", in S.S. Demidov et al. (eds.) *Amphora - Festschrift für Hans Wussing zu seinem 65. Geburtstag*, Berlin, Birkhäuser.

MEYER, F.

1890   "Bericht über den gegenwärtigen Stand der Invariantentheorie", *JDMV* 1, 79-292.

MICHAEL, E.

1963   Review of Bourbaki (1939-), Book III, Ch. 1-2, *MR* 25, 4480.

MINIO, R.

1984   "An Interview with Michael Atiyah", *MI* 6(1), 9-19.

MINKOWSKI, H.

1905   "Peter Gustav Lejeune Dirichlet und seine Bedeutung für die heutige Mathematik", *JDMV* 14, 149-163.

MONNA, A.F.

1973   *Functional Analysis in Historical Perspective*, Utrecht, Oosthoek Publishing Company.

MONTGOMERY, D.

1963   "Oswald Veblen", *BAMS* 69, 26-36.

MOORE, E.H.

1902   "Projective Axioms of Geometry", *TAMS* 3, 142-158.

1902a   "A Definition of Abstract Groups", *TAMS* 3, 485-492.

1902b   "The Betweenes Assumption", *AMM* 9, 98-101.

1905   "On a Definition of Abstract Groups", *TAMS* 6, 179-180.

MOORE, E.H. *ET AL* (EDS.)

1896   *Mathematical Papers Read at the IMC held in Connection with the World's Columbian Exposition - Chicago 1893*, New York, Mac Millan.

MOORE, G.H.

1982   *Zermelo's Axiom of Choice - Its Origins, Development, and Influence*, New York, Springer.

1987   "A House Divided Against Itself: the Emergence of First-Order Logic as the Basis for Mathematics", in E.R. Phillips (ed.) *Studies in the History of Mathematics*, MAA Studies in Mathematics, 98-136.

1991   "Sixty Years After Gödel", *MI* 13 (3), 6-11.

1995   "The Axiomatization of Linear Algebra: 1875-1940", *HM* 22, 262-303.

MOORE, R.L.

1908   "Sets of Metrical Hypotheses for Geometry", *TAMS* 9, 487-512.

MOULINES, C.U. AND J.D. SNEED

1979   "Patrick Suppes' Philosophy of Physics", in R.J. Bogdan (ed.) *Patrick Suppes*, Dordrecht, Reidel.

MUELLER, I.

1981   *Philosophy of Mathematics and Deductive Structure in Euclid's "Elements"*, Cambridge, Ma., MIT Press.

MUMFORD, D.

1965   *Geometric Invariant Theory*, New York, Springer.

1976   "Hilbert's Fourteenth Problem - The Finite Generation of Sub-rings such as Rings of Invariants", in Browder (ed.) (1976), 431-444.

1991   "Zariski's Papers on the Foundations of Algebraic Geometry and on Linear Systems", in Parikh (1991), 204-214.

MUNROE, H.E.

1958   Review of Bourbaki (1939-), Book V, Ch. 5 (1956), *BAMS* 64, 105-106.

NAGATA, M.

1959   "On the 14th Problem of Hilbert", *AJM* 81, 766-772.

NAGEL, E.

1939   "The Formation of Modern Conceptions of Formal Logic in the Development of Geometry", *Osiris* 7, 142-224.

NAKAYAMA, T.

1943   "Note on Free Topological Groups", *Proc. Imp. Acad. Tokyo* 19, 471-475.

NETTO, E.

1882   *Substitutionentheorie und ihre Anwendung auf die Algebra*, Berlin.

NEUMANN, O.

1981   "Über die Anstöße zu Kummers Schöpfung der 'Idealen Complexen Zahlen'" in J.W. Dauben (ed.) *Mathematical Perspectives. Essays on Mathematics and its HIstorical Development (Biermann Festschrift)*, San Francisco- London, Academic Press, 179-199.

1997   "Die Entwicklung der Galois-Theorie zwischen Arithmetik und Topologie (1850 bis 1960), *AHES* 50, 291-329.

NICHOLSON, J.

1993   "The Development and Understanding of the Concept of Quotient Group", *HM* 20, 68-88.

NOETHER, E.

GA   *Emmy Noether: Gesammelte Abhandlungen*, ed. by N. Jacobson, New York, Springer (1983).

1908   "Über die Bildung des Formensystems der ternären biquadratischen Formen, *JRAM*
        134, 23-90. (*GA*, 31-99.)
1916   "Die allgemeinsten Bereiche aus ganzen transzendenten Zahlen", *MA* 77, 103-128.
        (*GA*, 195-220.)
1921   "Idealtheorie in Ringbereichen", *MA* 83, 24-66. (*GA*, 354-396.)
1926   "Abstrakter Aufbau der Idealtheorie in algebraischen Zahl- und Funktionenkörper",
        MA 96, 26-61. (*GA*, 493-528.)
1929   "Hyperkomplexe Grössen und Darstellungstheorie", *MZ* 30, 641-692. (*GA* 563-614.)
1933   "Nichtkommutative Algebren", *MZ* 37, 514-541. (*GA*, 642-669.)

NOETHER, E. AND J. CAVAILLÈS (EDS.)
1937   *Briefwechsel Cantor-Dedekind*, Paris, Hermann.

NOETHER, E. AND W. SCHMEIDLER.
1920   "Moduln in nichtkommutativen Bereichen, insbesondere aus Differenzial- und Differ-
        enzausdrücken", *MZ* 8, 1-35. (In Noether *GA*, 318-352.)

NOETHER, M.
1887   "Ueber den Fundamentalsatz der Theorie der algebraischen Functionen", *MA* 30, 410-
        417.

NOVY, L.
1973   *Origins of Modern Algebra* (English trans. by J. Taver), Leyden, Noordhof Interna-
        tional Publications.

ORE, O.
1931   "Some Recent Developments in Algebra", *BAMS* 37, 537-548.
1931a   "Linear Equations in non-commutative Fields", *AM* 32, 463-477.
1932   "Formale Theorie der linearen Differentialgleichungen (Teil I)", *JRAM* 167, 221-34.
1932a   "Id. (Zweiter Teil)", *JRAM* 168, 233-52.
1933a   "Theory of non-commutative Polynomials", *AM* 34, 480-508.
1935   "On the foundations of Abstract Algebra, I", *AM* 36, 406-437.
1936   "On the foundations of Abstract Algebra, II", *AM* 37, 265-292.
1936a   "Direct Descompositions", *DMJ* 2, 581-596.
1936b   "On the Decomposition Theorems of Abstract Algebra", *PICM* Oslo, Vol. 1, 297-307.
1936c   *L'algèbre abstraite*, Paris, Hermann.
1937   "On the Theorem of Jordan-Hölder", *TAMS* 41, 247-269.
1937a   "Structures and Group Theory I", *DMJ* 3, 149-174.
1938   "Structures and Group Theory II", *DMJ* 4, 247-269.
1938a   "On the Application of Structure Theory to Groups", *BAMS* 44, 801-806.
1942   "Theory of Equivalence Relations", *DMJ* 9, 573-627.
1943   "Some Studies on Closure Relations", *DMJ* 10, 761-785.
1943a   "Combinations of Closure Relations", *AJM* 44, 514-533.
1943b   "Chains in Partially Ordered Sets", *BAMS* 49, 558-566.
1944   "Galois Connexions", *TAMS* 55, 493-513,
1948   *Number Theory and its History*, New York, McGraw-Hill,
1957   *Niels Henrik Abel, Mathematician Extraordinary*, New York, Chelsea.
1958   *Cardano, the Gambling Scholar*, Princeton, Princeton University Press.

OSTROWSKI, A.
1917   "Über einige Lösungen der Funktionalgleichung $\phi(x)\phi(y) = \phi(xy)$", *Acta Mathematica* 61, 271-284.

PAREIGIS, B.
1970   *Categories and Functors*, New York, Academic Press.

PARIKH, C.
1991   *The Unreal Life of Oscar Zariski*, San Diego, Academic Press.

PARSHALL, K.H.
1984   "Eliakim Hastings Moore and the Founding of a Mathematical Community in America, 1892-1902", *Annals of Science* 41, 313-333.
1985   "Joseph H.M. Wedderburn and the Structure Theory of Algebras", *AHES* 32, 223-349.
1989   "Towards a History of Nineteenth-Century Invariant Theory", in D.E. Rowe & J. McCleary (eds.) (1989), Vol. 1, 157-206.
1990   "The One-Hundredth Anniversary of the Death of Invariant Theory?", *MI* 12, 10-16.

PARSHALL, K.H. AND D.E. ROWE
1994   *The Emergence of the American Mathematical Research Community, 1876-1900: J.J. Sylvester, Felix Klein, and E.H. Moore*, (History of Mathematics, Vol. 8) Providence, AMS/LMS.

PASCH, M.
1882   *Vorlesungen über neuere Geometry*, Leipzig, Teubner.

PECKHAUS, V.
1990   *Hilbertprogramm und Kritische Philosophie. Des Göttinger Modell interdisziplinärer Zusammenarbeit zwischen Mathematik und Philosophie*, Göttingen, Vandenhoeck & Ruprecht.
1994   "Wozu Algebra der Logik? Ernst Schröders Suche nach einer universalen Theorie der Verknüpfungen", *ML* 4, 357-381.

PIAGET, J.
1971   *Structuralism*, London, Routledge and Keagan Paul. (Enlgish transl. by Chaninah Maschler of *Le Structuralism*, Paris, PUF, 1968.)
1973   "Comments on Mathematical Education", in Howson (ed.) (1973), 79-87.

PIAZZA, P.
1999   "Egor Ivanovich Zolotarev and the Theory of Ideal Numbers for Algebraic Number Fields", *Rendiconti Circ. Mat. di Palermo, Ser. II*, Suppl. 61, 123-150.

PLA I CARRERA, J.
1993   "Dedekind y la Teoría de Conjuntos", *ML* 3, 215-305.

PURKERT, W.
1971   "Zur Genesis des abstrakten Körperbegriffs", *NTM* 10, (1), 23-37 & (2), 8-20.
1976   "Ein Manuskript Dedekinds über Galois-Theorie", *NTM* 13, 1-16.
2002   "Grundzüge der Mengenlehre - Historischer Kontext und Rezeption", in *Felix Hausdorff. Gesammelte Werke*, Vol. 2, Springer Verlag (Forthcoming).

PUTNAM, A.
1979   "A Biographical Note", in I. Kaplanski (ed.) *Saunders Mac Lane - Selected Papers*, New York, Springer.

QUENEAU, R.

1962 "Bourbaki et les Mathématiques de Demain", *Critique* 18, 3-18.

REID, C.

1970 *Hilbert*, Berlin/New York, Springer.

REMAK, R.

1911 "Über die Zerlegung der endlichen Gruppen in direkt unzerlegbare Faktoren", *JRAM* 139, 293-308.

1924 "Über die Zerlegung der endlichen Gruppen in direkt unzerlegbare Faktoren", *JRAM* 153, 131-140.

REMMERT, V.

1995 "Zur Mathematikgeschichte in Freiburg. Alfred Loewy (1873-1935): Jähes Ende späten Glanzes", *Freiburger Universitätsblätter* 129 (3), 81-102.

RESNIK, M.

1974 "The Frege-Hilbert Controversy", *Philosophy and Phenomenological Research* 34, 386-403.

1981 "Mathematics as a Science of Patterns: Ontology and Reference", *Nous* 15, 529-550.

1982 "Mathematics as a Science of Patterns: Epistemology", *Nous* 16, 95-105.

RIBENBOIM, P.

1988 *The Book of Prime Number Records*, New York, Springer.

RICHARDSON, A.R.

1940 "Abstract Algebra", *The Mathematical Gazette* 24, 15-24.

1945 "Grupoids and their Automorphisms", *Proc. LMS (2)* 48, 83-111.

ROSENBERG, A.

1960 Review of Bourbaki (1939-), Book II, Ch. 8, *BAMS* 66, 16-19.

ROTA, G.C.

1997 "The Many Lives of Lattice Theory", *Notices AMS* 44 (11), 1440-1445.

ROTTLÄNDER, A.

1928 "Nachweis der Existenz nicht-isomorpher Gruppen von gleicher Situation der Unter-gruppen", *MZ* 28, 641-653.

ROWE, D.E.

1988 "Gauss, Dirichlet, and the Law of Biquadratic Reciprocity", *MI* 10 (2), 13-25.

1989 "Klein, Hilbert, and the Göttingen Mathematical Tradition", *Osiris* 5, 186- 213.

1993 Review of Gillies (ed.) (1992), *HM* 20, 320-322.

1993a "David Hilbert und seine mathematische Welt", Unpublished manuscript.

1994 "The Philosophical Views of Klein and Hilbert", in Sasaki et al. (eds.) (1994), 187-202.

ROWE, D.E. AND J. MCCLEARY (EDS.)

1989 *The History of Modern Mathematics*, 2 Vols., San Diego, Academic Press.

ROWE, D.E. AND E. KNOBLOCH (EDS.)

1994 *The History of Modern Mathematics*, Vol. 3, Boston, Academic Press.

RÜDENBERG, L. AND H. ZASSENHAUS (EDS.)

1973 *Hermann Minkowski - Briefe an David Hilbert*, Berlin/New York, Springer.

ŠAFAREVICH, I.R.

1974 *Basic Algebraic Geometry* (English trans. by K.A. Hirsh), New York, Springer.

SAMUEL, P.
1948   "On Universal Mappings and Free Topological Groups", *BAMS* 54, 591-598.
1972   Review of Bourbaki (1939-), Book II, Ch. 1-3 (1970), *MR* 43, #2.

SASAKI, CH. ET AL. (EDS.)
1994   *The Intersection of History and Mathematics*, (Science Networks, Historical Studies:
          Vol. 15), Basel, Boston, Berlin, Birkhäuser.

SCANLAN, M.
1991   "Who were the American Postulate Theorists?", *JSL* 50, 981-1002.

SCHARLAU, W.
1981   *Richard Dedekind 1831/1931*, Braunschweig/Wiesbaden, F. Vieweg and Sohn
1982   "Unveröffentliche algebraische Arbeiten Richard Dedekinds aus seiner Göttinger Zeit
          1855-1858", *AHES* 23, 335-355.
1989   *Rudolf Lipschitz- Briefwechsel mit Cantor, Dedekind, Helmholtz, Kronecker und
          Weierstrass und anderen*, Braunschweig/Wiesbaden: Vieweg.
1999   "Emmy Noether's Contributions to the Theory of Algebras", in Teicher (ed.) 39-55.

SCHAPPACHER, N.
1998   "On the History of Hilbert's Twelfth Problem, A Comedy of Errors", in: *Matériaux pour
          l'histoire des mathématiques au XXe siècle* (Actes du colloque à la mémoire de Jean
          Dieudonné - Nice, 1996), Société Mathématique de France 3 , 243-273.

SCHAPPACHER, N. AND K. VOLKERT
1997   "Heinrich Weber; un mathématicien à Strasbourg, 1895-1913", *L'Ouvert* (Journal de
          l'A.P.M.E.P. d'Alsace et de l'I.R.E.M. de Strasbourg) 89, 1-18 (preprint).

SCHIMMACK, R.
1903   "Ueber die axiomatische Begründung der Vektoraddition", *GN* 1903, 317- 325.

SCHMIDT, E.
1933   "Zu Hilberts Grundlegung der Geometrie", in Hilbert *GA* Vol. 2, 404-414.

SCHMIDT, O.
1912   "Über die Zerlegung endlicher Gruppen in direkte unzerlegbare Faktoren", *Sitzungsber.
          d. Physico-mathem. Gesellsch. an St. Wladimir-Universitat in* Kiew, 1-6.
1913   "Sur les produits directs", *B. de la Soc. math. de France* 41, 161-164.
1928   "Über unendlichen Gruppen mit endlicher Kette", *MZ* 29, 34-41.

SCHMIDT, R.
1994   *Subgroup Lattices of Groups* (de Gruyter Expositions in Mathematics, Vol. 14), Berlin,
          Walter de Gruyter & Co.

SCHOENEBERG, B.
1981   "Heinrich Weber", *DSB* 12, 202-203.

SCHOLZ, E. (ED.)
1990   *Geschichte der Algebra - eine Einführung*, Mannheim/Wien/Zürich, BI- Wissen-
          schaftsverlag.

SCHÖNEMANN, W.
1846   "Grundzüge einer allgemeinen Theorie der höhern Congruenzen, deren Modul eine
          reelle Primzahl ist", *JRAM*

SCHREIER, O.
1928   "Über den Jordan-Hölderschen Satz", *AMSHU* 6, 300-302.

SCHRÖDER, E.
1877   *Der Operationkreis der Logikkalküls*, Leipzig.
1890   *Vorlesungen über die Algebra der Logik*, 3 vols., Leipzig.
SCHUR, F.
1901   "Über die Grundlagen der Geometrie", *MA* 55, 265-292.
1903   "Über die Zusammensetzung von Vektoren", *ZMP* 49, 352-361.
SCRIBA, C.J.
1961   Review of Bourbaki (1960), *MR* 22, #4620.
SEBASTIÃO E SILVA, J.
1945   "Sugli automorfismi di un sistema matematico cualunque", *Comm. Pontif. Acad. Sci.* 9, 327-57.
1985   "On Automorphisms of Arbitrary Mathematical Systems" (English trans. by A.J. Franco de Oliveira), *HPL* 6, 91-116.
SEGRE, M.
1994   "Peano's Axioms in their Historical Context", *AHES* 48, 201-342.
SERRE, J.P.
1955   "Faisceaux algébriques coherents", *AM* 61, 197-278.
SERRET, J.
1849   *Cours d'algébre supérieure* (2d. ed. 1854, 3d. ed. 1866), Paris.
SHAPIRO, S.
1983   "Mathematics and Reality", *PS* 50, 523-548.
SIEGMUND-SCHULTZE, R.
1981   "Der Strukturwandel in der Mathematik um die Wende vom 19. zum 20. Jahrhundert, untersucht am Beispiel der Entstehung der ersten Begriffsbildungen der Funktionalanalysis", *NTM* 18 (1), 4-20.
1998   *Mathematiker auf der Flucht vor Hitler. Quellen und Studien zur Emigration einer Wissenschaft*, (Dokumente zur Geschichte der Mathematik, Band 10) Braunschweig/Weisbaden, Vieweg.
1998a   "Eliakim Hastings Moore's General Analysis", *AHES* 52, 51-90.
SILVESTRI, R.
1979   "Simple Groups of Finite Order in the Nineteenth Century", *AHES* 20, 313-356.
SINACEUR, H.
1991   See Benis-Sinaceur (1991).
SMILEY, H.F.
1944   "An Application of Lattice Theory to Quasi-groups", *BAMS* 50, 782-786.
SMITH, D.E. (ED.)
1929   *A Source Book in Mathematics*, 2 Vols., New York, Dover.
SMITH, H.J.S.
1965   *Report on the Theory of Numbers* (originally published in six parts as a Report of the British Assn.), Chelsea, New York.
SOBOLEV, S.L.
1973   "Some Questions of Mathematical Education in the USSR" in Howson (ed.) (1973), 181-193.

SONO, M.
  1917-8   "On Congruences", *Mem. Coll. Sci. Kyoto* 2, 203-26; 3, 113-149, 189- 197.
SPALT, D.
  1985   "Nach Bourbakis Fortschrittmodell", Review of Dieudonné (ed.) (1978), *Frankfurter Allgemeine Zeitung* Nr. 280 (3 Dez. 1985), L 14-15.
  1987   *Vom Mythos der mathematischen Vernunft*, Darmstadt, Wissenschaftliche Buchgesellschaft.
SPOHN, W.G.
  1961   "Can Mathematics be Saved?", *Notices of the AMS* 16, 890-894.
SPRINGER, T.A.
  1977   *Invariant Theory*, Berlin/Heidelberg, Springer.
STEGMÜLLER, W.
  1979   *The Structuralist View of Theories: A Possible Analogue to the Bourbaki Programme in Physical Sciences*, Berlin/New York, Springer.
STEINITZ, E.
  1910   "Algebraische Theorie der Körper", *JRAM* 137, 167-309. (2d. ed. (1930), edited by H. Hasse and R. Baer, Berlin/Leipzig.)
STONE, H.M.
  1932   *Linear Transformations in Hilbert Space and their Application to Analysis*, New York, AMS Coll. Publ., Vol.XV.
  1934   "Boolean Algebras and their Application to Topology", *PNAS* 20, 197-202.
  1935   "Subsumption of the Theory of Boolean Algebras under the theory of Rings", *PNAS* 21, 103 -115.
  1936   "The Theory of Representations for Boolean Algebras", *TAMS* 40, 37-111.
  1937   "Applications of the Theory of Boolean Rings to General Topology", *TAMS* 41, 321-364.
  1938   "The Representation of Boolean Algebras", *BAMS* 44, 807-816.
  1970   Remarks on Mac Lane (1970), in F.E. Browder (ed.) (1970), 236-241.
STUDY, E.
  1933   *Einleitung in die Theorie der Invarianten*, Braunschweig, Vieweg.
SUPPES, F.
  1969   *Studies in the Methodology and Foundations of Science*, Dordrecht, Reidel.
SUZUKI, M.
  1956   *Structure of Groups and the Structure of the Lattice of Subgroups*, Berlin, Springer.
TAMARI, D.
  1978   "Algebra: Its Place in Mathematics and Its Role, Past and Present for Mankind. A Historical and Critical Essay", *Eleuteria* (September 1978), 187- 201.
TARSKI, A.
  1949   *Cardinal Algebras*, Oxford/New York, Oxford University Press.
TARWATER, D. (ED.)
  1977   *The Bicentennial Tribute to American Mathematics, 1776-1976*, Providence, The Mathematical Association of America.
TATE, J.
  1976   "Problem of the General Reciprocity Law", in F.E. Browder (ed.) (1976), 311-322.

TAZZIOLI, R.
1994   "Rudolf Lipschitz's work on differential geometry and mechanics", in Rowe & Knobloch (eds.) (1994), 113-138.

TEICHER,, M.
1999   *The Heritage of Emmy Noether*, Israel Mathematical Conference Proceedings.

THOM, R.
1971   "Modern Mathematics: an educational and philosophical error?", *American Scientist* 59, 695-699.
1973   "Modern Mathematics: does it exist?", in Howson (ed.) (1973), 194-209.

TOEPELL, M.M.
1986   *Über die Entstehung von David Hilberts ,Grundlagen der Geometrie"*, Göttingen, Vandenhoeck & Ruprecht.

TOEPLITZ, O.
1922   "Der Algebraiker Hilbert", *Die Naturwissenschaften* 10, 73-77.

TOLLMIEN, C.
1990   "Emmy Noether, 1882-1935", *Göttinger Jahrbuch* 38, 153-219.
1991   "Die Habilitation von Emmy Noether an der Universität Göttingen", *NTM* 28, 13-32.

TORRETI, R.
1978   *Philosophy of Geometry from Riemann to Poincaré*, Dordrecht, Reidel.

TOTH, I.
1987   "Nicolas Bourbaki S.A.", *Schriftenreihe der Universität Regensburg* 14, 138-153.

TOTI-RIGATELLI, L.
1989   *La Mente Algebrica - Storia dello Sviluppo della Teoria di Galois nel XIX Secolo*, Varese, Bramante Editrice.
1996   *Evariste Galois, 1811-1832* (Vita Mathematica: Vol. 12), Berlin, Birkhäuser (Enlgish transl. by John Denton of *Evariste Galois, matematica sulle barricate*, Firenze, RCS Sansoni Editore, 1993).

TRICOMI, F.G.
1981   "Giuseppe Veronese", *DSB* 11, 623.

TYMOCZKO, T. (ED.)
1985   *New Directions in the Philosophy of Mathematics*, Boston, Birkhäuser.

VAUGHT, R.L.
1974   "Model Theory before 1945", in L. Henkin et al. (eds.) *Proceedings of the Tarski Symposium*, Providence, AMS, 152-172.

VEBLEN, O.
1904   "A System of Axioms for Geometry", *TAMS* 5, 343-384.
1905   "Definition in Terms of Order Alone in the Linear Continuum and in Well- Ordered Sets", *TAMS* 6, 165-171.
1906   "The Square Root and Relations of Order", *TAMS* 7, 197-199.

VEBLEN, O. AND J.H.M. WEDDERBURN
1907   "Non-Desarguesian and Non-Pascalian Geometries", *TAMS* 8, 379-388.

VERONESE, G.
1891   *Fondamenti di geometria a piu dimensioni e a piu specie di unitá rettilinee, esposti in forma elementare*, Padova, Tipografia del Seminario.

WAERDEN, B.L. VAN DER

1930 *Moderne Algebra*, 2 vols., Berlin, Springer. (English trans. of the 2d. ed. - Vol. I by Fred Blum (1949), Vol. II by T. J. Benac (1950), New York, Frederic Ungar Publishing.)

1933 "Nachwort zu Hilberts algebraischen Arbeiten", in Hilbert *GA* Vol. 2, 401- 403.

1935 "Emmy Noether- Obituary", *MA* 111, 469-74. (Repr. in Dick (ed.) (1971), 100-111.)

1966 "Die Algebra seit Galois", *JDMV* 68, 155-165.

1970 "The Foundations of Algebraic Geometry from Severi to André Weil", *AHES* 7, 171-180.

1972 "Die Galois-Theorie von Heinrich Weber bis Emil Artin", *AHES* 9, 240-248.

1975 "On the Sources of my Book *Moderne Algebra*", *HM* 2, 31-40.

1985 *A History of Algebra-From al-Kharizmi to Emmy Noether*, Berlin/New York, Springer.

WALLMAN, H.

1938 "Lattices and Topological Spaces", *AM* 39, 112-126.

WARD, M.

1939 "A Characterization of Dedekind Structures", *BAMS* 45, 448-451.

WEAVER, G.

1993 "Model Theory", in I. Grattan-Guinness (ed.) *Companion Encyclopaedia for the History of Mathematics*, London, Routledge, 670-679.

WEBER, H.

1891 *Elliptische Funktionen und algebraische Zahlen*, Braunschweig.

1893 "Die allgemeinen Grundlagen der Galois'schen Gleichungstheorie", *MA* 43, 521-549.

1895 *Lehrbuch der Algebra*, Vol. 1, (2d. ed. 1898) [Vol. 2, 1896 (2d. ed. 1899); Vol 3, 1908], Braunschweig.

WEBER, H. AND J. WELLSTEIN

1903 *Enzyklopädie der elementar Mathematik*, Leipzig, Teubner.

WEDDERBURN, J.H.M.

1909 "On the Direct Product in the Theory of Finite Groups", *AM* 10, 173-176.

WEIL, A.

1937 "Sur les espaces à structure uniforme et sur la topologie générale", *Actual. sci. et ind.* 551, 3-40. (Repr. in Weil (1978), 147-183.)

1938 "Science Française", in Weil (1978), 232-235.

1946 *Foundations of Algebraic Geometry*, New York, AMS Coll. Publ. (Vol. XXIX).

1950 "Number Theory and Algebraic Geometry", in *PICM* Cambridge - Mass., Vol. 2, 90-100.

1954 "Abstract versus Classical Algebraic Geometry", in *PICM* Amsterdam, Vol. 3, 550-558.

1978 *Collected Papers*, Vol. 1, New York, Springer.

1980 "History of Mathematics: Why and How?" in *PICM* Helsinki (1978), Vol. 1, 228-237.

1992 *The Apprenticeship of a Mathematician*, Boston, Birkhäuser. (English trans. by J. Gage of *Souvenirs d'apprentissage* (1991), Basel, Birkhäuser.)

WEINTRAUB, E.R. AND P. MIROWSKI

1994 "The Pure and the Applied: Bourbakism Comes to Mathematical Economics", *SIC* 7, 245-272.

WEYL, H.

1935 "In Memoriam Emmy Noether", in Dick (ed.) (1981), 112-152.

1939 *The Classical Groups: Their Invariants and Representations*, Princeton, Princeton University Press.

1944 "David Hilbert and his Mathematical Work", *BAMS* 50, 612-654.

WILDER, R.L.

1952 *Introduction to the Foundations of Mathematics*, New York, Wiley, (2nd ed. 1965.)

1976 "Robert Lee Moore, 1882-1974", *BAMS* 82, 417-427.

WITHEHEAD, A.N.

1898 *A Treatise on Universal Algebra, with Applications*, Vol. 1, Cambridge.

WUSSING, H.

1984 *The Genesis of the Abstract Group Concept. A Contribution to the History of the Origin of Abstract Group Theory*, Cambridge, Ma., MIT Press. (English trans. by Abe Shenitzer of *Die Genesis des abstrakten Gruppenbegriffs* (1969), Berlin.)

1989 *Vorlesungen zur Geschichte der Mathematik*, 2d. ed., Berlin, VEB Deutscher Verlag der Wissenschaften.

YONEDA, N.

1954 "On the Homology Theory of Modules", *J. Fac. Sci. Univ. Tokyo - Sect. I*, 7, 193-227.

YOUNG, D.M.

1997 "Garrett Birkhoff and Applied Mathematics", *Notices AMS* 44 (11), 1446-49.

ZARISKI, O.

1935 *Algebraic Surfaces*, Berlin, Springer.

1950 "The Fundamental Ideas of Abstract Algebraic Geometry", in *PICM* Cambridge - Mass., Vol. 1, 77-89.

1954 "Interprétations algébrico-géometriques du quatorzième problème de Hilbert", *Bulletin des Sciences Mathématiques* 78, 155-168.

1972 *Collected Works*, Vol. 1, "Foundations of Algebraic Geometry and Resolution of Singularities", ed. by H. Hironaka and D. Mumford, Cambridge, Mass., MIT Press.

ZARISKI, O. AND P. SAMUEL

1958 *Commutative Algebra*, Princeton, N.J., Van Nostrand.

ZASSENHAUS, H.

1934 "Zum Satz von Jordan-Hölder-Schreier", *AMSHU* 10, 106-108.

1973 "Zur Vorgeschichte des Zahlberichts", in Rüdenberg & Zassenhaus (eds.) (1973), 22-26.

ZINCKE, H. (SOMMER)

1916 "Erinnerungen an Richard Dedekind", *Braunschweigisches Magazin* 7 (Juli), 73-81.

ZOLOTAREV, E.I.

1880 "Sur la théorie des nombres complexes", *JMPA* 16, 51-83; 129-166.

Author Index

A

B

Subject Index

A